Academic Press Rapid Manuscript Reproduction

Proceedings of the Jubilee Symposium
of the Waite Agricultural Research Institute
held at Glen Osmond, South Australia,
on November 5-6, 1974

Trace Elements in Soil-Plant-Animal Systems

Trace Elements in Soil-Plant-Animal Systems

Edited by

D.J.D. Nicholas

Department of Agricultural Biochemistry and Soil Science
Waite Agricultural Research Institute
The University of Adelaide
Glen Osmond, South Australia

Adrian R. Egan

Department of Agronomy
Waite Agricultural Research Institute
The University of Adelaide
Glen Osmond, South Australia

Academic Press, Inc., New York San Francisco London 1975
A Subsidiary of Harcourt Brace Jovanovich, Publishers

ACADEMIC PRESS, INC.
111 Fifth Avenue, New York, New York 10003

United Kingdom Edition published by
ACADEMIC PRESS, INC. (LONDON) LTD.
24/28 Oval Road, London NW1

Library of Congress Cataloging in Publication Data
Main entry under title:

Trace elements in soil-plant-animal systems.

Proceedings of the Jubilee Symposium of the Waite Agricultural Research Institute held at Glen Osmond, South Australia, on Nov. 5-6, 1974.
Bibliography: p.
Includes index.
1. Trace element metabolism–Congresses.
2. Traces elements in soils–Congresses. I. Nicholas, David James Donald. II. Egan, Adrian R. III. Adelaide. University. Waite Agricultural Research Institute, Glen Osmond.
QP531.T7 574.1'33 75-17612
ISBN 0–12–518150–7

PRINTED IN THE UNITED STATES OF AMERICA

Contents

Preface

Over a period of some fifty years the study of trace elements in soils, plants, and animals has evolved from a recognition of their adverse effects or their essentiality for living systems under field and laboratory conditions, to the more basic studies of the physics, chemistry, and biochemistry of their availability to various organisms and their functions in cell metabolism. Our knowledge is, of course, far from complete. Most models on which our present understanding of these complex interactions is based are probably grossly oversimplified. The chain of events whereby trace elements are released from various rock and soil sources, taken up by living organisms, translocated to their sites of action, and function in various metabolic events is not well understood. The trace elements occur in different chemical compounds having varying solubilities, depending upon the chemical and physical environment, and they differ in the type of coordination compounds formed and in their mobility within soils and living organisms. The scope for interactions at each stage of release, uptake, and function is, of course, great.

As a consequence, studies on trace elements in soil, plant, and animal systems have brought together scientists of many disciplines. It seemed both appropriate and timely to assemble, in the two-day Jubilee Symposium, scientists from various disciplines involved in different aspects of research on trace elements. Their contributions encompass reviews of an historical nature and those of a more specialized kind, dealing with current research work on some of the blind-spots in our knowledge. In this way the Organizing Committee of the Symposium has aimed to present wide-ranging contributions which together provide information and current concepts on trace elements in relation to soil–plant–animal systems.

Acknowledgments

We are grateful for the financial support of the Alf Hannaford Research Bequest; for assistance with the preprints and in various other ways by the Bank of Adelaide, the University Information Office, the CSIRO Divisions of Soils and Nutritional Biochemistry, and the South Australian Department of Agriculture; and for the support given by staff of the Waite Institute and by the individual contributors to the Symposium. We thank particularly the members of the Organizing Committee, Mr. K.P. Barley, Mr. K.M. Doull, Dr. P.B. Nicholls, Dr. J.M. Oades, Mr. D.J. Reuter, and Dr. K.G. Tiller, and their subcommittees for their advice and assistance with the preparation of the program.

It is a pleasure to acknowledge the dedicated assistance and professional skill of Miss Janet Peavy in secretarial organization of the Symposium and in all aspects of preparation of this book.

Peter Waite
1834-1922

The Waite Institute Jubilee

Peter Waite came to Australia at the age of 25 years from Kirkcaldy, Fifeshire, Scotland, in 1859. Within a few years he took up pastoral land in the north of South Australia and was outstandingly successful in his ventures. In 1874 he purchased the property at Urrbrae, Adelaide, now known as the Waite Agricultural Research Institute.

In 1913 Peter Waite intimated to the Council of the University of Adelaide his desire to establish an agricultural research institute by making a gift of his land, and from that time until his death in 1922 he took a personal interest in the scheme. The total area of land which he gave to the University was 125 hectares plus a considerable endowment.

The University of Adelaide then proceeded to establish the Waite Agricultural Research Institute and in November 1924 research commenced under the Foundation Director, Professor A.E.V. Richardson, a distinguished graduate of the University. Professor Richardson expressed the hope at that time that the Waite Institute might become "the Rothamsted of the southern hemisphere." When he resigned in 1938 to become the Chief Executive Officer of CSIRO, the Waite Institute was already fulfilling this aspiration as a result of its research on "the plant and soil problems relating to agriculture and on insect and fungus pests which affect production." The continued development of the Institute was markedly influenced by Professor J.A. Prescott, FRS, Foundation Professor of Agricultural Chemistry, who became the second director of the Waite Institute in 1938.

The research program during these years included studies on trace elements, plant breeding, insect ecology, agricultural climatology, plant physiology, plant diseases, and the nature of viruses, as well as crop and pasture production. However, trace element research was probably more influential in establishing the Waite Institute's international standing than any other single area of work. The pioneer work in 1928 of G. Samuel and C.S. Piper, who found that "road-take-all" of oats was due to manganese deficiency, was outstanding. This research was the forerunner of other work in the Waite Institute, in CSIRO, and in the South Australian Department of Agriculture, as well as in other states. It established Australia as a world leader in the field of trace element studies in soil-plant-animal systems, the subject of this Symposium.

It is noteworthy that the interdisciplinary character of the early trace element work at the Institute is sustained by the staff of the six departments to the present day through their interest in an holistic approach to agricultural research.

Nineteen seventy four marks the fiftieth year of the establishment of the Waite Agricultural Research Institute and this Symposium was held to honor Peter Waite's generosity and foresight and those pioneer workers who did so much to influence the development of scientific and biological research throughout Australia and indeed contributed much to agricultural production in many parts of the world.

J. P. Quirk
Director

INTRODUCTION TO THE SYMPOSIUM

Professor G.M. Badger, Pres.A.A.
Vice-Chancellor, The University of Adelaide

Nineteen-seventy-four is an important year in our history. It marks the centenary of the establishment of the University of Adelaide and the fiftieth anniversary of the creation of the Waite Agricultural Research Institute. In those fifty years the Institute has attracted many able and distinguished scientists and its work has won world acclaim. This acclaim has perhaps never been more enthusiastic than in the response to the work on trace elements. It is therefore a fitting celebration of those fifty years that we have gathered here some of the world authorities on this subject.

The Jubilee Symposium will deal with all areas of knowledge of the chemistry and biology of trace elements. Some of the papers will be nominally of an historical nature, but the common thread running through them seems forward-looking and strongly oriented towards the future application of knowledge.

I hope, therefore, that my function here, in presenting an historical preface to the Symposium, will not label me as being inappropriately anachronistic. It is, however, a pleasure to review, if only briefly, the events which led to the establishment of the Institute and to honour the foresight and generosity of Peter Waite.

It began on 3 October 1913, when Peter Waite wrote a letter to his friend, Sir Samuel Way, the Chancellor of the University. He pointed out that the State had "done notably in all branches of agriculture and the allied arts and, largely without scientific direction and education". He added that in the future "competition threatens to be so keen that we must equip our people in the best possible manner". In this letter he offered the University the Urrbrae Estate of 134 acres, 67 acres of which was to be preserved as a public park.

In 1915 Peter Waite purchased Netherby Estate (112 acres) and Claremont (53 acres) adjoining Urrbrae, making a total of 299 acres to be given to the University, in addition to 114 acres to be handed over to the Government for an agricultural high school. He clearly had great faith in the benefits to be derived from agricultural science. In another letter to the Chancellor, on 21 May 1915, he wrote:

There are in South Australia large areas of proved agricultural land with a rainfall suitable for Closer Settlement. In recent years many of the large holdings have been divided up, and also some of the smaller, and it is reasonable to suppose that this process of subdivision will be continued as scientific discoveries render possible the more intensive cultivation of the soil, which will thus be able to sustain a larger population. Help is also required from scientific research to satisfactorily deal with the large area of somewhat inferior Mallee wheat land, as well as to discover what grasses, herbs, and shrubs may be acclimatized to advantage in the vast tracts of Pastoral country, which have but a small and erratic rainfall.

The University Council set up a sub-committee "to inquire into steps that should be taken by the University in order to give practical effect to the conditions attaching to the Peter Waite Land Endowment" and, as might be expected, Arthur Perkins, who was Director of Agriculture and (for a time) simultaneously the Principal of Roseworthy Agricultural College, was a member. Perkins wrote a position paper for the sub-committee and it is interesting to read what he had to say.

After considering the obligation which the gift imposed upon the University he went on:

The question may be raised as to whether there is room in South Australia for an institution of this kind; as to whether it will not be confined to the mere duplication of work at present in the hands of various Government institutions.

On this point I have no hesitation in being very emphatic; the field for agricultural investigation work in a new country such as ours is so vast, as to offer ample scope for more workers than all the institutions in the Commonwealth are likely to attract in our times. It is rather the lack of workers than of material that is to be feared. At the present time we are very largely dependent for what may be termed the scientific basis of Agriculture on principles established under totally different climatic and economic conditions in Europe and America: and it is often in the light of misleading principles that attempts are made to improve local practice. Hence, apart altogether from excursions into the unknown, there is a wide field of work in the way of

> *confirmation or correction of what are supposed to be the basic principles of a great national industry.*

He went on to consider the direction which a University agricultural investigation station (as he called it) might take. He referred to the work of Joseph Henry Gilbert and John Bennett Lawes and discussed the possible appointment of a Professor of Agriculture.

> *I foresee therefore very little scholastic work for the future Professor of Agriculture: true, he may have the handling of that infinitesimal minority, who are attracted to the subject for the subject's sake, and not because of the career to which University training is the passport ... I recognise however that there is one factor that may set at naught my present anticipations: and that is an altogether unprecedented increase in our population. In such perhaps improbable circumstances it is easy to conceive that the call for professional men may be such as to create a demand which the Professor of Agriculture alone will be able to supply.*

Perkins mentioned the other sciences which he thought should be represented on the staff and he added:

> *I am prepared however to hazard the suggestion that the early appointment of a Chemist would in my opinion be absolutely essential.*

With the death of Mr. and Mrs. Peter Waite in 1922, the estate came into the possession of the University and with it a number of shares in Elder, Smith and Co. Ltd. (of which Peter Waite was, for many years, the Chairman), which realized £58,450. In May 1923 the University Council resolved that the proposed Institute would be called 'The Waite Agricultural Research Institute', and the first appointments to the staff were made in 1924 and 1925. A.E.V. Richardson was appointed the first Director and the Waite Professor of Agriculture, and J.A. Prescott was appointed the Waite Professor of Agricultural Chemistry. Richardson had been born in Adelaide and was a graduate of Roseworthy and the University, and at the time of his appointment he was Superintendent of Agriculture in Victoria. It is also of some interest that he was a classical scholar who wrote odes in Latin. Prescott, a graduate from Manchester who had also been trained as a soil chemist at Rothamsted, was working in Egypt at the time of his appointment. These two men were to lay the foundations of the Waite Agricultural Research Institute we now know.

It so happens that these two men were also responsible for creating the long and mutually beneficial working relationship with the Council for Scientific and Industrial Research, which was established by the Australian Government just as the Waite Institute was finding its feet. The CSIR was under the direction of an Executive of three members, with an Advisory Council. The first three members of the Executive were G.A. Julius, W.J. Newbigin and A.C.D. Rivett; but Newbigin died in 1927 and was succeeded by A.E.V. Richardson. He later became Deputy Chief Executive Officer, and in 1946 he became the Chief Executive Officer. He remained in this position until his retirement in 1949. Prescott was also destined to be closely associated with CSIR because he was appointed Chief of the Division of Soils, the headquarters of which remained in the main building of the Waite Institute for the next 33 years. This close association of CSIR with the Institute is illustrated by the fact that, in 1940, there were 21 CSIR scientists and 24 University scientists working together at the Institute. It is pleasing to note that there is still close collaboration between CSIRO (as it is now called) and the University on what has become known as the Waite Campus.

The catalogue of the scientific achievements of the Institute is impressive. In the early days there was work on River Murray soil problems, fungus diseases in crops, and tomato wilt. The first soils map of Australia, covering the 3 million square miles, was prepared; and there was work on many insect pests. The Institute's work on trace element deficiencies was most impressive, complementing that in the CSIR Divisions of Soils and of Nutritional Biochemistry, in the University of Western Australia, and elsewhere.

The areas of trace element deficiency and of deleteriously high trace element availability may be predictable from regional geochemical data, but three major areas need to be held in mind in our consideration of the subject.

First, there is an incalculable loss of biological and economic efficiency associated with trace element disorders not identifiable from frank deficiency signs. These may be continuous and marginal, or they may be seasonally transient and may have their effects on plants and animals at a particular phase of the biological cycle.

Secondly, with increasing intensity of production, applications of ever-increasing levels of fertilizers can result in the production of deficiencies which have not been previously experienced. This problem is encountered in many countries having a high intensity of cultivation and animal production, and there is no simple solution. Often the imbalances and inadequacies are compounded by interactions between a number of elements in a way which can be unique to a particular biological system. Moreover, the 'displacement' of disorders occurs when one remedial action results in induction of a new disorder, or when waste materials of poor trace element balance re-enter biological systems.

And thirdly, the quest for new higher yields in crop plants, and the drive for the introduction of crops not previously included in traditional rotations, may result in mineral disorders because of their differing abilities to draw specific trace elements from the soil.

Such problems call for solutions through biological knowledge which will permit identification of the nature of the disorder long before it affects the productivity of the crop or animal system; or better still, the prescribing of necessary remedial action before new industry commences in a given area. In this way we may hope to move towards efficient innovation without the risk of failure of a new enterprise on a minor nutrient inadequacy or imbalance. In this Symposium results will be presented which will permit us to reduce the time interval for solution of the new problems of this nature which may arise. In other words, this Symposium is designed to review the past, to assess current work, and to encourage future work.

TRACE ELEMENTS IN AUSTRALIAN CROP AND PASTURE PRODUCTION, 1924-1974

C.M. Donald* and J.A. Prescott†

Waite Agricultural Research Institute
The University of Adelaide
Glen Osmond, South Australia 5064

INTRODUCTION

No country has derived greater benefit from trace elements than Australia. Three principal factors have been involved — the infertility of so many of our soils, the use of trace elements in extensive land development and the emphasis on these nutrients in our plant investigations over several decades.

The use of chemical fertilizers in Australia began in 1882 with the experimental application of superphosphate to wheat at Roseworthy College, South Australia, by the first principal, J.D. Custance; he wrote that superphosphate is "the cheapest and best manure for the wheat crop". By 1924, when the Waite Institute was established, Australian superphosphate consumption had reached 420,000 tonnes per annum, mainly on the wheat crop, while other fertilizers (N and K) were used only in small quantity, principally on sugarcane and horticultural crops.

Trace elements were unknown as fertilizers in Australia, and indeed few of these elements had been proved to be essential for plants. The need for several elements in minute amounts had been demonstrated in the case of microorganisms, particularly by workers at the Pasteur Institute [for example, Zn for *Aspergillus* by Raulin (1869)], but only Mn (McHargue 1922) and B (Warrington 1923) had been clearly demonstrated

---

*Emeritus Professor of Agriculture
†Emeritus Professor of Agricultural Chemistry

to be essential trace elements for higher plants. The need for Zn was unresolved. The outlook at this time towards these elements was as 'stimulants' rather than as nutrients, as illustrated in the review of Cu, Zn, As, B and Mn by Brenchley (1927): "... some inorganic poisons act as stimulants when present in small amounts, the stimulating concentrations varying both with the poison used and the plants on which they re-act."

It was with this background from overseas that Australian workers became involved in trace element problems in the mid-twenties, and it is of interest to recall the circumstances of the first field responses in Australia to each of the plant micronutrients.

THE FIRST FIELD RESPONSES

In 1924 A.J. Perkins, then Director of Agriculture in South Australia and a member of the Council of the University of Adelaide, drew the attention of J.A. Prescott, Professor of Agricultural Chemistry at the newly established Waite Institute, to a disease of oats on volcanic ash soil in the Mount Gambier district. This disease was known as 'road take-all' for the singular reason that it was always more severe in those parts of the paddocks bordering the road. A strip of rye was commonly grown in place of oats along the roadside because rye was not susceptible to the disease.

G. Samuel and C.S. Piper of the Waite Institute investigated this problem and found that the disease also occurred on heavy rendzina soils near Penola, where it was known as 'the grubs'. By 1927 they had firmly established that the disease was due to Mn deficiency (the *Dorrfleckenkrankheit* or 'grey speck disease' of Holland and Germany), and had secured an increase in the grain yield of oats from nil to 2960 kg/hectare with 84 kg/hectare manganese sulphate (Fig. 1). They recognized that the occurrence of the disease along roadsides was due to the dust blown from the limestone roads; this raised soil pH, shown in the laboratory to reduce the availability of the Mn (Samuel & Piper 1928). In the following year, they successfully produced grey speck disease in water culture, the first laboratory demonstration of a known micronutrient deficiency disease (Samuel & Piper 1929). Indeed they achieved such an effective removal of Mn that several species died before flowering.

An important step at this time was the publication by Prescott (1931) of his monograph on the soils, vegetation and climate of Australia, adding greatly to our understanding of

Fig. 1. The first trace element response in Australia. Experimental plots at Mount Gambier, South Australia, 1927. On right, untreated plot with most of the plants dead; path in centre; on left, plot with 84 kg/hectare manganese sulphate. [*From Samuel & Piper 1928*]

of the Australian environment. The involvement of the Waite Institute in mineral nutrition problems was further stimulated through a cooperative research programme promoted by the first Director, Dr. A.E.V. Richardson. In 1927 the University of Adelaide, the Council for Scientific and Industrial Research, the Empire Marketing Board and (later) the Carnegie Corporation of New York agreed to finance investigations at the Institute on the mineral content of pastures. Several programmes on major and trace elements were conducted under these funds between 1930 and 1950.

The next trace element response in Australia was recorded on citrus trees. Pittman & Owen (1936), working on a disease of citrus in Western Australia, suspected a nutritional disorder. In view of results reported on 'citrus mottle leaf' (Chandler *et al.* 1933) in the U.S.A., they decided to apply zinc sulphate to affected trees and quickly gained a marked improvement in tree health. Workers in other states soon followed and Zn is now widely used on Australian citrus.

In the winter of 1933, J.D. Atkinson of the Department of Scientific and Industrial Research, New Zealand, conducted a survey of an obscure disease of apples known as 'corky-pit', in which lesions occur in the flesh of apples; he concluded (Atkinson 1935) "that the disease may be due to mineral deficiency" and accordingly administered individually 16 elements in solution to affected trees. Fifteen of them had no effect, but B reduced the incidence of corky-pit from 98% to 3%. "This result," wrote Atkinson cautiously, "suggests that corky-pit is materially influenced by the addition of boron." Carne & Martin (1937), in Tasmania, immediately applied B successfully for the like condition of apples in Australia known as 'internal cork', a disease first described by Cobb in New South Wales in 1892.

It will be noted that these first recorded deficiencies of three trace elements in Australia were all related to the cure of a plant disease — of grey speck of oats, mottle-leaf of citrus and internal cork of apples. But from the mid-thirties onwards, trace elements progressively came to be regarded in Australia as fertilizers in the hands of agronomists, permitting successful crop and pasture production where previously such crops and pastures could not be grown.

In 1935 the Division of Animal Health and Nutrition of CSIR extended its work on coast disease of sheep to the calcareous aeolian sands (65% $CaCO_3$; pH 8.5) at Robe, South Australia (Fig. 2). Mr. R. Dawson, on whose property the work was being done, pointed out that, in addition to disibili-

Fig. 2. Coastal calcareous aeolian dunes at Robe, South Australia, on which Cu and Co deficiencies lead to 'coast disease' of sheep and on which plants show Cu and Zn responses.

ties he experienced in maintaining animals, his cropping was severely limited by the failure of various crop and pasture plants. Oats and barley produced no grain, while wheat even made scant vegetative growth. Only rye could be grown successfully. Lucerne and all other sown legumes failed completely.

The poor growth of such unrelated species indicated a non-pathogenic disorder and so studies were begun both in the field and in pot culture with various fertilizer treatments. It was established that there was an acute deficiency of Cu for the growth of most species (Riceman & Donald 1938; Piper 1938); subsequently it was shown that there was a less marked but significant deficiency of Zn. With these trace elements, wheat, oats, barley and lucerne could all be grown successfully.

Fig. 3. Oats on Cu-deficient calcareous sand near Kingston, South Australia. Left: superphosphate only, showing failure to produce panicles. Right: superphosphate plus copper sulphate. Each sheaf is from 0.5 m^2. *[From Riceman* et al. *1940]*

It was on this soil type that for the first time a trace element response was recorded by a pasture community. A mixture of *Lolium rigidum* and *Medicago lupulina* was sown in 1937, with and without copper sulphate. At the end of the second season the pasture on the no-Cu section had reverted to *Bromus madritensis* and *Lagurus ovatus*, two inferior Mediterranean annual grasses, whereas the plus-Cu pasture was still largely made up of the sown species (Riceman *et al*. 1940).

Following these field responses, Piper confirmed in water culture the essentiality of Cu for plant growth (demonstrated for tomatoes by Sommer, 1931) and showed that as little as 3 μg Cu/ℓ produced a considerable increase in the growth of oats compared with a Cu-deficient culture (Piper 1942). Fig. 4 shows the striking response by oats to increasing levels of Cu — first a response in vegetative growth, then the production of panicles and grain, and finally a depression in growth at high levels of Cu.

In the following years the emphasis in fertilizer studies in Australia moved strongly towards the attainment of satisfactory growth by pasture legumes in 'problem' areas. One such problem site was the so-called 'ironstone soil' on the property of Mr. Norman Brookman of Meadows in the Adelaide Hills, with whom the Waite Institute conducted cooperative experiments for more than 12 years. Subterranean clover sown with superphosphate on this lateritic podzolic soil was extremely stunted and pale in colour. Mr. Brookman observed, however, that the clover grew well wherever a tree stump had burned; and accordingly he put down an experiment with nil, superphosphate and superphosphate-plus-ash treatments. The

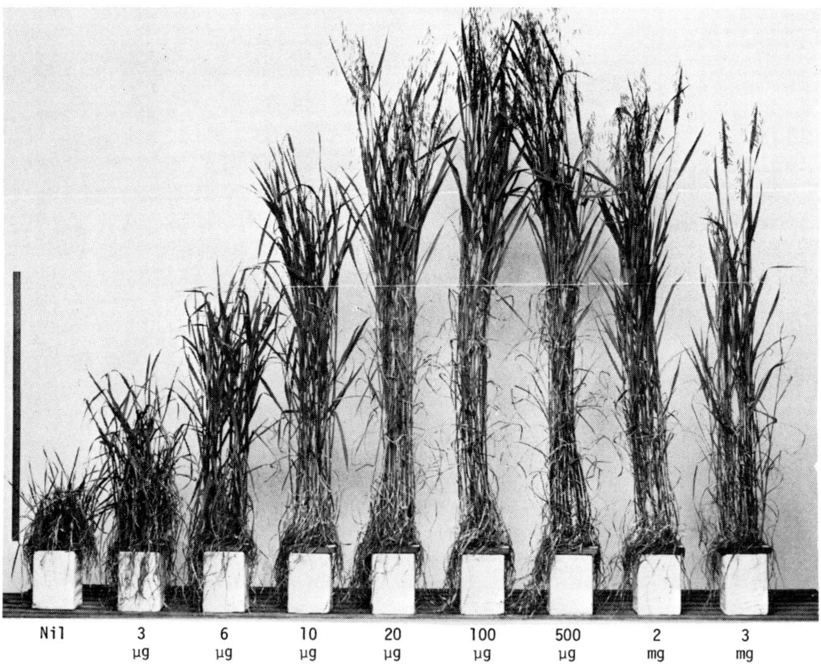

Fig. 4. The influence of Cu on the growth of oats in water culture, Waite Institute, 1939. The concentration of Cu per litre *is shown for each culture (Piper 1942).*

response to the ash was spectacular, with vigorous growth of clover. A.J. Anderson, continuing these studies under the Joint Mineral Investigations, decided to seek an account of the mineral status of the defective clover, and spectrographic analyses by A.C. Oertel showed that all plant samples from untreated areas had low levels of Mo compared to those of the vigorous, ash-treated plants. Molybdenum had been shown in 1939 to be essential for the growth of higher plants (Arnon & Stout 1939), and accordingly Anderson decided to include ammonium molybdate among his fertilizer treatments — and the yield of clovers was increased from 314 to 3340 kg/hectare, a truly remarkable response (Anderson 1942).

The comments on these events 30 years later by Stout, co-discoverer of the essentiality of Mo, are worthy of note (Stout 1972). "Professor Arnon and I wondered," he wrote, "whether Mo might have implications for practical agricultural situations." But after searching in vain for plant symptoms of Mo deficiency in California, "we concluded that the amounts of Mo essential for plant nutrition were so very, very small that nature by itself would not be able to clear any soil environment sufficiently to result in a demonstrable Mo deficiency in the field ... that the essentiality of Mo for the growth of higher plants was probably destined to remain forever as a laboratory curiosity ... until 1942 when A.J. Anderson, in a series of beautifully conducted field experiments

Fig. 5. Characteristic responses by subterranean clover on an N-, P- and Mo-deficient soil. Left: No fertilizer, with dark, P-deficient foliage. Centre: Superphosphate only, with pale N-deficient foliage. Right: Superphosphate + sodium molybdate. On a podzolic soil from the southern tablelands of New South Wales.

on some low producing hill-land near Adelaide, South Australia, showed that Mo was the cause of the difficulty".

Thus by 1942 Mn, Zn, B, Cu and Mo deficiencies had been shown to occur on Australian soils. The history of Fe deficiency in Australia is much more diffuse, with many early but unpublished records of deficiencies in citrus, stone fruit, vines, pineapples, tomatoes and other horticultural crops, mainly on solonized brown soils, the more calcareous red-brown earths and grey soils of heavy texture. Baxter (1957) gave the first published account of Fe deficiency in a wide range of trees and crops on brown solonized soils of the irrigation areas of northern Victoria. He emphasized the difficulty of treatment and even proposed avoidance by planting less susceptible species.

The next deficiency — that of Co — was recorded in 1960, only one year after its essentiality for *Rhizobium* had been established in the United States (Ahmed & Evans 1959). One of us (C.M.D.) was at the University of California, Berkeley, in 1959 and was shown by C.M. Johnson water cultures of lucerne growing normally with either N or inoculation plus Co. Johnson remarked, "Now that cobalt is known to be essential for nitrogen fixation, it's up to Australians to find a plant response in the field — because that's where it will be." Such was our reputation — the hungry country!

At that time the late J.K. Powrie was perplexed by the intractability for pasture growth of the soils of the Bangham Scrub, lying between Bordertown and Naracoorte, South Australia. It was indeed a hungry soil, "a grey siliceous sand (pH 5.8) underlain at 60 cm by a yellow sandy clay, capped by a layer of ferruginous concretions ... deficient in P, Zn, Cu and Mo". Immediately news of the essentiality of Co came to hand, Powrie decided to test it on this recalcitrant soil; he gained a 30% increase in the yield of subterranean clover over that of plots receiving P, S, Ca, Zn, Cu and Mo, the first plant response to Co in the field (Powrie 1960). A few years later a similar response by subterranean clover to Co was recorded on a coarse siliceous sand at Muchea, north of Perth, Western Australia (Ozanne *et al.* 1963).

These then are the records of the early responses to each of the trace elements in Australia. In the following pages we seek to assess their relationship to our soils and to the productivity of Australian agriculture.

AUSTRALIAN SOILS AND THEIR TRACE ELEMENT DEFICIENCIES

Australia has vast areas of soils of great antiquity, soils which have undergone such long-continued and deep weathering and leaching as to become impoverished of all nutrients. Acute deficiencies of P and N are almost universal, together with marked deficiencies of many other nutrients.

There are several factors which have contributed to this widespread infertility of our soils, in particular the geomorphic history of the continent (Stephens & Donald 1958). Above all, Australia has long been a vast plain with limited uplift; in an area of over 7.7 million km^2 the highest point is a mere 2230 m, and the Great Dividing Range is the only major topographic feature. As a consequence, dissection has been slow and shallow. Great areas of Tertiary landscape have been preserved, sometimes as substantial areas of peneplain, sometimes as dissected remnants. The extensive lateritic podzolic soils of south-west Western Australia, the Adelaide Hills and parts of the Great Divide all lie on ancient Tertiary surfaces, as do also the lateritic red earths of the monsoonal tropics and subtropics.

The lateritic soils, commonly truncated to the lateritic crust so that ironstone gravel is characteristic of the surface horizon, have usually been podzolized under current climatic conditions; they at first proved very unattractive for development but are now extensively sown to leguminous pastures, commonly with dependence on P, Mo, Cu and Zn. The lateritic red earths have shown responses to Mo, B and Fe by vegetable crops in Queensland, but the trace element status of most of these soils in the north is not known because of the absence of cropping or sown pastures.

Even where the dissection of the Tertiary landscape has been more extensive and deeper, and has generated new surfaces, it has not necessarily exposed basement rocks for juvenile soil formation. On the contrary, the slow dissection of the gentle landscape has been into zones of deep weathering and leaching, developed in advance of the exposure of the new surface. Thus soils formed in Pleistocene times were commonly on strongly weathered lower horizons of earlier profiles, and have the nutrient status of older soils. They include extensive, trace element-deficient podzolic and solodic soils, such as the Riverine Plain of north-east Victoria and its associated hill country, and the Salmon Gums region of Western Australia. Thorp (1949) has remarked: "New soils from old are, in this instance, poor indeed."

The detritus removed by dissection and deposited elsewhere was itself heavily impoverished. The siliceous sands of coastal and subcoastal plains formed in this way are again commonly deficient in trace elements, responses to Cu and Zn being widely prevalent. The plains of the upper South-East of South Australia and of the Esperance region of Western

Australia are characteristic of these trace element-deficient siliceous plains.

The ancient soils on Tertiary (mainly Pliocene) and Pleistocene surfaces thus embrace the lateritic podzolic soils, the lateritic red earths and most of the podzolic and solodic soils, all of extremely low fertility; they constitute perhaps a third of all Australian soils and two-thirds of the soils of our better rainfall agricultural regions, where most of our crop production and pasture development lies. They are all acutely deficient in P and N and often also in S and K,

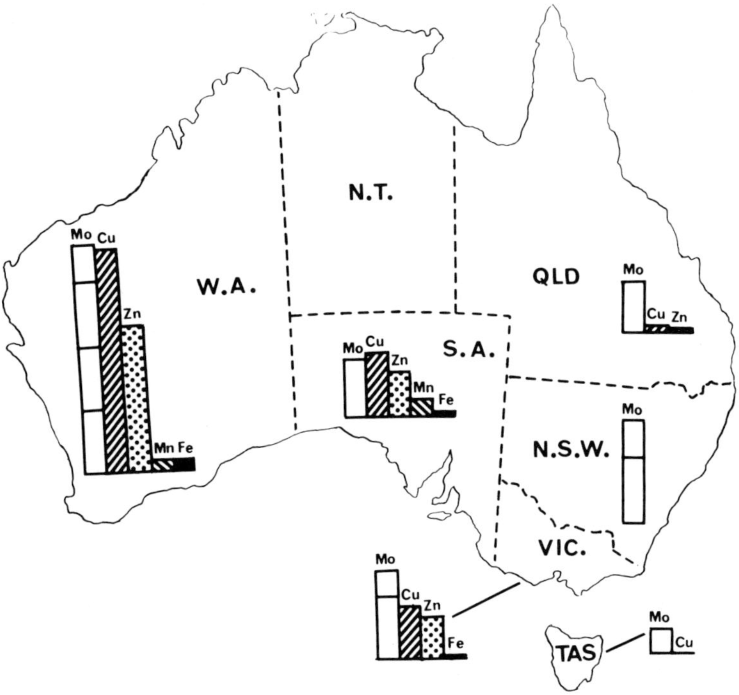

Fig. 6. The areas to which trace elements were applied annually in each State during the 5 years 1967/68 to 1971/72. Each full subdivision of the Mo columns is 100,000 hectares and the same scale applies to the other elements. The estimated annual areas for the whole of Australia, in 000's hectares, were Mo, 874; Cu, 556; Zn, 374; Mn, 46; Fe, 34 and B, 1. State areas of less than 5000 hectares are not shown in the histograms. The areas are approximations only, based on the purchases of trace element substances by fertilizer firms during this 5-year period and the estimated mean rate of application of the element.

as well as the trace elements.

The heavy leaching of our soils in the late Tertiary period doubtless washed great quantities of nutrients into the sea (Wild 1961). Profiles of a number of Australian soils lying on granite parent material show values for soil P much lower than those of the rocks on which they lie and it is predictable that the same is true of the trace elements. Further, since Pliocene times there have been several periods of heavy leaching in Australia associated with climatic fluctuations (Jessup & Wright 1971). These have undoubtedly contributed exponentially to nutrient losses from ancient profiles (Conway 1945), so that the nutrient depletion of the soil mantle is often far more severe than might seem consistent with the present-day climate. Nevertheless, continued loss under present-day conditions is a significant process of high rainfall areas in some soils — in those developed on coarse-textured sedimentary rock, in the strongly aggregated krasnozems and in aeolian and fluviatile sands, all of which show great porosity and associated nutrient depletion. Leaching may have been further intensified in many soils, notably solodized solonetz and solonized brown (mallee) soils, by the continuous solonizing influence of cyclic salt characteristic of the Australian rainfall, but this process is not well understood.

Beadle (1962) has emphasized a further factor in fertility loss, that many of the sedimentary rocks from which Australian soils are derived underwent sorting and leaching by wind and water under sandsheet conditions during the several arid periods of the Pleistocene era. This led to a loss of mineral-rich particles before the coarse fractions were reconsolidated into rock, which was again weathered to give acutely mineral-deficient soils. He also points to the progressive loss of nutrients which can occur when plant parts are blown or washed away or when surface soil, enriched by plant growth, is eroded in a sandsheet environment.

A further major geomorphic feature, due to Australia's low latitude (11°-43°S) and lack of any great mountain mass, is the limited extent of Pleistocene glaciation, confined to the south-east corner of the mainland and Tasmania. There was little exposure of new surfaces and no appreciable deposition of morainic material or loess, a situation in marked contrast to that in North America and Europe. Additionally, youthful vulcanism is confined to a relatively narrow zone extending along the eastern seaboard and for a short distance inland. These occurrences include volcanic ash and basalt of Tertiary to late-Recent age; all volcanoes are now extinct.

Thus the low relief and lack of uplift within the contin-

TABLE 1

Trace element responses on Australian soils by various crop, pasture and forest species

This table summarizes the responses to trace elements recorded on Australian soils. Some warrant correction, some do not, but they give a broad picture of the incidence of deficiencies. A 'nil' record does not necessarily mean that the crop in question has shown no response on a particular soil, since it may never have been grown on that soil because of climatic or soil limitations other than trace element deficiency, or may never have been specifically tested.

Soil^φ	Pasture* Legumes	Pasture Grasses	Grain† Legumes	Cereals	Potatoes	Linseed or Flax	Tobacco	Turnips Swedes Beets	Stone & Pome Fruits	Citrus	Vines	Pine-apples	Crucifers and Cucurbits	Tomatoes	Other Vege-tables	Pines
Alluvial soils	Cu Mo Mn			Mo						Zn B					B	
Calcareous aeolian sands (Fe, all spp.)	Cu Zn Mn		Cu Mn	Cu Zn Mn						Cu Zn Mn			Mo		Mn B	
Acid swamp soils	B Fe	Cu Fe	Cu	Cu Fe	Cu			Cu					Mo	Cu	Cu Mn B	
Podzols & ground-water podzols	Cu Zn B Mo Fe Co		Co Mn	Zn Fe Mo					Cu Zn Mn B				Mo		B	Zn
Lateritic podzolic soils	Co Zn Mo Mn B	Zn	Cu Mo Mn	Cu Zn Mo Mn		Zn			Cu Zn Mn	Cu			Mo			Zn
Grey-brown, brown, red, yellow & meadow (gley) podzolic soils ∇	Cu Zn Mo B Mn Co	Zn	Cu Mo B Mn	Cu Mn Mo		Mo	Cu Mo	B	Cu Zn Mn B Fe	Cu Zn Mn Fe	Cu B Fe		Mo B	B Fe	Mo Mn B	Zn Cu
Krasnozems	Cu Zn Mo B		Mo	Mo				B	B			Fe				
Lateritic red earth								B				Fe	Mo B			
Terra rossa	Cu Zn Mn			Mn							Mn					
Black earths	Mo		Zn	Zn		Zn		B	Mn Fe				Mo		Mn B	
Rendzina & ground-water rendzina	Cu Zn Mn	Zn Mn		Mn												
Fen soils (Fe, all spp.)	Cu Zn Mn	Zn	Cu Mn	Cu	Cu Mn			Cu Mn						Cu		
Solodized solonetz and solods	Cu Zn B Mo Mn Co			Cu Zn Mn												Zn
Solonized brown soils	Zn			Mn Zn					Cu Zn Fe	Cu Zn Fe Mn	Fe					
Red-brown earths	Mo			Mo				B	Zn Fe	Zn Mn Fe	Zn		Mo		B	
Grey calcareous soils	Zn Mn		Mn	Mn		Zn			Mn							
Grey & brown soils of heavy texture	Zn			Zn Cu		Zn			Fe	Zn	Zn					

φBased on C.G. Stephens, "A Manual of Australian Soils", 1962

*Includes lucerne

†Includes peas, lupins, soybeans, french beans

∇Includes yellow earths

ent, the slow and shallow dissection of Tertiary surfaces, the prolonged leaching of ancient profiles and formation of 'new soils from old', the loss of finer fractions due to wind and water sorting, and the limited development of new surfaces or new materials through glaciation or vulcanism, all contributed to the poverty of so many of our soils.

The soil groups so far discussed have pH values ranging from 4.5 to 6.5, but there are some groups of high lime status and pH, notable the coastal calcareous aeolian sands, with up to 80% $CaCO_3$ (pH 8.5), the calcareous solonized brown soils (mallee soils) of uncertain origin, occurring in semi-arid southern areas, and various shallow soils over limestone. The high lime status and high pH of these soils limit the availability of several of the trace elements — for example, Mn on rendzina and calcareous soils.

It may be asked to what extent these deficiencies are sufficiently consistent within any soil group to permit confident predictions from one location to another. Some soil groups show a highly consistent response pattern but, conversely deficiencies occur across apparently unrelated soil types (Table 1). Wherever calcareous aeolian sands are found, it is predictable that they will show deficiencies of Cu and (for animals) Co, with a marginal Zn status. It is likely also that any lateritic podzolic soils will provide insufficient Mo for N fixation by legumes and that acid swamp soils will show Cu response. Another widespread deficiency within soil groups is that of Zn in grey and brown soils of heavy texture and in black earths; conversely, Cu deficiency has not been recorded in the red-brown earths. In more general terms, one or more deficiencies may be found in most podzolic soils. There is also the possibility of Mo deficiency in any strongly acid soil and of Mn deficiency in any alkaline soil. Yet Mn response has been recorded also on acid soils, such as the lateritic podzolic soils of Kangaroo Island, South Australia, presumably due to an absolute deficiency of Mn (Carter & Wigg 1963).

While the soil group has limitations in indicating probable fertilizer responses, common genesis may have a notable influence on mineral deficiency. In the South-East of South Australia, there are calcareous sand dunes, groundwater rendzinas, ridges of deep sandy podzolic soils, solodized solonetz and terra rossa soils (Blackburn 1964) all sharing a more or less common set of deficiencies involving P, Cu, Zn, Mn and Co (Tiver 1955). All these soils have been formed during the gradual seaward retreat of the shoreline of a Pliocene-Pleistocene coast and all are derived, by different processes of varying durations, from calcareous dunes or from underlying and closely related Miocene limestone (Stephens 1951). Only their common genetic factors — parent material, leaching and shallow groundwater movement — explain the like patterns of trace element deficiencies on soils of such different texture, depth and natural fertility.

Conversely, trace element responses within Great Soil Groups may show strong dependence on parent material. Thus among the podzolic soils of West Gippsland, the grey loams on Silurian mudstones and sandstones show strong Mo response, whereas the sandy loams on granite show none (Hosking 1957).

It will be seen that, while we have gained some appreciation of the factors governing the mineral status of Australian soils, our understanding is still far from complete. It especially remains to gain a fuller picture of the chemistry and availability of trace elements and, as discussed by other contributors, to develop quantitative tests of soils and plants as reliable indices of the occurrence and intensity of mineral deficiencies.

TRACE ELEMENT RESPONSES IN CROPPING AND FORESTRY

Although most Australian cereals are still grown without trace elements, commercial responses to Cu, Zn, Mo and Mn have been recorded over substantial sectors of our cereal areas. The most important deficiencies (see Table 2 for references) are as follows:

- the podzolic, lateritic podzolic, yellow earths and calcareous sands of the south-west of Western Australia : Cu and Zn responses by wheat, oats and barley;
- included within these Western Australian soils are certain sandy yellow earths and grey-brown lateritic podzolic soils : Mo response by wheat, oats and barley in addition to the responses to Cu and Zn;
- the black earths of Queensland and New South Wales : Zn response by wheat, barley, oats, maize and sorghum;
- the podzolic soils of coastal New South Wales : Mo response by maize;
- the grey soils of the Wimmera district of Victoria : Zn response by wheat;
- the siliceous sands, calcareous sands, solodized solonetz, solodic and lateritic podzolic soils of Eyre Peninsula, Murray Mallee and the South-East of South Australia : Cu response by wheat, oats and barley;
- the calcareous sands of Yorke Peninsula and Eyre Peninsula, South Australia : Mn response by barley, oats and wheat.

The red-brown earths are the only important soil group for

cereal production in Australia without any present use of trace elements. In some soils, especially in parts of Western Australia and South Australia, economic production is not possible without supplying these micronutrients, which contribute both directly to cereal growth and also to the vigorous development of ley legumes and the associated input of N. Other field crops as well as cereals are affected on these soils; for example, the responsive crops on the black earths include cotton, linseed, soybeans and sunflowers (all to Zn), and on the lateritic podzolic soils of Western Australia, lupins (Cu, Mn) and flax (Zn). Other important crop responses are those to Cu by potatoes on swamp and fen soils and to B by root crops on podzolic soils and lateritic red earths.

TABLE 2

Responses to trace elements by cereals in Australia

Place	Soil	Cereal	Yield (kg grain/hectare)			Reference
Stokes, Eyre Peninsula, S.Aust.	Solodic soils (Mean of 5 expts)	Wheat		-Cu 2200	+Cu 2908	King (1974)
Yorke Peninsula, S.Aust.	Calcareous soils (Mean of 5 expts)	Barley		-Mn 855	+Mn 1715	Reuter *et al.* (1973)
Central wheat belt, W.Aust.	Sandplain soils (sandy and gravelly lateritic soils) (Mean of 15 expts)	Wheat		-Cu	+Cu	Gartrell (1969)
			-Zn	618	934	
			+Zn	470	994	
Northern Esperance Plain, W.Aust.	Solonized soil, with ironstone gravel in the profile	Wheat		-Cu 968	+Cu 1599	Gartrell (1969)
Nhill, Wimmera District, Vic.	Grey and brown soils of heavy texture	Wheat		-Zn 1955	+Zn 2587	Millikan (1938)
Gutha and Bodallin, W.Aust.	Sandy and lateritic podzolic soils (Mean of 4 expts)	Wheat (3) Oats (1)		-Mo 824	+Mo 970	Gartrell (1966)
Taree, coastal N.S.W.	Podzolic soil	Maize*		-Mo 2.2 (kg of tops/m row)	+Mo 2.5	Noonan (1953)
Darling Downs, Qld.	Black earth	Maize		-Zn 1700	+Zn 2998	Duncan (1967)

*No grain data available

Despite these well established responses, there is concern that many growers do not yet make adequate use of trace elements in crop production. Only about 10% of the wheat of the Wimmera region of Victoria receives Zn each year, though application to each second or third crop is recommended on these heavy soils (Jessop & Tuohy 1973). Similarly in South Australia there is a great deal of uncorrected Cu and Mn

Fig. 7. The upper picture shows the stunted growth of vines on deep sandy podzolic soil at Gin Gin, Western Australia. In the lower picture the vines have received soil dressings and sprays containing Cu compounds. [Photos: Department of Agriculture, Western Australia]

deficiency of cereals (King & Alston, this volume).

Among the horticultural crops, numerous responses have been recorded to individual elements. Many deficiencies are corrected only as they appear or re-appear, but regular treatment is adopted in particular districts on an annual or longer term basis, usually by soil application or spraying; affected fruit crops include citrus (Zn, Cu, B, Mn), apples (B, Cu, Zn), grapes (Zn, Cu), stonefruits (Zn, B) and pineapples (Fe, Zn, B, Cu). Among the vegetable crops the most notable responses are by crucifers and cucurbits to Mo; some of these crops were formerly of marginal worth on various acid soils (acid swamps, podzolic, lateritic podzolic, lateritic red earths) because only partial control of such severe disorders as whiptail of crucifers was attainable with organic fertilizers. Boron deficiency occurs in cucurbits, beetroot and crucifers, while various other deficiencies are shown in Table 1.

One of the earliest trace element responses in Australia was to Zn in pine plantations (Kessell & Stoate 1938) on the lateritic podzolic soils of the Darling Range and the podzolic sands of the coastal plain of Western Australia, where new plantations were commonly "a spectacular failure" (Stoate 1950). Experiments with 30 elements in 1936 showed the pines to be seriously deficient in P and Zn; trees treated with Zn nails when 9 years old in 1937 had attained a height of 6.5 m by 1948, compared to 3.5 m for the control pines. Phosphorus and zinc have now become almost standard fertilizers for pines in Western Australia, South Australia and parts of Victoria, using a single aerial spray of zinc sulphate at 3 years or a soil application at planting. Copper (Ruiter 1969), manganese, boron, zinc, iron and molybdenum (M. Raupach and C.G. Stephens, personal communication) have each also given responses, though often slight or complex; one South Australian firm is now using Cu on second rotation sites. Meanwhile, more precise foliar analysis is urgently needed (Raupach, this volume), especially in relation to interacting responses between elements. As the major elements, including N and K as well as P, become more widely used in forestry, so it may be that other trace element deficiencies will become commercially significant.

THE ROLE OF TRACE ELEMENTS IN LAND DEVELOPMENT

Australia has made use of phosphatic fertilizers and pasture legumes in raising the fertility of poor soils to a degree unparalleled elsewhere in the world. Through the input of P and N by fertilized leguminous pastures, more than 20 million hectares of infertile soils have been made productive. Each kg of P results in the addition to the soil by pasture legumes of some 8 kg N (e.g. Donald & Williams 1954), and thus we can estimate from our fertilizer usage the input of N by rhozobial fixation; it is currently about two million tonnes N per annum, added as organic matter, a huge contribution. Increased pasture productivity and the associated fertility lift have been the principal factors both in the postwar increase in stock numbers and in raising Australian wheat yields from 860 kg/hectare in 1941-50 to 1230 kg/hectare in 1961-70 (Donald 1965).

It is within this context that trace elements have played such a significant role in the postwar II development of 'new country', of great areas of infertile virgin land.

There lay in South Australia a tract of several million hectares of undeveloped mallee broombush and mallee heath on solodized solonetz soils and deep sands of low mineral status. This region was known as 'The 90-Mile Desert', not because of aridity (most of it receives over 430 mm per annum), but because of the lack of development and the sparsity of settlement. The carrying capacity of the scrub was rated at 1 sheep to 5 hectares but, except in more favoured outliers, clearing and the use of clover and superphosphate gave no economic increase. And so the region remained uncleared, unpopulated.

In 1945, Riceman demonstrated that the failure of subterranean clover on the Laffer sand, a solonetzic soil lying south of Keith on the 90-Mile Plain (a more appropriate name), was due to a triple deficiency of P, Cu and Zn, with marked responses to each of the trace elements (Table 3).

TABLE 3

Responses to Cu and Zn on Laffer sand on the 90-Mile Plain (Riceman 1948)

	Sub-Clover (kg/hectare)	Lucerne (kg/hectare)
No fertilizer	187	62
Superphosphate	625	937
+ Cu	875	2000
+ Cu + Zn	3250	2000

There was speedy realization that trace elements might provide the key to the development of this infertile but reasonably watered area and an enterprising assurance company (the AMP Society) launched a huge scheme involving the acquisition of some 320,000 hectares and its development into sheep holdings. The vegetation lent itself to new and sophisticated methods of clearing, using a 250 m loop of heavy anchor chain drawn by two 105 kW tractors, 'scrub-bashing' at the rate of 40 hectares/hr. When dry the scrub was burned, several thousand hectares at a time, and ploughed by the 'big ploughs', with their eight 750 mm stump-jump discs. A pasture mixture based on sub-clover was broadcast with 200 kg/hectare of superphosphate and 7 kg/hectare each of copper and zinc sulphates, and then covered with the anchor chain. Here was a new phase in Australian agriculture made possible by the final input — trace elements.

Meanwhile, the potential of Anderson's finding of Mo deficiency in the Adelaide Hills and of Mo response on an 'ironstone' soil at Cressy, Tasmania, in the following year (Fricke 1943,1944) lay unrealized. But in the mid-forties Anderson (1948) showed that Mo deficiency was widespread on the southern tablelands of New South Wales, part of the peneplain of the Great Divide. Podzolic soils of sedimentary, granitic and basaltic origin all showed responses to Mo. Within a further few years, remarkable results were also recorded on the hill country of the central Dividing Range of Victoria. This area, much of it long cleared, carried only low-yielding native grasses; clovers sown with superphosphate gave poor growth or totally failed. Here, on podzolic and solodic soils on Ordovician and Silurian shales, sandstones and mudstones (pH 4.9-6.2), Newman (1955) recorded increases in the yield of subterranean clover from about 1000 kg/hectare with superphosphate to over 5000 kg/hectare with superphosphate plus 140 g sodium molybdate per hectare (Fig. 8). Important responses were also recorded on podzolic soils on granite, on heavy grey soils on basalt, and on krasnozems. It is estimated that the correction of Mo deficiency is necessary for the development of pastures on some 4 million hectares of land in Victoria — and much of it has already received its few grams of Mo and shown an increase in carrying capacity from 1 to 8 sheep per hectare! (Newman 1962). Little wonder that one atom of Mo can, as Stout has calculated (1972), bring together 100×10^6 carbon atoms in carbohydrates and proteins!

From 1950 onwards land development proceeded apace; chain-

Fig. 8. A strong Mo response by subterranean clover on a podzolized soil on granite at Carlsruhe, 80 km north-west of Melbourne, Victoria. Molybdenum applied 1956, photographed in 1957. Left of white peg: Superphosphate only. Right: Superphosphate + 140 g/hectare sodium molybdate. [*Photo: Department of Agriculture, Victoria*]

ing, sometimes supplemented by a 1.5 m steel ball, logging by a pair of tractors hitched to each end of a 12 m log and bull-dozing enabled any class of heath, scrub or forest to be cleared. Some development depended only on legumes and super-phosphate, much of it needed trace elements (Tiver 1958). Molybdenum, copper and zinc were the principal nutrients, with a lesser role for Mn and, rarely, for B. And always the input of N by legumes was critical, whether made possible by P, S, Cu, Zn, Mn or B in clover metabolism and the provision of energy to the *Rhizobium*, or whether depending on Mo for the N fixation itself.

The greatest areas of 'new land' were in Western Australia, but there were huge developments also in South Australia (Eyre Peninsula, Fleurieu Peninsula, Kangaroo Island and the South-East), Victoria (Central Highlands, Gippsland, the Otway Ranges, Portland), New South Wales (mainly Mo on the Great Divide) and Tasmania. Queensland was the state with the least 'pasture improvement' because the development of suitable legumes for tropical and subtropical areas was still in progress, but research had shown that leguminous pastures could be developed on certain humic gleys of the coastal lowlands (the Wallum) if Cu, Zn, Mo and B (as well as P, K, S and Ca!) were applied.

The south-west of Western Australia warrants particular mention. This region of 25 million hectares, with an inland boundary from Geraldton to Esperance, has a Mediterranean climate and a rainfall ranging from 1000 mm down to 250 mm at the inland limit of cereal cultivation and sown pastures; it produces most of the crops and livestock of the State. At the time of European settlement it carried forest, woodland and heath on extensive sandy soils of very low fertility. The native vegetation was of limited value for livestock, a situation aggravated by widespread poison plants. The region scarcely seemed to merit the appraisal published in 1718 in Amsterdam of its southern part, discovered by Pieter Nuyts in 1627: "Nuyts Land, being in the fifth climate between 34° and 36° of latitude, ought to be like other countries so situated, one of the most habitable, most rich and most fertile parts of the world" (Flinders 1814). On the contrary the progress of the early settlement, founded in such unpromising terrain in 1829, excites both respect and wonder.

The first work on trace elements in Western Australia was on the use of Zn on citrus and on pines, discussed earlier. Then, in a notable score of papers, Teakle (1939-1945) and Dunne (1946-1950) and their colleagues reported the widespread occurrence of Cu, Zn and Mn deficiencies variously affecting both clover and crops (wheat, oats, potatoes, vines). By 1958

Fig. 9. Clearing and developing new land in the 1950's and 1960's, based on pasture legumes, superphosphate and trace elements.

(a) Clearing forest in the south-west of Western Australia with ball and chain. Two tractors pulling the 8-tonne chain and 5-tonne armour-plate steel ball can fell 25-40 hectares per day, preparatory to burning and ploughing a year later. [Photo: Aust. News & Information Bureau]

(b) Virgin heath country on the 90-Mile Plain near Keith, South Australia. This type of vegetation was cleared with the chain alone. [Photo: CSIRO]

(c) The same country as in (b), after chaining, burning, ploughing and seeding to Phalaris tuberosa *and* Trifolium subterraneum, *with superphosphate and copper and zinc sulphates. [Photo: CSIRO]*

a general recommendation was issued for the use of Cu and Zn on cereals and pastures on all sandy and gravelly (lateritic) soils of the entire south-west (Toms 1958): "In most cases it will result in marked improvement in crop yields and pasture growth, and even where no spectacular plant growth responses are obtained, there will usually be improvements in animal health." The recommended initial rate per hectare was 200 kg of superphosphate containing 1.5 kg each of Cu and Zn, with no re-application of the trace elements for at least 5 years except on acutely Cu-deficient limestone soils.

These findings meant not only that crop yields could be substantially increased, but that extensive 'light lands' which previously defeated the settler could now be developed for pasture and crop production, including both uncleared land on thousands of existing farms and large areas of virgin land, especially of scrubplain. Some of the scrubplain, sandy soils on laterite or clays, carried low heath which could be ploughed even without burning, while some the land cleared from forest involved heavy clearing costs. By 1960 no less than half a million hectares were being developed each year.

The use of Mo in Western Australia was strangely delayed; trials with Mo on subclover in 1944 were almost wholly negative (Teakle 1944), possibly because of the very dry season and the lack at that time of well adapted subclover cultivars for many districts. Despite some localized responses (Dunne 1950; Fitzpatrick 1957), it was not until the early sixties, 20 years after Mo responses had been recorded in eastern Australia, that Mo deficiency of clover was shown to be widespread on many soil types in Western Australia; within a few years, responses were recorded also in wheat and oats (Gartrell 1966) and further trials have led to recommendations (Gartrell & Glencross 1968) that Mo be applied, in addition to Cu and Zn, on all sandy and gravelly soils in large sectors of the south-west province (Fig. 10). Within a few years, over 70% of trace element mixtures in Western Australia contained Mo (Williams & Andrew 1970).

The whole region, except for a narrow coastal fringe, lies on the Western Australian granite shield. The triple deficiencies of Cu, Zn and Mo are mainly in acid sandy or gravelly soils from laterite, over granitic gneiss; the western coastal soils, deficient in Cu and Zn but not in Mo, are aeolian calcareous or leached siliceous sands overlying limestone, neutral to alkaline and sometimes Mn deficient; the southern coastal soils are mainly sand over clay on Jurassic sediments. The western wheat belt soils (the central zone in Fig. 10, not usually deficient in Mo) are grey and brown loams and sands which, in contrast to the more acid soils to the east and west, are generally of neutral reaction.

It is estimated (J.W. Gartrell, personal communication) that no less than 8 million hectares, constituting a third of the entire south-west of Western Australia, has now received trace elements, usually Cu and Zn, but also including Mo over vast areas and Mn on a few soils. Here undoubtedly lies the most extensive trace element-deficient province of the world — now becoming "a rich, habitable and fertile land" and a notable tribute to the Western Australian Department of Agriculture.

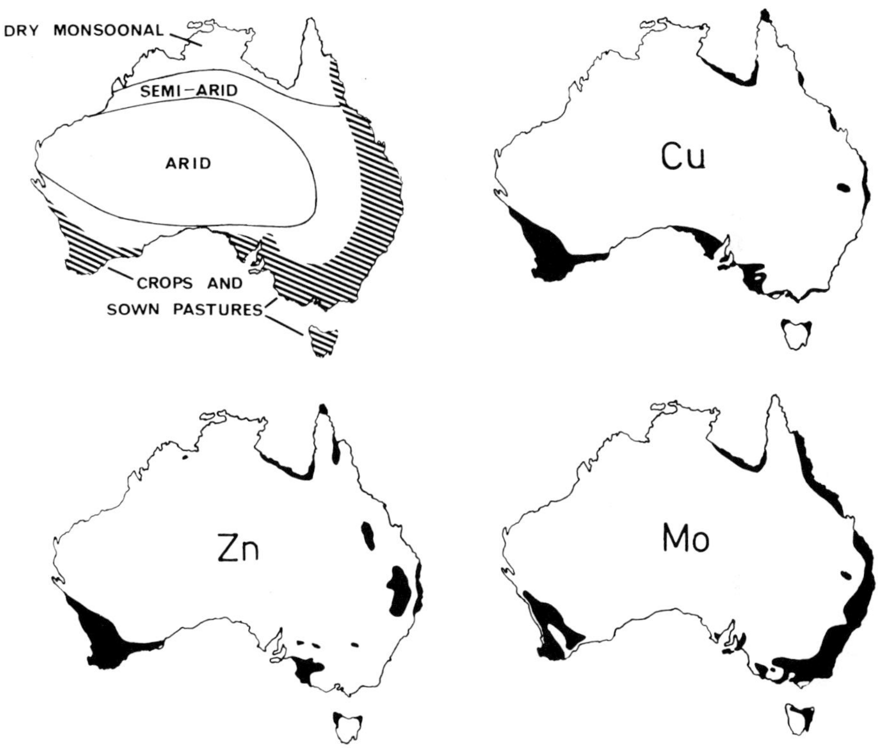

Fig. 10. The distribution of Cu, Zn and Mo deficiencies in Australia. Widespread responses to each of the elements are recorded within the black areas. A comparison with the upper left map serves to indicate the considerable proportion of Australia's better watered lands which suffer one or more trace element deficiencies. The infrequency of recorded deficiencies in the dry monsoon zone is almost certainly due to the low incidence of cropping and pasture development, and thus of agronomic trials in the zone, rather than to the absence of deficiencies.

No precise account is possible of the area of 'new' land developed in Australia since 1945 from virgin forest, heath or scrub, or converted from low-carrying native grassland to leguminous pasture through the catalysis of trace elements, but it may well exceed 15 million hectares. Trace elements have been a critical input in Australia's land development and economic progress.

MOLYBDENUM IN AUSTRALIAN AGRICULTURE

Molybdenum has become of great importance in Australian pasture and land development and for field and horticultural crops; it is therefore of interest to follow our progressive understanding of its role as a fertilizer.

Early studies showed that the deficiency of Mo on the lateritic podzolic soils of the Adelaide Hills inhibited N fixation by the *Rhizobium* of the legume nodule, so that the response to Mo application was virtually a response to a non-limiting supply of N (Anderson & Thomas 1946). On the other hand, the level of Mo was sufficient for nitrate reduction in the plants themselves, so that there was no response to Mo by non-legumes on these soils. The whole emphasis in the use of Mo in pasture and land development in all states has been on the attainment of the satisfactory growth of clovers and other pasture legumes.

Nevertheless early responses were also recorded by non-legumes, including crucifers (Waring *et al.* 1947), cucurbits (Wilson 1948*a*), lettuce (Wilson 1948*b*), flax (Millikan 1948) and oats, in which Mo cured the Blue Chaff Disease of Tasmania and gave the first recorded response by a cereal (Fricke 1947). In terms of plant families, known Australian deficiencies of Mo affect Leguminosae (for N fixation), Cruciferae, Compositae, Cucurbitaceae, Solanaceae, Linaceae, Gramineae and Leguminosae (for the host plant). On the basis of the study by Johnson *et al.* (1952), these last two groups indicate an acute soil deficiency.

Piper (1940), following Arnon & Stout's evidence on tomatoes (1939), showed in water culture the essentiality of Mo for oats, with little effect of low Mo on the biomass, but with a reduction of 60% in grain yield. Nevertheless Mo deficiency of cereals in the field was of limited importance in Australia for several decades, being confined to oats in Tasmania (Fricke 1947) and maize in parts of coastal New South Wales (Noonan 1953). It was not known to limit yield in any of our main cereal areas until, as noted earlier, Gartrell reported substantial responses by wheat and oats in Western

Australia. Molybdenum-treated wheat was deeper green in colour than the untreated crop and matured 7-10 days earlier (Gartrell 1966).

It now seems clear that we must suspect the possibility of Mo deficiency for both legumes and non-legumes on all our acid soils. Furthermore, marginal Mo deficiency may be compounded when N is used as fertilizer in cereal production (Freney & Lipsett 1965). Typical results are shown in Table 4, where N gives a response only in the presence of applied Mo. These interactions are due to the damaging effect of nitrate accumulation in the Mo-deficient plant and the role of Mo as a component of nitrate reductase (Evans & Hall 1955).

TABLE 4

Interaction of N (75 kg/hectare) and Mo (0.17 kg/hectare) on wheat on an alluvial soil at Canberra (kg/hectare grain) — Lipsett & Simpson (1971)

	No N	Plus N
No Mo	930	851
Plus Mo	896	1097

The depression in the grain yield of wheat commonly recorded with N application in Australia has been ascribed to the premature depletion of the water supply by the fertilized crop and indeed there is strong evidence that this is the principal factor (Fischer & Kohn 1966). Nevertheless these workers have written that "there is some evidence that high nitrogen has special effects on the crop, independent of the effect on dry matter production", though they did not implicate Mo in their speculations at that time. It is possible that the poorer earlier responses by wheat to Mo in Western Australia may have been due to the lower N level of the soils prior to the widespread use of legumes in pasture leys, and there are also indications that Mo may have a role, when N levels are high, even in our moderately fertile red-brown earths, the most important wheat soils of the eastern states (Piper & de Vries 1964; Freney & Lipsett 1965).

In 1943 Stephens & Oertel, working on an acid, Mo-deficient soil in Tasmania, showed that the application of lime led to the release of plant-available Mo, so that at pH 8, white clover grew equally well with or without Mo. Thus clover on an acid soil, deficient in P, N and Mo for rhizobial activity, will grow well if fertilized with P together with any one of lime or N or Mo (Anderson & Oertel 1946). But of these three fertilizers, Mo is by far the least costly.

All the responses to Mo in South Australia, New South Wales and Victoria have been on acid soils, with low availability of the soil Mo. This is still the prognosis — that Mo deficiency is unlikely on alkaline soils. But in Western Australia there are at least two recorde of Mo response on soils of high pH: a calcareous sand of pH 8.2, and a loam containing 70% calcium carbonate and of pH 8.6 (Dunne & Jones 1948). There are also occasional records of Mo response on black earths, generally of neutral to alkaline reaction. In addition, some acid soils do not respond to lime to the degree attainable through Mo application (Newman 1955). It follows that the resources of Mo available for liberation through the raising of pH vary considerably. Commercially it is much cheaper and easier to apply Mo than lime, but future experience may show variation in residual values according to the differential capacity of the soil to 'fix' native or applied Mo.

The full requirement by clover for Mo for N fixation (a relatively high level) on an Mo-deficient soil can be met by pre-soaking the seed in a 0.1% solution of sodium molybdate (Donald & Spencer 1951). But again this has not proved a worthwhile procedure in pasture development because deficiencies can so easily be met with molybdenized superphosphate. It was therefore unexpected that the Mo status of seed became the basis of control of Mo deficiency in maize. Weir & Hudson (1966) had noted the erratic incidence of Mo deficiency in maize crops in coastal New South Wales and postulated that the Mo status of the seed might be a controlling factor. Tests showed that there were no Mo-deficiency symptoms in seedlings when the Mo of the grain was > 0.08 ppm but that severe symptoms were recorded when the Mo was < 0.02 ppm. They further showed that when Mo is applied as a fertilizer to the female parent, the hybrid seed is of adequate Mo status to avoid deficiency in the next generation. Accordingly all registered growers of certified hybrid maize in New South Wales are now required to maintain an adequate Mo level in the grain they sell. A general raising of Mo level in much of the Western Australian wheat grain is presumably also to be expected.

The contribution by Mo to Australian agriculture is thus very considerable, with its role in legume growth and land development, cereal yield and horticultural production. And it seems likely to increase substantially further.

THE FUTURE OF TRACE ELEMENTS IN AUSTRALIA

The definition of the geographic and soil boundaries of trace element deficiencies in Australia is far from complete.

It will involve not only new records of occurrences, such as the recently announced deficiencies of Cu and Zn in the cattle country of Yorke Peninsula, Queensland (Isbell & Gillman 1973; Jones 1973), but also detailed studies of the relationship of soil type to nutrient deficiency within particular districts (King & Alston, this volume). It may also be that the recent further records of Co responses by legumes [by lucerne on a podzol and by subterranean clover on a solodized solonetz soil (Victorian Department of Agriculture) and by lupins on a podzol (Western Australian Department of Agriculture)] are a prelude to Co becoming a commercial fertilizer for plant production as well as for animal health.

In addition to this definition of deficiencies as related to soils, we may expect further findings in regard to particular species and agricultural practices. The use of Mo to raise the Mo status of maize seed, the recent identification of 'split seed' of sweet lupins as Mn deficiency (Gartrell 1974) and the possible use of Mo in enhancing N responses by cereals are examples of new roles for these nutrients.

Agronomists in all states emphasize the need for fuller definition of optimal rates and frequency of application of trace elements, including residual effects — first because any unnecessary application of these expensive compounds is wasteful, and secondly because excess amounts of trace elements may be toxic to plants (it is possible that Zn application has been supra-optimal on some light soils) or to animals (as when excess Mo in the pasture depresses the Cu status of the animal).

There has been a progressive tendency to reduce the rate and frequency of application of Cu, Zn and Mo. The 'standard' application of 7 kg/hectare of copper sulphate, emerging from Riceman's studies on the 90-Mile Plain, is now regarded in Victoria as adequate for 5-7 years, in South Australia for at least 12 years and in Western Australia perhaps even for a century, depending on soil type and land use. These views are based on the non-response to further applications, on the limited removal of Cu in animal and plant products and, in a study at Newdegate, Western Australia, on the accountability of the applied Cu in the soil surface horizons and the greater value of earlier than current applications (Reuter, this volume; J.W. Gartrell, personal communication). The residual value of Zn is enhanced by the Zn content of superphosphate and no further application may be needed on light soils, depending on the Zn content of the rock phosphate (Walkley 1940; Fitzpatrick 1962; Ozanne *et al.* 1965). Molybdenum also seems to be effective on most soils for 10 or more years, while on the other hand, Mn and Fe compounds have little residual value. But despite these generalizations there is inadequate information on the soil chemistry of these elements and on quantitative as well as observed residual values.

A much debated issue has been the breeding of crop and pasture plant cultivars tolerant of low levels of supply of trace elements. Striking differences between genera, species and cultivars are widely recorded and there seems little doubt that breeding for tolerance could successfully be undertaken, at least for some micronutrients. However Cu, Zn and Mo give such long-term responses that breeding may not be warranted, while in the case of elements important for grazing animals as well as for plants, continued application may be needed. Perhaps only for Mn deficiencies may breeding be worthwhile as a means of avoiding or reducing the annual treatment of crops.

Finally, the most pressing need in the use of trace elements in Australian agriculture is for effective tests of deficiency. The individual land holder may be guided by local field trials and farmer experience; or he may acquire an appreciation of particular deficiency symptoms. But field trials cannot cover all situations, and plant symptoms are unreliable in marginal or complex situations. As discussed in later papers, there is urgent need for the fuller development of tissue testing and soil analysis as reliable guides to trace element deficiencies.

ACKNOWLEDGEMENTS

We are greatly indebted to the State Departments of Agriculture and to many colleagues at the Waite Institute and in CSIRO for information and discussion regarding trace element deficiencies in Australia. The section on Australian soils draws substantially on the writings of Dr. C.G. Stephens and we express our particular appreciation of his constructive comments.

REFERENCES

AHMED, S. & EVANS, H.J. (1959) - *Biochem. biophys. Res. Commun. 1* : 271

ANDERSON, A.J. (1942) - *J. Aust. Inst. agric. Sci. 8* : 73

ANDERSON, A.J. (1948) - *J. Aust. Inst. agric. Sci. 14* : 28

ANDERSON, A.J. & OERTEL, A.C. (1946) - Coun. sci. ind. Res. Bull. 198, p. 25

ANDERSON, A.J. & THOMAS, M.P. (1946) - Coun. sci. ind. Res.

Bull. 198, p. 7
ARNON, D.I. & STOUT, P.R. (1939) - *Pl. Physiol. 14* : 599
ATKINSON, J.D. (1935) - *N.Z. Jl Sci. Technol. 16* : 316
BAXTER, P. (1957) - *J. Agric. Vict. 55* : 11
BEADLE, N.C.W. (1962) - *Aust. J. agric. Res. 13* : 434
BLACKBURN, G. (1964) - CSIRO Soils & Land Use Series No. 45
BRENCHLEY, W.E. (1927) - "Inorganic Plants Poisons and Stimulants". Cambridge University Press : Cambridge
CARNE, W.M. & MARTIN, D. (1937) - *J. Coun. sci. ind. Res. Aust. 10* : 47
CARTER, E.D. & WIGG, P.M. (1963) - *J. Agric. S. Aust. 66* : 464
CHANDLER, W.H., HOAGLAND, D.R. & HIBBERD, P.L. (1933) - *Proc. Am. Soc. hort. Sci. 30* : 70
CONWAY, E.J. (1945) - *Am. J. Sci. 243* : 583
DONALD, C.M. (1965) - *Aust. J. Sci. 27* : 187
DONALD, C.M. & SPENCER, D. (1951) - *Aust. J. agric. Res. 2* : 295
DONALD, C.M. & WILLIAMS, C.H. (1954) - *Aust. J. agric. Res. 5* : 664
DUNCAN, O.W. (1967) - *Qd J. Agric. Anim. Sci. 24* : 293
DUNNE, T.C. (1950) - *J. Dep. Agric. West. Aust. (2nd Ser.) 27* : 118
DUNNE, T.C. & colleagues (1946-1950) - Various papers, *J. Dep. Agric. West. Aust. (2nd Ser.) Vols. 23-27*
DUNNE, T.C. & JONES, L.T. (1948) - *J. Dep. Agric. West. Aust. (2nd Ser.) 25* : 412
EVANS, H.J. & HALL, N.S. (1955) - *Science 122* : 922
FITZPATRICK, E.N. (1957) - *J. Agric. West. Aust. (3rd Ser.) 6* : 65
FITZPATRICK, E.N. (1962) - *J. Agric. West. Aust. (4th Ser.) 3* : 323
FISCHER, R.A. & KOHN, G.D. (1966) - *Aust. J. agric. Res. 17* : 281
FLINDERS, MATTHEW (1814) - "A Voyage to Terra Australis". London
FRENEY, J.R. & LIPSETT, J. (1965) - *Nature, Lond. 205* : 616
FRICKE, E.F. (1943) - *Tasm. J. Agric. 14* : 69
FRICKE, E.F. (1944) - *Tasm. J. Agric. 15* : 65
FRICKE, E.F. (1947) - *J. Aust. Inst. agric. Sci. 13* : 75
GARTRELL, J.W. (1966) - *Nature, Lond. 209* : 1050
GARTRELL, J.W. (1969) - West. Aust. Dep. Agric. Tech. Bull. No. 3
GARTRELL, J.W. (1974) - *In* "Our Land" *7* : 12. Published by CSBP & Farmers Ltd., Western Australia
GARTRELL, J.W. & GLENCROSS, R.N. (1968) - *J. Agric. West. Aust. (4th Ser.) 9* : 517
HOSKING, W.J. (1957) - *J. Agric. Vict. 55* : 205

ISBELL, R.F. & GILLMAN, G.P. (1973) - *Aust. J. exp. Agric. Anim. Husb. 13* : 81
JESSOP, R.S. & TUOHY, C.L. (1973) - *J. Agric. Vict. 71* : 273
JESSUP, R.W. & WRIGHT, M.J. (1971) - *Geoderma 6* : 275
JOHNSON, C.M., PEARSON, G.A. & STOUT, P.R. (1952) - *Pl. Soil 4* : 178
JONES, R.K. (1973) - *Aust. J. exp. Agric. Anim. Husb. 13* : 89
KESSELL, S.L. & STOATE, T.N. (1938) - Forests Department, Western Australia, Bull. 50
KING, P.M. (1974) - Thesis, University of Adelaide
LIPSETT, J. & SIMPSON, J.R. (1971) - *J. Aust. Inst. agric. Sci. 37* : 348
McHARGUE, J.S. (1922) - *J. Am. chem. Soc. 44* : 1592
MILLIKAN, C.R. (1938) - *J. Agric. Vict. 36* : 409
MILLIKAN, C.R. (1948) - *J. Agric. Vict. 46* : 511
NEWMAN, R.J. (1955) - *J. Agric. Vict. 53* : 451
NEWMAN, R.J. (1962) - Grassld Soc. Vict. Bull. 1
NOONAN, J.B. (1953) - *Agric. Gaz. N.S.W. 64* : 422
OZANNE, P.G., GREENWOOD, E.A.N. & SHAW, T.C. (1963) - *Aust. J. agric. Res. 14* : 39
OZANNE, P.G., SHAW, T.C. & KIRTON, D.J. (1965) - *Aust. J. exp. Agric. Anim. Husb. 5* : 29
PIPER, C.S. (1938) - Coun. sci. ind. Res. Aust. Pamphlet 78
PIPER, C.S. (1940) - *Emp. J. exp. Agric. 8* : 199
PIPER, C.S. (1942) - *J. agric. Res. 32* : 143
PIPER, C.S. & DE VRIES, J. (1964) - *Aust. J. agric. Res. 15* : 234
PITTMAN, H.A. & OWEN, R.C. (1936) - *J. Dep. Agric. West. Aust. 13* : 137
POWRIE, J.K. (1960) - *Aust. J. Sci. 23* : 198
PRESCOTT, J.A. (1931) - Coun. sci. ind. Res. Bull. 52
RAULIN, J. (1869) - *Ann. Sci. Naturel. (5th Ser.) Botany 11* : 139
REUTER, D.J., HEARD, T.G. & ALSTON, A.M. (1973) - *Aust. J. exp. Agric. Anim. Husb. 13* : 434
RICEMAN, D.S. (1948) - *J. Coun. sci. ind. Res. Aust. 21* : 229
RICEMAN, D.S. & DONALD, C.M. (1938) - Coun. sci. ind. Res. Aust. Pamphlet 78
RICEMAN, D.S., DONALD, C.M. & EVANS, S.T. (1940) - Coun. sci. ind. Res. Aust. Pamphlet 96
RUITER, J.H. (1969) - *Pl. Soil 31* : 197
SAMUEL, G. & PIPER, C.S. (1928) - *J. Dep. Agric. S. Aust. 31* : 696, 789
SAMUEL, G. & PIPER, C.S. (1929) - *Ann. appl. Biol. 16* : 493
SOMMER, A.L. (1931) - *Pl Physiol. 6* : 339
STEPHENS, C.G. (1951) - Br. Commonw. Spec. Conf. Agric., Aust., 1949, p. 51. HMSO : London

STEPHENS, C.G. & DONALD, C.M. (1958) - *Adv. Agron. 10* : 167
STEPHENS, C.G. & OERTEL, A.C. (1943) - *J. Coun. sci. ind. Res. Aust. 16* : 69
STOATE, T.N. (1950) - Aust. Timber & Forestry Bureau, Canberra, Bull. 30
STOUT, P.R. (1972) - *In* "Micronutrients in Agriculture" (J.J. Mortvedt, P.M. Giordano & W.L. Lindsay, eds.), p. 1 Soil Science Society of America Inc. : Madison, Wisconsin
TEAKLE, L.J.H. (1944) - *J. Dep. Agric. West. Aust. 21* : 335
TEAKLE, L.J.H. & colleagues (1939-1945) - Various papers, *J. Dep. Agric. West. Aust. (2nd Ser.) Vols. 16-22*
THORP, J. (1949) - *Bull. geog. Soc. Am. 60* : 1517
TIVER, N.S. (1955) - *J. Dep. Agric. S. Aust. 59* : 100
TIVER, N.S. (1958) - *J. Dep. Agric. S. Aust. 62* : 117
TOMS, J. (1958) - *J. Agric. West. Aust. (3rd Ser.) 7* : 197
WALKLEY, A. (1940) - *J. Coun. sci. ind. Res. Aust. 13* : 255
WARING, C.J., SHIRLOW, N.S. & WILSON, R.D. (1947) - *J. Aust. Inst. agric. Sci. 13* : 187
WARRINGTON, K. (1923) - *Ann. Bot. 37* : 629
WEIR, R.G. & HUDSON, A. (1966) - *Aust. J. exp. Agric. Anim. Husb. 6* : 35
WILD, A. (1961) - *Aust. J. agric. Res. 12* : 286
WILLIAMS, C.H. & ANDREW, C.S. (1970) - *In* "Australian Grasslands" (R.M. Moore, ed.), p. 321. Australian National University Press : Canberra
WILSON, R.D. (1948*a*) - *J. Aust. Inst. agric. Sci. 14* : 118
WILSON, R.D. (1948*b*) - *J. Aust. Inst. agric. Sci. 14* : 180

TRACE ELEMENTS IN ANIMAL PRODUCTION

H.J. Lee

Division of Nutritional Biochemistry, CSIRO
Adelaide, South Australia 5000

INTRODUCTION

Traces of certain heavy metals were recognized in plant and animal tissues early in the Nineteenth Century, and about 50 years later biological importance was attached to some of these and potential toxicity to others.

Little productive interest was taken in the possible nutritional importance of any of these elements, except Fe, until the essential role of Cu in haemoglobin formation was announced in the 1920's.

During the following decade, specific importance to higher animals was assigned, also, to Zn, Mn and Co and, in the 1950's, to Mo and Se.

It is against this background that I have attempted to present the Australian and, particularly, the South Australian contribution to the elucidation of problems concerning trace elements in animal production.

Because of restricted space it has proved possible to make general reference only to much of the historical material; specific references not included may be found in key publications quoted in this Historical Preface, in the comprehensive treatment of the subject by Underwood (1971) or, in some instances, elsewhere in these Proceedings.

COPPER

Deficiency

The first reported response to Cu in grazing ruminants was probably that of Neal, Becker and Shealy in 'salt sick' cattle in Florida in 1931.

Enzootic Ataxia in Lambs

In 1932, Bennetts described an ataxic condition affecting lambs in Western Australia, the principal feature of which was nervous degeneration in the spinal cord. He attributed the cause to either dietary deficiency or a cumulative toxic agent and identified the pathology with that of similar diseases in Europe, South America and other parts of Australia.

By 1935 Bennetts had concluded that the disease was of toxic origin and developed an hypothesis that accumulated Pb was the cause; major prophylactic success attended the provision of ammonium chloride to ewes as a de-leading agent during gestation.

However, the plumbism hypothesis was disproven when Bennetts & Chapman (1937) reported that Pb concentrations in affected and unaffected animals were similar, that chemically pure ammonium chloride had no preventive effect and that there was less Cu in the tissues of affected animals. The implication that Cu had been a beneficial impurity in the commercial ammonium chloride originally used was proved when, in circumstances which gave rise to 100% ataxia in the progeny of untreated ewes, 15 mg Cu/day given to comparable ewes during gestation completely protected their lambs and assured their normal development.

This important discovery of the role of Cu in preventing enzootic ataxia was very soon found to apply elsewhere to similar demyelinating diseases.

Wool Production

While Bennetts had referred to the abnormal, "stringy" appearance of the fleeces of ewes which produced ataxic lambs, South Australian studies demonstrated that, with an increasing degree of Cu deficiency, crimp formation deteriorated and wool production diminished progressively (Marston & Lee 1948*a*). The earliest sign of deficiency proved to be the failure of pigmentation in black fleeces (Lee 1951).

The characteristic failure of crimp has been encountered in Merino sheep in Queensland (Lee & Moule 1947), New South Wales (McClymont 1957) and Victoria (Hewitt 1941) and, in South Australia, also in crossbreds (Marston & Lee 1948*b*) and several English breeds (Lee 1956); the lesion has been used to map the Cu-deficient areas in South Australia, by plotting the origins of affected clips (Lee 1951).

Requirements

Disability of some sort is ordinarily suffered by sheep which graze pastures containing less than about 5 ppm Cu, and the daily requirement has been considered to lie between 5 and 10 mg/day (Marston & Lee 1948*a*; Beck 1962). However, Dick (1954) has deduced experimentally that the minimal requirement for a mature sheep is likely to be only about 1 mg/day in the absence of complicating factors, and it would seem appropriate, therefore, to substitute "Cu-responsive condition" for "Cu deficiency" in most situations where the latter term has been applied.

Correction of Copper Deficiency

Various methods have been devised in Australia to provide adequate Cu supplements by including copper sulphate in drenches, licks, drinking water or phosphatic fertilizers. The last of these has commonly been the method of choice, particularly in those areas where pastures, also, respond to Cu. Whatever the method, over-zealous supplementation has, from time to time, induced Cu poisoning.

"Falling Disease" in Cattle

In an area in Western Australia where enzootic ataxia had commonly occurred, Bennetts & Hall (1939) described a disease of dairy cows characterized by sudden death in seemingly normal animals. Very low Cu concentrations in liver (6 ppm) and pasture (2 ppm) were an invariable feature and subsequent investigations (Bennetts *et al.* 1948) demonstrated that "falling disease" could be completely prevented by Cu supplementation. The essential lesion was progressive myocardial atrophy with replacement fibrosis, the genesis of which remains unknown.

General Considerations

As the most severe depletion of Cu does not invariably result in the most obvious signs of deficiency, it has been contended that uncomplicated deficiency is not responsible for all Cu-responsive conditions. Interference with Cu metabolism by Mo and sulphate (*vide infra*) is certainly one complicating factor and others have been suggested, e.g. the possibility that the usual post-natal fall in vitamin B_{12} in the serum of lambs might precipitate enzootic ataxia has been studied and disproven (H.J. Lee, unpublished).

Currently, in South Australia, the biochemistry of Cu in relation to myelination or demyelination in lambs is being studied intensively (G.R. Russell & R.M. Smith, personal communication), while the continuing investigation (E.G. Cleary, personal communication) of the relationship of Cu to elastin and collagen maturation might conceivably throw light on the remaining puzzle of "falling disease".

Chronic Poisoning

This condition was first described about 100 years ago in sheep which grazed European vineyards sprayed with Cu fungicides and, more recently, by Boughton & Hardy (1934) in Texan sheep with access to licks containing copper sulphate.

In Australia, Albiston recorded the occurrence of haemolytic ("toxaemic") jaundice of unknown aetiology in sheep in Victoria in 1929 and the characteristic jaundice was subsequently reported to be indistinguishable from that which developed in sheep known to have been poisoned by copper sulphate used as a molluscicide (Albiston *et al.* 1940).

The specifically Australian contribution to the study of chronic poisoning, however, has been the hard-won realization that pastures *not* contaminated by Cu could produce the same haemolytic crisis in three circumstances — (1) where Cu-rich soils induce extremely high Cu concentrations in pasture plants, (2) where pastures of normal Cu content but with low concentrations of Mo lead to excessive accumulation of Cu in the liver and (3) where liver cells damaged by hepatotoxic alkaloids in plants (e.g. *Heliotropium europaeum*) develop an avidity for Cu.

These investigations have been summarized by Bull (1949) and in the final report of the Investigation Committee on Toxaemic Jaundice in Sheep (Anon. 1956).

Haemolytic jaundice due to one or other of these circumstances has occurred in New South Wales, Victoria, South Australia and Western Australia.

COBALT

Deficiency

From as far back as the Eighteenth Century, when the first accounts appeared in Great Britain, wasting diseases affecting ruminants in restricted areas have been described in many countries. In all of them, anaemia and progressive loss of appetite led to death in a state resembling hunger oedema unless the affected animals were moved to healthy pastures.

The possibility that an unspecified soil and plant deficiency might be responsible for "bush sickness" in New Zealand was put forward in about 1900 and supported later by observations that affected sheep improved if dosed with suspensions of soil from sound areas. The New Zealand workers adopted an Fe-deficiency theory to account for the anaemia and the associated wasting disease, but the undoubted success achieved by providing licks containing naturally occurring hydrated ferric oxide was marred by the later realization that not all limonite deposits were equally effective. An excellent account of the Iron Starvation Theory has been presented by Hopkirk & Patterson (1954).

In South Australia, "coast disease" of sheep confined to the coastal littoral, first referred to by Professor Ralph Tate in 1883, clearly resembled similar wasting maladies in Western Australia, Britain, New Zealand, Kenya and Florida. Thomas conducted a geochemical survey of affected areas on Kangaroo Island, South Australia, and drew attention to the highly calcareous nature of the soils, composed largely of comminuted marine shell fragments which, he postulated, would lack heavy metals and available phosphate.

Bone fragility was a frequent feature of coast disease and the first experimental work undertaken in 1929 on Kangaroo Island was designed by Lines to test an aphosphorosis hypothesis. The result was negative and a re-appraisal of the situation followed under the guidance of the Coast Disease Committee set up in 1933, comprising Dr. J.A. Gilruth, Sir Charles Martin, Dr. H.R. Marston, Mr. C.G. Dickinson and Mr. R.H.F. Macindoe (all deceased) with Dr. L.B. Bull and Professor J.A. Prescott. At the same site, internal parasites and toxic plants were disposed of as possible causes and Lines then examined one of the remaining possibilities by treating sheep with a modified Jensen trace element mixture. His work was hampered by unpredictable fluctuations in the severity of the disease so he transferred sheep and fodder to the laboratory to continue his intensive observations. His results were equivocal until Thomas, having read that polycythemia could be induced in rats by excess Co and knowing that coasty soils probably lacked Co, suggested that a specific deficiency of this element might be responsible for the anaemia and other features of coast disease. Lines accepted the suggestion and in 1934 dosed his coasty sheep with 1 mg Co daily; the response in clinical appearance, appetite, body weight and haemoglobin concentration was prompt and dramatic.

These early considerations and investigations, which also revealed that coast disease was complicated by Cu deficiency, have been covered by Marston *et al.* (1938).

During the period referred to, there had been parallel activity in Western Australia. There, in 1933, Filmer described Denmark wasting disease ("enzootic marasmus") in sheep and cattle, reported the curative properties of limonite and expressed the opinion that the virtue of this material was more likely to reside in an unknown mineral impurity rather than in Fe itself. Filmer & Underwood (1934) demonstrated that the effective agent was indeed in the Fe-free extract of curative limonite and not in the extracted residue. Preliminary fractionation led to the suggestion that the active ingredient might be Ni.

Underwood visited the Division of Animal Nutrition in Adelaide in November, 1934, and described the Western Australian experience. The South Australian discovery was made known to him and he soon confirmed that Co was the essential element lacking in the Western Australian situation also. He graciously withheld publication of his findings (Underwood & Filmer 1935), which carried a complete acknowledgement of the South Australian priority, until that had been claimed as a general statement (Marston 1935) and as a brief account of the experimental evidence (Lines 1935).

In historical context it may be stated that New Zealand workers first devised a practical therapeutic treatment for Co deficiency but assigned to it an incorrect explanation; Thomas made the first suggestion that Co might be an element essential to ruminants; Lines first demonstrated a response to Co in deficient sheep; Marston first announced this discovery (at ANZAAS in Melbourne in 1935) while Underwood and Filmer made the first logical and methodical approach to a solution which was bound to succeed.

Murnane (in Marston *et al.* 1938) had carried out a survey in 1934 which showed that coast disease occurred in South Australia wherever calcareous soils of aeolian origin predominated. Later, other soils derived from this parent material were found to be deficient (Lee 1951); on these derived soils, deficiency was not invariably acute but fluctuated in intensity from year to year (Lee 1949).

On a world basis, soils of great diversity have been found to be Co deficient. The essentiality of Co to ruminants was quickly found to apply throughout the world to many hitherto obscure wasting diseases.

Prevention of Cobalt Deficiency

The minimum oral requirement of Co by sheep — no benefit is derived from parenteral Co — has been found to be about 0.08 mg/day, and practical control measures to provide at least this amount have been adapted to a diversity of husbandry practices.

Manurial application of Co salts to pastures was adopted in New Zealand and elsewhere but did not find ready acceptance on less productive pastures in South Australia. To meet the situation there, a novel method of continuous Co supplementation was devised, stemming from an initial suggestion by a visiting Californian, Professor P.R. Stout. This was the development of a dense, heavy pellet containing Co_3O_4 which, when lodged in the reticulum, slowly releases sufficient Co to meet the animal's requirements for several years (Dewey *et al.* 1958).

Vitamin B_{12} and Cobalt Deficiency

Early in the South Australian investigations the similarity between the blood dyscrasia in Co-deficient sheep and in human pernicious anaemia patients was recognized (Marston 1949), but attempts to correct the condition in sheep with "potent liver extracts" failed and for almost 20 years the physiological function of Co remained unknown. Then, in 1948, the announcement (Rickes *et al.* 1948; Smith 1948) that Co was a constituent of vitamin B_{12} (anti-pernicious anaemia factor) re-kindled interest.

A few mg of this new vitamin were made available to South Australian workers but the small amounts sufficient to control pernicious anaemia failed to elicit any response when injected into Co-deficient sheep (Marston & Lee 1952). Then, in a paper describing similar findings at Cornell, Becker & Smith (1951) announced in an addendum, "Data more recently obtained ... show that the amount of vitamin B_{12} injected in these lambs is suboptimum. Larger amounts of the vitamin do result in a positive response in Co-deficient lambs." This major advance was confirmed (Smith *et al.* 1951; Marston & Lee 1952; Marston & Smith 1952) and the requirement of sheep for vitamin B_{12} determined (Marston 1970).

The intensive studies on the nature of the biochemical involvement of vitamin B_{12} in the the metabolism of sheep, which make it clear that the sole function of Co is to ensure the adequate microbial synthesis of the vitamin in the rumen, have been reviewed elsewhere in this Symposium.

Phalaris Staggers

A staggers syndrome in sheep and cattle grazing pastures in which *Phalaris tuberosa* predominated was first described by McDonald (1942) in South Australia; affected animals suffered irreversible and commonly fatal nervous degeneration. Other *Phalaris* spp. have proved similarly toxic and between 1948 and 1969 an identical disease involving *P. tuberosa* has been reported in New South Wales (Southcott 1956), Western Australia (Watson 1956), Argentina (Gaggino *et al.* 1963), California (Dunbar, personal communication), New Zealand (Le Souef 1948; Milne 1955), and South Africa (Wessells 1961); *P. minor* in South Africa (Wessells 1961); *P. arundinacea* in New Zealand (Simpson *et al.* 1969) and *P. tuberosa* × *P. arundinacea* in Florida (Ruelke & McCall 1961), Rhodesia (Anon. 1960) and South Africa (van der Merwe 1959; Wessells 1961).

It was discovered in South Australia that oral Co supplements completely prevent phalaris staggers (Lee & Kuchel 1953) and this has been universally confirmed. Cobalt exerts its preventive influence only within the rumen but seemingly without involving vitamin B_{12} (Lee *et al.* 1957).

The presence of toxic tryptamine alkaloids in phalaris was reported (Culvenor *et al.* 1964) and the suggestion advanced (Gallagher *et al.* 1966) that these were responsible for staggers (as well as for acute forms of phalaris poisoning not preventable by Co). This view has not been supported by Oram (1970) nor by our own (unpublished) experience that staggers in sheep may occur on *P. tuberosa* pastures containing negligible concentrations of alkaloids, while infusion of alkaloids into the Co-deficient rumen has not induced the disease.

Both the agent responsible for staggers and the manner in which Co acts remain unknown.

MOLYBDENUM

The realization that Mo was of significance to grazing animals came with the discovery that high levels of Mo, sometimes exceeding 100 ppm, in Somerset pastures were responsible for "teart", a grave scouring disorder of cattle (Ferguson *et al.* 1938). Industrial contamination has precipitated an identical disease in localized areas in Britain while naturally occurring molybdenosis has been reported in Ireland, Europe, Scandinavia, U.S.A. and New Zealand.

A specific biochemical role involving xanthine oxidase has been attributed to Mo (Richert & Westerfield 1953) but its importance in animal production lies primarily in its relationship to Cu metabolism.

In Australia, Dick & Bull (1945) first observed that a relatively high intake of Mo by cattle markedly reduced liver Cu content. They reported the limitation imposed by Mo on Cu storage in sheep and drew attention to the possibility that very low Mo concentrations in pasture might lead to excessive accumulation of Cu from pasture of normal Cu content.

This first indication that milligram quantities of Mo were intimately involved in Cu metabolism led to the detailed study of the quantitative relationship between Cu and Mo and the manner in which this was influenced by sulphate.

Dick (1954) showed that the accumulation of Cu diminishes as the intake of Mo increases and that interaction is controlled by sulphate. As sulphate ingestion increases, so does the rate of Mo excretion; with small concentrations of sulphate in pasture the effect of relatively high concentrations of Mo on Cu metabolism is small, while sufficient sulphate will result in reduced Cu storage with even small amounts of Mo. These important discoveries have improved the understanding of both Cu deficiency and toxicity in grazing animals, and the mechanisms of the interactions involved are currently being examined further.

Not everywhere in Australia has it been accepted that interaction with Mo and sulphate is responsible for all manifestations of Cu deficiency (Beck 1962; Harvey *et al.* 1963), but the probability remains that the interaction is of significance in most of them.

MANGANESE

The essentiality of Mn for adequate testicular and ovarian development and function in the rat was demonstrated in the U.S.A. as early as 1931 and, subsequently, Mn involvement with growth and bone formation was shown in several species. The discovery that Mn was essential for the prevention in poultry of perosis ("slipped tendon") proved to be of major commercial importance.

More recently there has been a spate of claims and denials that a conditioned Mn deficiency was responsible for loss of productivity in cattle, particularly in Holland, but the debate has probably been concluded by Hartman's statement during discussion at the International Symposium on Trace Element Metabolism in Animals in 1970, "In reply to Professor Underwood's remark on conditioned Mn deficiency in the Netherlands, it can now be clearly stated that the symptoms described in the publications dealing with this work must be due to factors other than Mn. Following an extensive re-investigation of these cases it has been found that there were no signs of Mn deficiency even when the rations contained as little as 25-30 ppm in the dry matter of the ration. These earlier reports of the existence of a conditioned Mn deficiency in cattle must therefore be regarded as incorrect."

In Australia, Dick (1956) has shown that in certain circumstances Mn may block the limiting effect of Mo *plus* sulphate on Cu storage by sheep but that, on high protein rations, Mn may enhance the limitation imposed by Mo. Mylrea (1958), however, could find no effect of Mn on the Mo-sulphate-Cu interaction in cattle under his experimental conditions.

In South Australia, Egan (1972) has provided evidence that Mn supplements have sometimes increased the number of lambs born to a flock of Dorset Horn ewes in which reproductive performance had declined. The practical importance of these Australian observations is not clear but they do point to the necessity for continuing awareness of the possible necessity for Mn supplements for grazing ruminants.

ZINC

The presence of Zn in plant and animal tissues had been recognized for about a hundred years but its long-suspected nutritional importance was not proven until 1934 when the Wisconsin group, adding to their successes during that period of intense activity, showed it to be essential to the rat. Since then, experimentally induced Zn deficiency has been demonstrated in many animals and in poultry, several biochemical processes have proved to be Zn dependent and a few reports of Zn deficiency in grazing ruminants have appeared.

Early experience in Australia led workers generally to conclude that the minimum concentration of Zn present even in Zn-deficient plants was sufficient to satisfy the requirements of grazing animals. However, Underwood and Somers (1969) have shown that ram lambs fed a diet containing only 2.4 ppm Zn developed severe clinical Zn deficiency, with impaired testicular development and a complete cessation of spermatogenesis. When supplementation raised dietary Zn to 17.4 ppm, the lambs grew normally and showed no signs of deficiency, but testicular development and function remained significantly less than in similar lambs whose diet contained 32.4 ppm Zn. In these experimental lambs, Zn deficiency was accompanied by impaired protein utilization (Somers & Underwood 1969).

More recently, in South Australia, Egan (1972) has reported that significantly more lambs were produced by Dorset Horn ewes given 140 mg Zn weekly, than by untreated ewes. This flock grazed pastures in which Zn concentration varied, during a two-year period, between 12 and 45 ppm.

The importance of Zn deficiency in grazing animals in Australia appears to remain uncertain; a high proportion of the pastures we have sampled over the years contained less than 20 ppm, many of them appreciably less than 10 ppm. There was no indication of impaired reproductive performance in sheep in the many areas sampled but perhaps the possibility that subclinical Zn deficiency exists should be examined.

FLUORINE

Chronic fluorosis in grazing animals, recognized as a

disease entity in 1931, has been widely encountered throughout the world. The condition has resulted from industrial contamination of pastures, the ingestion of phosphatic supplements rich in F or the consumption of underground waters of high F content.

In Australia, naturally occurring mottling of dental enamel (one of the recognized signs of fluorosis) was first described in human teeth in 1937 and in sheep in 1945 (*vide* Seddon 1945), both situations arising in Queensland from the consumption of artesian bore waters of high F content. These waters may contain as much as 15 ppm F at the borehead and, as a result of evaporation, as much as 40-50 ppm in distant parts of the surface drains. The distribution of Queensland bore waters according to their F content has been mapped and the effects on Merino sheep of exposure to fluorided waters have been studied (Harvey 1952,1953; Harvey & Moule 1954).

In South Australia, the increasing use of rock phosphate licks led, in 1933, to the investigation of the effects on Merino sheep of the F ingested in such supplements and, also, in fluoride-rich drinking water (Peirce 1938,1952,1954,1959).

These comprehensive Australian studies confirmed the development in mature sheep of exostoses of the long bones and mandibles and the abnormal development and wear of teeth erupted by young sheep exposed to F excess. Reproductive performance was not diminished by F intoxication, but wool production in mature sheep with faulty dentition and the growth rate of their lambs were adversely affected by the inability of the ewes to graze and masticate adequately.

SELENIUM

Toxicity

The toxicity to animals of an excess of Se compounds in edible plants (e.g. *Astralagus* spp.) was recognized in the U.S.A. about 40 years ago, although the dire effects of the consumption of such plants had long been known. Subsequently, Se toxicity has been reported in many countries.

In Australia, naturally occurring selenosis has been described in two situations in Queensland, one in which the Se concentration in the soil is low and, in the other, abnormally high (Knott *et al.* 1958; McCray & Hurwood 1963). In the former, restricted to part of Cape York Peninsula, horses may be poisoned by the accumulator plant *Merinda reticulata*, which sometimes contains more than 1000 ppm Se, and the toxic hazard posed has restricted the expansion of cattle raising.

The second situation applies near Richmond, where the surface soil is derived from an outcrop of the very extensive Tambo Formation. In this area a native legume (*Neptunia amplexicaulis*) may accumulate in excess of 4000 ppm Se, while grasses, with as much as 263 ppm, have proved toxic to horses and sheep.

Deficiency

The demonstration in the U.S.A. in 1957 that Se supplements would prevent several disease entities in deficient rats and chickens was followed promptly by the recognition that nutritional muscular dystrophy ("white muscle disease", WMD) in lambs and calves could be prevented by Se in Oregon (Muth *et al.* 1958) and in New Zealand (McLean *et al.* 1959). Subsequently, WMD and other Se-responsive conditions have been widely reported.

For the control of Se deficiency in ruminants, periodic drenches or injections with sodium selenite or selenate have been variously advocated and the topdressing of deficient pastures with selenite considered. In Australia, WMD has occurred most commonly in Western Australia (Gardiner 1961), but also in New South Wales (Walker *et al.* 1961), Victoria (Christie 1962) and South Australia (Godwin *et al.* 1970). A significant contribution from the CSIRO Division of Nutritional Biochemistry has been the development of a heavy ruminal pellet consisting of powdered elemental Se (5%) mixed and pressed with Fe filings (Kuchel & Buckley 1969). This device raises the Se concentration in the blood of sheep by about 0.1 ppm — a level sufficient in itself to satisfy all nutritional requirements — and maintains this advantage for several years; it raises the blood Se status of untreated lambs from treated ewes sufficiently to obviate the necessity to treat the lambs prior to weaning; it will prevent or correct WMD in lambs, and as many as four Se pellets do not harm the sheep or raise the Se concentration in the tissues to unacceptable levels (unpublished data). Larger pellets of similar composition will raise the blood Se levels in cattle where these have been found to be extremely low (Dodson & Judson 1973).

OTHER TRACE ELEMENTS AND INTERACTIONS

Cadmium

Interest in Cd has centred on its undoubted toxicity in excess and on the importance of its interactions with other elements. Cadmium may act as a Zn antimetabolite (Cotzias *et al.* 1961), depress the storage of Cu in the liver of sheep (Mills & Dalgarno 1972) or damage testicular and ovarian tissues; the last effect may be reduced by Se (*vide* Gunn & Gould 1967).

A recently completed study at the Division of Nutritional Biochemistry, initiated in view of the possibility of environmental pollution with Cd, has examined the effect of Cd-Se-Cu interactions on the storage of these elements in various tissues in sheep. Concentrations of Se and Cd appear not to have been influenced markedly by interaction and the significance of changes in Cu concentration is currently being assessed (Lee & Jones, unpublished).

Chromium

Apart from its known toxicity in excess, interest in the physiological importance of Cr stems from the demonstration of its involvement in carbohydrate metabolism (Schwarz & Mertz 1959). Deficiency has not been reported in livestock in Australia but an examination of Australian wheats has been undertaken at the Division of Nutritional Biochemistry to provide a preliminary assessment of the possibility of Cr deficiency in human and animal nutrition (Jones, unpublished).

Other Elements

Evidence presented at the Second International Symposium on Trace Element Metabolism in Animals, Wisconsin, 1973, increases the credibility of an essential physiological role for several additional trace elements — Sn, V, F, Si, Pb and Ni. However, responses to these elements were elicted only in laboratory animals maintained for more than one generation under stringent experimental conditions and the likelihood of their practical importance in animal production is remote.

Interactions

Apart from the interactions already mentioned, it has been shown or claimed that others exist between Co and Cu, Co and Se, Zn and Cu, Se and Cu, and Se and Hg. The practical significance of these has not in every case been established but it seems possible that it is in this area that future work with trace elements may prove of importance to animal production.

REFERENCES

ALBISTON, H.E., BULL, L.B., DICK, A.T. & KEAST, J.C. (1940) - *Aust. vet. J. 16* : 233
ANON. (1956) - *Aust. vet. J. 32* : 229
ANON. (1960) - Annual Report, Grasslands Research Station, Southern Rhodesia, p. 35
BECK, A.B. (1962) - *Aust. J. exp. Agric. Anim. Husb. 2* : 40
BECKER, D.E. & SMITH, S.E. (1951) - *J. Nutr. 43* : 87
BENNETTS, H.W., BECK, A.B. & HARLEY, R. (1948) - *Aust. vet. J. 24* : 237
BENNETTS, H.W. & CHAPMAN, F.E. (1937) - *Aust. vet. J. 13* : 138
BENNETTS, H.W. & HALL, H.T.B. (1939) - *Aust. vet. J. 15* : 152
BOUGHTON, I.B. & HARDY, W.T. (1934) - Tex. Agric. Exp. Sta. Bull. No. 499, p. 1
BULL, L.B. (1949) - Br. Commonw. Spec. Conf. Agric., Aust., p. 300. HMSO : London
CHRISTIE, B. (1962) - *Aust. vet. J. 38* : 392
COTZIAS, G.C., BORG, D.C. & SELLECK, B. (1961) - *Am. J. Physiol. 201* : 927
CULVENOR, C.C.J., DAL BON, R. & SMITH, L.W. (1964) - *Aust. J. Chem. 17* : 1301
DEWEY, D.W., LEE, H.J. & MARSTON, H.R. (1958) - *Nature, Lond. 181* : 1367
DICK, A.T. (1954) - *Aust. J. agric. Res. 5* : 511
DICK, A.T. (1956) - *Soil Sci. 81* : 229
DICK, A.T. & BULL, L.B. (1945) - *Aust. vet. J. 21* : 70
DODSON, M.E. & JUDSON, G.J. (1973) - *Aust. vet. J. 49* : 320
EGAN, A.R. (1972) - *Aust. J. exp. Agric. Anim. Husb. 12* : 131
FERGUSON, W.S., LEWIS, A.H. & WATSON, S.J. (1938) - *Nature, Lond. 141* : 553
FILMER, J.F. & UNDERWOOD, E.J. (1934) - *Aust. vet. J. 10* : 83
GAGGINO, P.O., CARRILLO, B.J. & FRONTERA, A.R. (1963) - INTA Publ. Tech. No. 3, Balcarce, Argentina, p. 1
GALLAGHER, C.H., KOCH, J.H. & HOFFMAN, H. (1966) - *Aust. vet. J. 42* : 279
GARDINER, M.R. (1961) - *J. Agric. West. Aust. (4th Ser.) 2* : 497
GODWIN, K.O., KUCHEL, R.E. & BUCKLEY, R.A. (1970) - *Aust. J. exp. Agric. Anim. Husb. 10* : 672
GUNN, S.A. & GOULD, T.C. (1967) - *In* "Selenium in Biomedicine" (O.H. Muth, ed.), p. 395. Avi Publishing Co. Inc. : Westport, Conn.
HARVEY, J.M. (1952) - *Qd J. agric. Sci. 9* : 47
HARVEY, J.M. (1953) - *Qd J. agric. Sci. 10* : 109
HARVEY, J.M., BEAMES, R.M., HEGARTY, A. & O'BRYAN, M.S. (1963) - *Qd J. agric. Sci. 20* : 137

HARVEY, J.M. & MOULE, G.R. (1954) - *Qd agric. J. 79* : 291
HEWITT, A.C.T. (1941) - *J. Dep. Agric. Vict. 39* : 279
HOPKIRK, C.S.M. & PATTERSON, J.B.E. (1954) - "The Story of Cobalt Deficiency in Animal Health", p. 1. Mond Nickel Co. : London
KNOTT, S.G., McCRAY, C.W.R. & HALL, W.T.K. (1958) - *Qd J. agric. Sci. 15* : 43
KUCHEL, R.E. & BUCKLEY, R.A. (1969) - *Aust. J. agric. Res. 20* : 1099
LEE, H.J. (1949) - Br. Commonw. Spec. Conf. Agric., Aust., p. 262. HMSO : London
LEE, H.J. (1951) - *J. Dep. Agric. S. Aust. 54* : 475, 527
LEE, H.J. (1956) - *J. agric. Sci., Camb. 47* : 218
LEE, H.J. & KUCHEL, R.E. (1953) - *Aust. J. agric. Res. 4* : 88
LEE, H.J., KUCHEL, R.E., GOOD, B.F. & TROWBRIDGE, R.F. (1957) - *Aust. J. agric. Res. 8* : 502
LEE, H.J. & MOULE, G.R. (1947) - *Aust. vet. J. 23* : 303
LE SOUEF, H.D. (1948) - *Aust. vet. J. 24* : 12
LINES, E.W.L. (1935) - *J. Coun. scient. ind. Res. Aust. 8* : 117
MARSTON, H.R. (1935) - *J. Coun. scient. ind. Res. Aust. 8* : 111
MARSTON, H.R. (1949) - Br. Commonw. Spec. Conf. Agric., Aust., p. 247. HMSO : London
MARSTON, H.R. (1970) - *Br. J. Nutr. 24* : 615
MARSTON, H.R. & LEE, H.J. (1948*a*) - *Aust. J. scient. Res. 1(B)* : 376
MARSTON, H.R. & LEE, H.J. (1948*b*) - *J. agric. Sci., Camb. 38* : 229
MARSTON, H.R. & LEE, H.J. (1952) - *Nature, Lond. 170* : 791
MARSTON, H.R. & SMITH, R.M. (1952) - *Nature, Lond. 170* : 793
MARSTON, H.R., THOMAS, R.G., MURNANE, D., LINES, E.W.L., McDONALD, I.W., MOORE, H.O. & BULL, L.B. (1938) - *Bull. Coun. scient. ind. Res., Melbourne 113* : 1
McCLYMONT, G.L. (1957) - *Agric. Gaz. N.S.W. 68* : 428
McCRAY, C.W.R. & HURWOOD, I.S. (1963) - *Qd J. agric. Sci. 20* : 475
McDONALD, I.W. (1942) - *Aust. vet. J. 18* : 182
McLEAN, J.W., THOMSON, G.G. & CLAXTON, J.H. (1959) - *N.Z. vet. J. 7* : 47
MILLS, C.F. & DALGARNO, A.C. (1972) - *Nature, Lond. 239* : 171
MILNE, J.A. (1955) - *N.Z. vet. J. 3* : 119
MUTH, O.H., OLDFIELD, J.E., REMMERT, L.F. & SCHUBERT, J.R. (1958) - *Science 128* : 1090
MYLREA, P.J. (1958) - *Aust. J. agric. Res. 9* : 373
ORAM, R.N. (1970) - *Proc. XI Int. Grassld Congr.*, p. 785
PEIRCE, A.W. (1938) - *Bull. Coun. scient. ind. Res., Melb.*

121 : 1
PEIRCE, A.W. (1952) - *Aust. J. agric. Res. 3* : 326
PEIRCE, A.W. (1954) - *Aust. J. agric. Res. 5* : 545
PEIRCE, A.W. (1959) - *Aust. J. agric. Res. 10* : 186
RICHERT, D.A. & WESTERFIELD, W.W. (1953) - *J. biol. Chem. 203* : 915
RICKES, E.L., BRINK, N.G., KONIUSZY, F.R., WOOD, T.R. & FOLKERS, K. (1948) - *Science 108* : 134
RUELKE, O.C. & McCALL, J.T. (1961) - *Agron. J. 53* : 406
SCHWARZ, K. & MERTZ, W. (1959) - *Archs Biochem. Biophys. 85* : 292
SEDDON, H.R. (1945) - *Aust. vet. J. 21* : 2
SIMPSON, B.H., JOLLY, R.D. & THOMAS, S.H.M. (1969) - *N.Z. vet. J. 17* : 240
SMITH, E.L. (1948) - *Nature, Lond. 162* : 144
SMITH, S.E., KOCH, B.A. & TURK, K.L. (1951) - *J. Nutr. 44* : 455
SOMERS, M. & UNDERWOOD, E.J. (1969) - *Aust. J. agric. Res. 20* : 899
SOUTHCOTT, W.H. (1956) - *Aust. vet. J. 32* : 225
UNDERWOOD, E.J. (1971) - "Trace Elements in Human and Animal Nutrition", 3rd Edn. Academic Press : New York and London
UNDERWOOD, E.J. & FILMER, J.F. (1935) - *Aust. vet. J. 11* : 84
UNDERWOOD, E.J. & SOMERS, M. (1969) - *Aust. J. agric. Res. 20* : 889
VAN DER MERWE, F.J. (1959) - *Farming in S. Afr. 35* : 44
WALKER, D.J., HARRIS, A.N.A., FARLEIGH, E.A., SETCHELL, B.P. & LITTLEJOHNS, I.R. (1961) - *Aust. vet. J. 37* : 172
WATSON, E.R. (1956) - *J. Aust. Inst. agric. Sci. 22* : 209
WESSELLS, C.C. (1961) - *Jl S. Afr. vet. med. Ass. 32* : 289

GEOCHEMISTRY AND MINERALOGY OF TRACE ELEMENTS

K. Norrish

Division of Soils, CSIRO
Glen Osmond, South Australia 5064

INTRODUCTION

Figure 1, plotted from the data of Mason (1952), shows the abundance of the lighter elements in crustal rocks. For elements with atomic numbers greater than 45, concentrations are low, and with the exception of Ba(56), La(57), Ce(58), Nd(60) and Pb(82), concentrations are less than 10 ppm.

As life evolved, nature used those elements that were available, so that of the first 30 elements, about 20 are essential for plant growth, and additional elements are essential for particular plants or animals. Helium, neon, argon, lithium, beryllium, scandium and titanium seem to be the only elements that are not required in some biological system. Above atomic number 30, there are probably only three essential elements, Se(34), Mo(42) and I(53), although some others such as Ga(31), As(33), Br(35) and Ba(56) may be required for

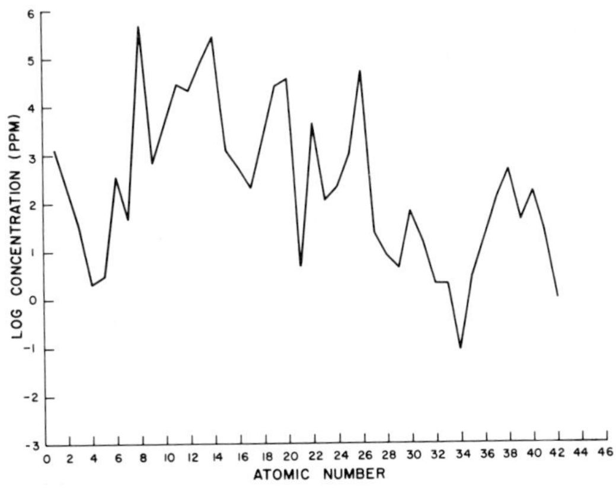

Fig. 1. Element concentrations in surface rocks (data from Mason 1952).

specific organisms.

For the purposes of this paper, those elements generally regarded as majors (H, C, N, O, Na, Mg, P, S, Cl, K, Ca) are excluded.

There are excellent reference books on geochemistry (Rankama & Sahama 1950; Mason 1952; Goldschmidt 1958; Day 1963; Degens 1965; Wedepohl 1969), and for rocks, the minerals that contain specific elements are known and the principles governing the distribution of elements between the various minerals are reasonably understood.

For soils, less is known concerning the geochemistry of the elements, and this applies particularly to trace elements. The concentrations of the various elements in soils in general, and in soils developed from specific parent materials, have been measured (Swaine 1951,1955; Vinogradov 1959; Swaine & Mitchell 1960; Oertel & Giles 1963; Mitchell 1964), but for many elements very little is known of their mineralogy or of the soil factors that control their distribution (see review by Mitchell 1964). If geochemical studies of the soils are to be meaningful in terms of agriculture, then we require information on the concentration, mineral form and availability of elements in the active soil constituents.

There are good reasons for the lack of knowledge concerning the mineralogy of trace elements in soils. In terms of nutrient reactions, the clay is the most important part of the soil but this is so fine grained that classical methods of mineral separation and identification cannot be used. X-ray diffraction techniques were developed to identify soil colloids, because of their small size, and these methods have been dramatically successful in their application to clay minerals. Although in principle all minerals in the soil could be identified in this way, X-ray diffraction methods are not particularly sensitive, and for many crystalline phases the limit of detection will be about 1%. In soils the sensitivity of X-ray diffraction analyses is often reduced by the mineral phases being disordered. For some trace elements there may be no crystalline phases involved; the element may be present in organic matter or in an adsorbed form.

Because of the lack of success of direct methods of identification, various other techniques have been used. Most of these can be roughly classified as solubility studies or synthesis experiments. In solubility studies the mineral form of the element is inferred from its solubility in water or some particular extracting medium. For a system as complex as a soil, simple solubility data give very little information on the nature of the solid phase. For this reason, reagents have been used to extract particular minerals, or remove the adsorbed phase from certain minerals, and although these methods have contributed to our understanding, rarely are the extractants as specific as intended, so that ambiguities arise in interpreting results.

The synthesis experiments usually involve test-tube type experiments where various elements are reacted with pure minerals or with soils. Very often the conditions of reaction are very far removed from field conditions and the relevance of the experiments becomes questionable. Often these indicate several reactions that may be possible in a soil, but they do not indicate which will be the important one.

Soil phosphorus is the element most studied by the indirect methods, and despite a very large number of publications on the subject the precise forms of native P, and the fate of fixed P, are unknown—see reviews by Wild (1950), Hemwall (1957) and Kardos (1964). After considering the literature on P fixation, Kardos rightly concludes that "the facts are not incontrovertible and the various mechanisms of fixation are aptly described in the literature as theoretical rather than dogmatic". Certainly the same could be said of many of the trace element studies.

Some recently developed analytical instruments are capable of analysing exceedingly small volumes, and such analyses are often sufficient to identify the mineral phases in soils. The electron probe micro-analyser is such an instrument. It uses a fine electron beam ($\simeq$ 0.5 μm diameter) to cause a small volume of specimen to emit X-rays. An X-ray spectrograph can be used to disperse the radiation and measure the intensity of selected wavelengths. In this way, volumes of several cubic microns can be analysed accurately for most elements ($Z = 9$-92). This instrument is mentioned particularly as results obtained with it are presented later.

Such techniques are now being used on soils and they should resolve many of the currently conflicting suggestions as to the mineralogy of trace elements.

Geochemistry of the Elements

The principles governing the distribution of elements in igneous rocks and minerals are reasonably understood (Mason 1952; Goldschmidt 1958). The chemical affinity of an element determines the rock type in which an element will be concentrated, while its ionic radius governs the position it can occupy in a mineral structure. Note that the latter requirement does not involve valence, and it results in chemically unrelated elements occupying the one structural position in a mineral, e.g. Li, Mg, Zn and Al in micas, and K, Ba and Pb in feldspars.

In sediments and weathered rocks, ionic radius is still an important factor, but physicochemical and colloidal factors are also important in governing the associations of elements. Oxidation-reduction potentials and ionic potentials (valence/radius) determine whether elements will behave as cations or anions and whether they will tend to remain in solution or be precipitated.

Biological influences are an additional factor in soils. Plants can translocate elements, while their exudates can complex and make soluble some of the more insoluble elements. These soil processes are not considered further as there are good reviews on the subject (Hodgson 1963; Mitchell 1964; Silanpää 1972). The following discussion is concerned with the nature of trace elements in soils, i.e. their mineralogy, rather than the processes by which they got there. Table 1 gives the average concentrations of elements for different rocks and soils. Individual rocks, and particularly soils, can show great variations in concentration and the averages are only approximate.

TABLE 1

*Concentration (ppm) of trace elements in rocks and soils**

Element	Earth's crust	Basic rocks	Acid Rocks	Sedimentary rocks	Soils
B	3	1-2	3	100	10-20
F	700	100	1000	100-1000	20-1000
V	110	200	50	100	100
Cr	200	2000	2	100-500	200
Mn	1000	2000	1000	1000	1000
Fe	50,000	100,000	25,000	35,000	30,000
Co	23	50	8	20	3
Ni	80	200-1000	10		40
Cu	45	150	10	10-100	2-50
Zn	65	100	60		60
As	2	1.5	1.5	12	1.10
Se	0.09	0.1	0.1		0.01
Br	3	2.5	2.5		6
Mo	1	2	2.5	2	2.5
I	0.3	0.3	0.3	0.3	5
Ba	400	300	800		500
Pb	16				12

*Data from Rankama & Sahama (1950), Mason (1952), Swaine (1955), Goldschmidt (1958), Vinogradov (1959), Wedepohl (1969)

BORON

It is now well known that soils formed from marine sediments are likely to contain more B than those formed from igneous rocks, and soils on an average have a higher content than rocks.

Tourmaline (3-4% B) is present in soils formed from acid rocks and metamorphosed sediments, but this very resistant mineral will not be a source of B for plants. Boron can substitute for tetrahedrally coordinated Si in some minerals, and it is likely that much of the B in rocks and soils is dispersed in the silicate minerals in this way. Such B would be available only over long periods.

Many workers (Sims & Bingham 1967,1968; Couch & Grim 1968; Jasmund & Lindner 1973) have studied B uptake by illite, other clay minerals and oxides in the laboratory. Some of the retained B is readily exchanged, but some is fixed, the amount increasing with time. Couch & Grim (1968) suggest that the exchangeable B is held electrostatically (anion exchange) at the surface, but from there it can diffuse into the crystal lattice occupying positions of tetrahedral coordination.

Often much of the B of soils is water soluble and possibly linked with organic matter (Vinogradov 1959).

FLUORINE

Topaz and fluorite are the only common rock minerals in which F is an essential constituent. In other common rock minerals, e.g. apatite, mica, etc., F replaces hydroxyl groups. This is possible because F has the same valence and radius (1.33Å) as OH (1.40Å).

Fluorapatites are common in soils that are not highly weathered, but this mineral would not be a ready source of F. J.T. Hutton (private communication) has studied soil formation from granite in an arid soil from near Alice Springs. In the granite, much of the F was present as fluorapatite but in the soil, fluorite was present in the clay fraction and occurred as a sharp layer, with 3% F, at the base of a diffuse carbonate horizon. This result suggests that fluorite is more soluble than calcite and would not persist in leached soils.

To what extent the structural F, replacing hydroxyls of clays, is available is not known, but it would not be a ready source of F. However, clays can exchange reversibly F and OH, probably at crystal edges (Hofmann *et al.* 1956; Hubner 1969), and F in this position would be available to plants.

VANADIUM

Most of the V of rocks does not occur as V minerals, but is present in low concentrations in other minerals replacing Fe, Ti, and even Al (in micas). As with rocks, V in the soil is generally present substituting for Fe, in the Fe oxides, and several investigators obtain a close correlation between the Fe and V contents of soils (Oertel & Prescott 1944; Turton *et al.* 1962; Taylor & Giles 1970). Soils high in Fe oxides, e.g. bauxites, can have high V contents, $\sim$ 500 ppm (Rankama & Sahama 1950).

Taylor & McKenzie (1966) found that in removing Mn oxides from the soil less than 10% of the soil V was extracted, whereas Le Riche & Weir (1963) in removing free Fe (and Mn) oxides removed 41% of the V. Although the two sets of data were obtained using different soils, they suggest that Fe oxides hold a reasonable fraction of the soil V.

The author has studied the weathered surface (soil?) of a V-bearing shale. Electron microprobe analyses gave results showing a close correlation ($r = 0.93$) between K and V (Fig. 2), suggesting that the latter was contained in a clay mica. Subsequent separation of the fine clay yielded a mixed-layer mica-montmorillonite containing 7% V_2O_5. In this clay the V is replacing Al, and similar clays are known from the United States (Hathaway 1959). Goethite separated from the above clay contained 3% V_2O_5 (see Table 4, sample 2).

Vanadium in silicates will only be released over geological

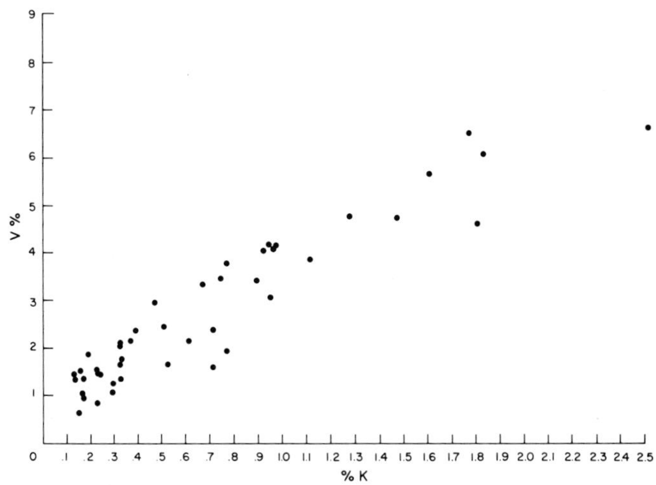

Fig. 2. Microprobe analyses showing the relationship between V and K in a weathered shale.

periods, but it seems likely that the fine grained and more mobile Fe oxides could supply V to plants.

CHROMIUM

In rocks, Cr is often present in chromite, $FeCr_2O_4$, or other spinel structures in which the Cr is substituting for Fe or Al. These minerals are exceedingly resistant to weathering and probably account for most of the Cr in soils. This is confirmed somewhat by the fact that procedures to extract Mn and Fe oxides remove less than 10% of the soil Cr (Le Riche & Weir 1963; Taylor & McKenzie 1966).

MANGANESE

In igneous rocks, Mn is usually present in silicates in the manganous form. During weathering the Mn oxidizes and separates out as Mn oxides in the soil. Although the black Mn concentrations in soils, as nodules and coatings, have long been recognized, it was not until 10 years ago (Tiller 1963; Taylor *et al.* 1964) that any such minerals were identified. Contrary to expectations, the chemically simple oxides and hydroxides do not occur in soils. Of the 28 samples examined by Taylor *et al.* (1964), lithiophorite and birnessite were the most common mineralogical forms, 10 each, but hollandite, 3, todorokite, 1, and pyrolusite, 1, were also identified. Despite mineralogical differences between the various oxides, there were similarities in their chemistry, particu-

TABLE 2

Chemical analyses of soil Mn oxides (Taylor et al. *1964)*

	Lithiophorite		Birnessite		Hollandites	
	mean (5)	range	mean (5)	range	mean (3)	range
Mn_3O_4	73.2	57-83	83.8	78-90	65.3	52-84
Al_2O_3	11.5	7-25	1.0	0.4-1.9	5.7	1-14
SiO_2	3.8		1.38		6.7	
Fe_2O_3	4.7	1-11	0.94	0.5-1.7	6.6	1-15
CoO	1.53	0.4-4.7	0.57	0.2-1.1	1.4	0.4-2.9
NiO	0.13	0.02-0.41	0.43	0.05-1.4	0.21	0.06-0.4
CaO	0.20		3.16		5.7	
MgO	0.25		1.13		1.0	
BaO	3.6	0.5-8	4.5	1.7-5.7	6.0	3-9.7
Li_2O	0.16		0.06		0.12	
Na_2O	0.20		0.09		0.14	
K_2O	0.32		0.22		0.37	

larly in the high Co, Ni and Ba contents (see Table 2). Figure 3 shows results obtained with the electron probe analyser on a laterite from Western Australia, and the association of Co and Ni with Mn is very clear. Often Mn concretions are also rich in Fe and the assumption has been that both elements are present in the one mineral. The studies of Taylor *et al.* (1964) showed that in nodules the Fe and Mn oxides were present as separate phases, and that the Fe contents of the Mn minerals were not high (Table 2). In Figure 3 the low Fe content in the Mn oxide mineral is obvious, and Brewer *et al.* (1973) have confirmed that the two elements usually segregate into separate phases. Taylor *et al.* (1964) found that lithiophorite occurred in acid to neutral soils, while birnessite was more common in alkaline horizons. Although they used the coarser fractions of Mn oxides, i.e. nodules, etc., the crystallite size of all the minerals was very small, averaging

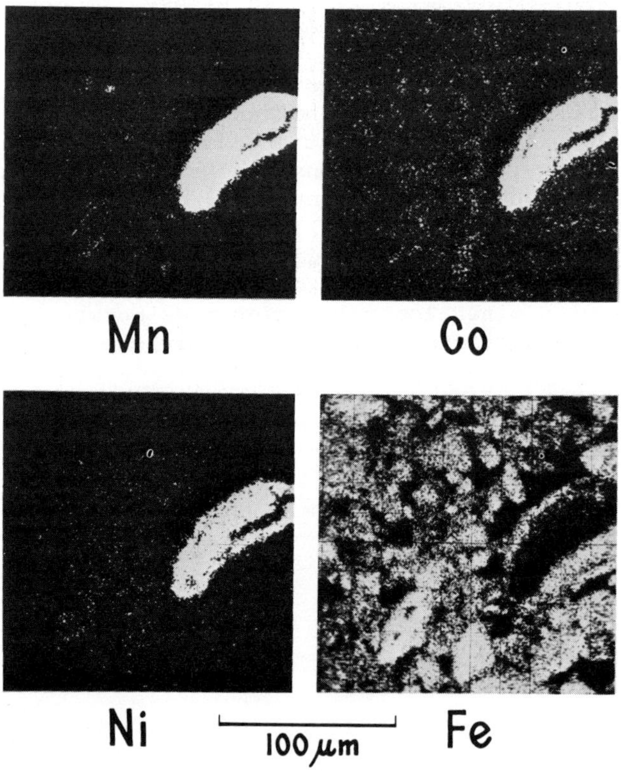

Fig. 3. Microprobe displays showing the association of Co and Ni with Mn, in a laterite from Western Australia.

about 0.02 μm for birnessites, 0.02-0.1 μm for hollandites and 0.1 μm for lithiophorites. This small particle size, and consequent high surface area, has important implications. Even though the Mn oxides are only a minor constituent of soils, they contribute significantly to the surface activity of the soil. With such a high surface area, elements adsorbed on the surface can be present in major concentrations in the minerals. Where the particle size is only around 0.02 μm, the distinction between surface adsorption and lattice replacement becomes blurred.

The availability of Mn to plants fits the known solubility of Mn oxides in that it is very dependent on oxidizing conditions and pH. Under acid conditions the Mn is sufficiently soluble, but deficiencies can occur in soils with a pH of 7-8. Even in alkaline soils, Mn is probably dissolved in small regions where reducing conditions exist. The very small particle size is important and enables the minerals to go into solution under local and temporary changes of pH and eH.

McKenzie (1971) has recently written an excellent review on Mn oxides in soils, and Jenne (1968) has reviewed the role of Mn and Fe oxides in fixing trace elements. The high content of other trace elements in Mn oxides is obviously significant and will be discussed later in this paper, and more fully in a subsequent paper by R.M. McKenzie.

Manganese, along with other heavy metals, is often concentrated in surface soils. This, and the fact that the elements are released by hydrogen peroxide treatment, is usually taken as good evidence that the elements are organically bound. Jenne (1968) has queried this interpretation, as peroxide will solubilize Mn oxides and any metal associated with them (Taylor *et al.* 1964). Electron microprobe analyses of some surface soils support his idea. Although Mn nodules were not recognizable, even under the microscope, in these soils, very small areas of Mn enrichment could be found using the microprobe. Table 3 shows the results of analysing points in which the Mn is so concentrated that it must be present as an oxide rather than in organic form. Because the Mn grains were always smaller than the area of analysis (several square microns), high amounts of Si and Al from adjacent minerals are found, and the measured Mn concentration depends on the size of the grain analysed, rather than the concentration of Mn in that phase. Because of this the results have been normalized to 70% MnO_2 and the associated elements have had their concentrations adjusted by the same factor. The results are the average of six or more point analyses. It is noteworthy that in these analyses, and also for those of Mn nodules, individual analyses made with the microprobe show very considerable

TABLE 3

Some trace element contents of the Mn-rich, and Fe-rich, points of surface soils

Sample	1		2		3	
	Mn rich	Fe rich	Mn rich	Fe rich	Mn rich	Fe rich
			Measured concentration, %			
MnO_2	8.7	.11	21.5	1.5	9.6	2.4
Fe_2O_3	5.6	11.4	25.3	42.4	11.1	14.8
			Normalized concentration, %*			
CoO	1.22	<0.1	.69	<.05	3.1	<.1
NiO	.15	.08	.11	.03	.62	.03
CuO	.12	.09	.04	<.02	<0.05	<0.05
ZnO	.69	.32	.04	.09	.11	.10
BaO	2.35	<0.2	.64	<0.1	3.2	<.2
PbO	2.2	<.15	.28	<.05	.37	<.15

* Normalized relative to 70% MnO_2 for the Mn-rich areas, and to 70% Fe_2O_3 for the Fe-rich areas.

1 Urrbrae loam. Red-brown earth developed on shale. Adelaide, South Australia.

2 Wollongbar clay loam. Krasnozem developed on basalt. Lismore, New South Wales.

3 Black earth developed on shale. Saddleworth, South Australia.

variations in trace elements contents. The results in Table 3 agree with those in Table 2, and show that Co, Ba and Pb are concentrated in the Mn oxides of these surface soils. The Fe oxides in the same soils had low contents of these elements. Although for Ni, Cu and Zn the Mn oxides had higher concentrations, the Fe oxides had comparable concentrations so that Fe oxides would account for more of the elements in the soil.

IRON

Of the elements considered here, Fe is by far the most abundant. In rocks, Fe occurs in both the ferrous and the ferric state, and in specifically Fe minerals as well as replacing Mg and Al in other minerals.

During weathering the Fe released from minerals may be removed from the soil, but usually much of it is retained as free oxides or in the clay minerals substituting for Al. Many of the black soils formed from basalt have most of their Fe in the clay mineral montmorillonite. Illites (clay micas) commonly contain several per cent Fe, but kaolin has little or no Fe in its structure. Iron in the soil that does not enter the clay minerals forms free Fe oxides. These have long been recognized because of their high pigmenting power. Goethite, α-FeOOH, and hematite, α-Fe_2O_3, are by far the commonest, but maghemite, γ-Fe_2O_3, is widespread and occurs particularly in ironstone nodules (Oades & Townsend 1963; Kojima 1964). Lepidocrocite, γ-FeOOH, is restricted to gleyed soils. Hematite gives soils their red colour, but even in many red soils goethite (yellow coloured) is the dominant Fe oxide.

Norrish & Taylor (1961) have shown that soil goethites are not pure Fe oxides, but contain up to 25 mole % of Al. Their studies showed also that soil goethites are very fine grained, as small as 0.02 μm, so that, like Mn oxides, they can contribute significantly to the surface activity of soils. Elec-

TABLE 4

Analyses of goethites

	Concentration, %			
	1	2	3	4
Fe_2O_3	73.9	88.5	81.3	73.7
Al_2O_3	8.68	-	0.30	2.96
TiO_2	1.53	0.7	0.02	0.01
V_2O_5	-	3.0	-	-
MnO	-	-	0.34	0.65
ZnO	-	2.15	-	2.93
CuO	-	0.10	-	-
MoO_3	-	1.42	-	-
P_2O_5	2.16	2.66	4.43	0.03
SO_3	0.7	-	0.08	-
SiO_2	1.02	-	0.30	5.02
Crystal size* Å	300	-	1000	500

* Crystal size estimated from X-R-D line broadening.

1 Soil goethite 'single crystals' from Queensland - see text. Electron microprobe analysis.

2 Goethite, < 0.5 μm, from the weathered surface of a V-rich shale. Analysis by difference, from results prior to and after removing the goethite. Ignited basis.

3 Goethite from Gossan, Cobar district, New South Wales. Electron microprobe analysis.

4 Goethite from near Broken Hill, New South Wales. Electron microprobe analysis.

tron micrographs of soil Fe oxides confirm that they occur as discreet crystals of the above size (Greenland *et al.* 1968).

Soil goethites are so fine that they cannot be analysed as pure separates. Table 4 gives some analyses of goethites and, although only sample 1 is a soil goethite, all have formed as surface weathering products. Sample 2 was analysed by extracting the goethite using sodium dithionite. The other three samples, despite their chemistry, appear to be pure goethite with no other phases when examined petrographically and by X-ray diffraction. Under the electron microprobe the samples are uniform and the analyses are the averages of many spots. Sample 1 consists of grains similar to those described by Brown & Stephen (1959). The grains are pure goethite replacing olivine crystals and, although they appear as single crystals under the petrological microscope, the grains are made up of very fine three-dimensionally oriented crystals. This specimen contains no unaltered olivine or layer silicates and appears to be more weathered than that studied by Brown & Stephen. Figure 4 shows a scanning electron image of an etched surface of such a goethite grain, and it shows clearly a crystallite size of several hundred Angstroms, which agrees with the X-ray diffraction measurement.

The complex chemistry of the goethites given in Table 4 is not unusual. The author has examined many others and it is very rare to obtain anything approaching pure Fe hydroxide. Almost without exception the crystallite size is small, and a variety of elements is present in appreciable amounts. To what extent these are present in the goethite lattice, or as

Fig. 4. Scanning electron micrograph of a soil goethite grain.

surface phases, is not known, but obviously goethites have the capacity to retain large amounts of the nutrient elements.

Despite the fact that many papers have been written on the amorphous nature of soil Fe oxides (Mackenzie 1949; Matsusaka & Sherman 1961), it is true that in most Australian soils these minerals are crystalline, though fine grained. Certainly some of the identifications of amorphous Fe oxides can be traced to insensitive or incorrect X-ray diffraction techniques. Taylor (1959) has shown that identifications of amorphous Fe oxides, based on differential thermal analysis (DTA), are often incorrect. Other Fe-bearing minerals that resist weathering occur in soils, e.g. chromite, garnets, ilmenite, etc., but these would have little significance agriculturally. Jarosite, a comparatively soluble mineral, has been identified in particular soils from Australia and overseas. In most soils it is probable that the free Fe oxides supply the required Fe for plant growth. Their very small particle size would be an aid in this respect, as in oxidized soils the minerals have a very low solubility and solution may be accomplished in micro-environments that are short lived.

Iron oxide minerals have not been investigated fully, but the indications are that these minerals have a very complex chemistry, like the Mn oxides, that involves other biologically important elements besides Fe.

COBALT

In igneous rocks there are no Co minerals; the Co is hidden in ferromagnesian minerals proxying for Fe. Cobalt does not enter the layer lattice silicate structures. Vinogradov (1959) suggested that Co was associated with Mn in soils but he presented no data to support the claim. For a long time it has been known that Co and other metals are high in marine Mn nodules, and Goldschmidt (1958) speculated on the possibility of a similar chemistry for the soil oxides. Oertel & Giles (1964), after observing a correlation between the elements in Queensland soils, suggested that Co was associated with Mn oxides. The data of Tables 2 and 3 show clearly that the soil Mn oxides have a very high Co content, and investigations by Taylor & McKenzie (1966) have shown clearly that most of the Co in soils is associated with the Mn oxides. Further studies by McKenzie (1967,1970) and Murray *et al.* (1968) have shown that both oxides separated from soils and those synthetically prepared absorb Co specifically from very dilute solutions. Adams *et al.* (1969) found that the uptake of Co fertilizer by plants was dependent on the Mn content of the soils, and Table 5 shows that relatively quantitative predictions can be made of Co uptake using pH and Mn content of the soil.

If Co is retained in Mn oxides, it seems possible that the Co/Mn ratio might be a measure of Co potential in the soil. Table 6 shows results obtained using data from the experiments of Adams *et al.* (1969). The Co content of plants grown in pots is correlated with Co/Mn ratio of the soil ($r = 0.84$ with no effect due to pH). When data from a larger group of soils (90) from different localities were tested in the above way, the results were less encouraging. But the data were obtained in different laboratories, under undoubtedly different conditions, and in different years, so the result is perhaps not surprising.

TABLE 5

Comparison of the actual and predicted uptake of applied Co by clover. Prediction using MnO content and pH of soil†.*

Great Soil Group	Site	Clay%	C%	MnO ppm	Co ppm	pH	Recovery of applied Co, μg per pot	
							Actual	Predicted
Prairie	3	41.4	4.1	5800	97	7.1	0.4	-0.3
	4	25.9	4.3	2100	51	6.3	0.6	0.6
	5	41.6	4.0	1570	49	7.1	0.01	0.1
	6	39.2	8.3	3500	78	6.3	0.05	0.3
	7	36.2	4.8	1100	46	6.5	-0.2	0.8
Gleyed podzolic	8	26.4	3.1	1200	16	5.6	-	-
	9	18.4	3.5	495	3.7	6.0	2.7	2.2
	10	9.2	2.4	500	5.2	6.2	2.9	1.8
	11	18.6	2.6	2700	33	5.8	2.1	0.9
	12	10.9	2.7	990	8.5	5.2	4.0	2.7
Humus podzolic	13	4.4	2.9	30	0.22	5.8	6.2	8.8
	14	4.9	3.5	40	0.45	5.5	9.9	9.4
	15	4.6	2.9	125	0.18	7.5	1.1	1.3
	16	5.2	3.5	40	0.35	5.2	7.1	11.3
	17	4.7	3.7	49	0.48	5.1	7.1	11.0
Yellow podzolic	18	16.2	3.8	350	1.3	5.5	5.3	3.7
	19	29.4	2.8	4100	8.3	5.4	0.9	1.0
	20	33.2	3.7	166	2.4	5.1	10.6	6.7
	21	28.9	4.5	3600	5.6	6.0	0.2	0.5
	22	25.2	3.9	100	0.9	5.2	15.5	7.8
Krasnozem	23	29.2	10.9	560	4.9	5.2	4.4	3.6
	24	42.9	9.1	3800	10.4	5.4	0.2	1.0
	25	38.4	8.3	6700	26.2	5.8	0.3	0.3
	26	36.4	5.7	21000	89	5.8	-0.2	-0.1
	27	25.9	5.8	4600	29.9	5.5	0.3	0.8

* Data from Adams *et al.* (1969).

† $\log(\text{Co} + 1) = 2.9023 - 0.369 (\log \text{MnO ppm}) - 0.235\ \text{pH}$.
Actual recovery of Co : S.D. between duplicates = 2 μg.
S.D. between actual and predicted recovery = 1.6 μg.

TABLE 6

Comparison of actual and predicted Co contents of clover grown in pots. No Co added.*

Site†	100(Co/Mn) for soil	Co content of plants (ppm)	
		Actual	Predicted#
8	1.36	0.57	0.50
9	0.75	0.14	0.26
10	1.05	0.12	0.38
11	1.21	0.37	0.44
12	0.86	0.45	0.30
13	0.73	0.23	0.25
14	1.1	0.48	0.40
15	0.14	0.05	0.02
16	0.88	0.30	0.31
17	1.0	0.54	0.36
18	0.37	0.18	0.11
19	0.20	0.10	0.04
20	1.63	0.66	0.60
21	0.15	0.06	0.02
22	0.90	0.40	0.32
23	0 87	0.10	0.31
24	0.27	0.05	0.07
25	0.39	0.06	0.12
26	0.42	0.14	0.13
27	0.65	0.16	0.22

† See Table 5.
# ppm = 39.14 (Co/Mn) - 0.0344.
* Data from Adams *et al.* (1969).

It is known that Co availability is increased in waterlogged soils, and, if it is in Mn oxides, it will have the same sensitivity to pH and Eh as Mn.

The author considers that all the above evidence shows that the active Co is associated with the Mn oxides in soils, and, in this respect, it is one of the trace elements whose behaviour is reasonably understood. However, it must be pointed out that various workers have presented data that are not in accord with this view. Gonzalez Garcia & Garcia Gomez (1964), analysing certain Spanish soils, obtained a high correlation of Co with Fe ($r = 0.84$)*. In some other soils they obtained

* The authors give $r = 0.96$, but using the data of their Table 1 a value of 0.84 is obtained.

a poorer correlation with Mn (r = 0.51).

Tiller & Hodgson (1962) in adsorption experiments have shown that Co is fixed by clay minerals, and Grimme (1968) in similar experiments has shown that Fe oxide fixes Mn, Co, Cu and Zn. Undoubtedly clays and Fe oxides can fix Co, and yet in most normal soils it is the Mn oxides, which are in much lower concentration, that hold the Co. Presumably the Co is adsorbed much more strongly on the Mn than the Fe oxides, and the latter may hold the Co only in the absence of Mn oxides. In soil solutions, many competing ions will be present and perhaps some of the other ions prevent the adsorption of Co by Fe oxides.

NICKEL

In igneous rocks some Ni may occur in sulphides, but most of it, like Co, is in ferromagnesians replacing Fe. During weathering, Ni can combine with silica, so that the weathered zone of a Ni ore deposit may consist of garnierites, minerals that are high in Ni ($\sim$ 10%) and are related to clay minerals. To what extent Ni is incorporated into clay minerals in normal soils is not known.

From Tables 2 and 3 it can be seen that Ni concentrations are high in Mn and Fe oxides, but the studies of Taylor & McKenzie (1966) show that for most soils less than 15% of the total Ni is extracted with the Mn oxides. Le Riche & Weir (1963) extracted less than 30% of the total Ni from the soils using methods to remove free Fe (and Mn) oxides.

It seems from the limited data available that only a fraction of the Ni of soils is carried in the oxides of Fe and Mn; nevertheless it may be that this is the fraction that is most available for plants.

COPPER

In igneous rocks Cu is present as sulphides; it does not enter into silicate minerals. Copper does not behave like a simple ion, and in weathering it tends to form hydroxy compounds and carbonates rather than enter clay mineral structures. These compounds of Cu can persist in very arid climates but would not be expected in agricultural soils.

Copper is known to accumulate in Mn nodules (Table 3), and McKenzie (1971) mentions that it is high in the soil Mn oxides but that only a small fraction of the total Cu in the soil is in this form. Copper is also known to be high in Fe oxides, and Le Riche & Weir (1963) removed an average of 50% of the soil Cu in treating the soil to remove Fe and Mn oxides. There is evidence that in some soils a large fraction of the Cu is combined with organic matter.

ZINC

Some Zn in igneous rocks is present in sulphides but most of it is in silicates in small amounts, as no independent Zn silicates occur. Zinc can be high in such minerals as ilmenite and magnetite that resist weathering in soils. It is known to accumulate in Mn and Fe oxide sediments, so it is likely that the same association might be present in soils. Table 4 shows that goethites can contain appreciable amounts of Zn. Taylor & McKenzie (1966) found that less than 15% of the total Zn was extracted in removing Mn oxides from soils. Against this, White (1957) extracted 45% in removing free Fe oxides from soils, and considered that an average of 35% was held within the clay mineral structures. Table 3 shows that the Mn and Fe oxides have comparable concentrations of Zn, and in this case Fe oxides, because of their abundance, will account for much more of the Zn. It is well known that Zn can enter layer lattice silicate structures and this is exemplified by the mineral sauconite, a montmorillonite containing approximately 30% ZnO.

It seems likely that, in soils, that fraction associated with the Fe and Mn oxides may be the most available to plants.

ARSENIC

Compared to igneous rocks, soils are enriched in As. In igneous rocks, As is present as arsenide in sulphides. In an oxidizing atmosphere the As will oxidize to arsenate, AsO_4^{3-}, and in this form its crystal chemistry resembles phosphates and vanadates, and sometimes silicates and sulphates.

It is well known that arsenate is retained by Fe oxides — ferric hydroxide is used as an antidote to As poisoning. Also marine Fe ores are high in As, containing about 500 ppm (Goldschmidt 1958). It is worth noting that sedimentary Mn ores are lower than this, about 70 ppm. Goldschmidt (1958) also reports that Fe-rich bauxites contain about 500 ppm As, an amount that is much higher than the level of white bauxites.

There is good evidence that arsenate in soils behaves like phosphate and is fixed by Fe oxides (Jacobs *et al.* 1970). However, the literature on phosphate retention in soils is very confused [see reviews by Wild (1950), Hemwall (1957) and Kardos (1964)]. It has been proposed that phosphate (and arsenate) forms insoluble crystalline compounds with Fe, Al and other oxides, and with clay minerals.

Recently Fordham & Norrish (1974) have published results which show clearly that, in the soil they studied, added arsenate was retained by goethite. They added a radioactive isotope of As (as arsenate) to clay, after which the particles that had adsorbed arsenate were identified under the electron microscope by autoradiography and analysed using the microprobe. Figure 5 shows some of their results.

While investigating the nature of As in a lateritic podzolic soil near Adelaide that was high in As, the author obtained the electron microprobe results shown in Figure 6. These analyses show a clear association of As with Fe (goethite), $r = 0.91$.

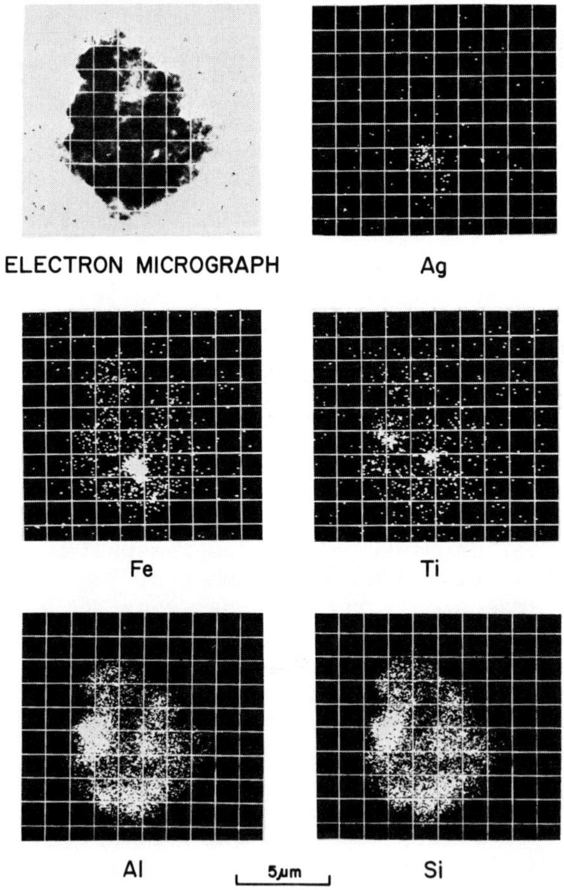

Fig. 5. Electron microscope and microprobe photos of a soil aggregate showing As uptake by Fe oxide. The region of As uptake is indicated by high Ag. [From Fordham & Norrish 1974).]

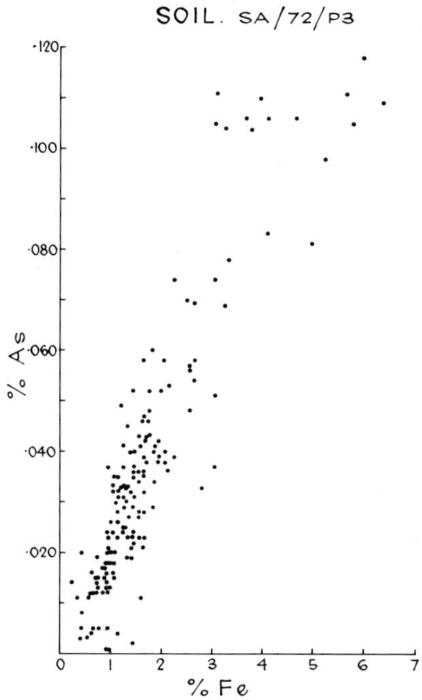

Fig. 6. Microprobe results showing the association of As with Fe in a lateritic podzolic soil from Mt. Torrens, South Australia.

Although adsorption experiments indicate that Al oxides and kaolin as well as Fe oxides are important in phosphate (and arsenate) retention (Muljadi *et al.* 1966; Hingston *et al.* 1972), the above evidence suggests that most of the soil As is held by Fe oxides, and this agrees with Goldschmidt's observations (see above).

In the numerous clay mineral analyses made in this laboratory, it is found that the Fe oxides, and particularly goethites, are very fine grained, ∿ 0.02 μm. Against this, gibbsite, which is the commonest Al hydroxide in soils, is relatively coarse, often appearing in the coarse clay or silt fractions. In these circumstances, even assuming an equal specific surface activity for the Fe and Al oxides, the greater surface area of the Fe oxides would make them much more active.

SELENIUM

Normal soils average about 0.01 ppm Se, but certain regions in the U.S., Canada, Colombia, Mexico and the Hawaiian Islands can have relatively high contents, 20 ppm.

In igneous rocks, Se replaces S in sulphides. During weathering, S is more readily oxidized and leached from the soil. The chemistry of Se in soil has been the subject of controversy as elemental Se, metal selenides, selenites and selenates have all been suggested as the likely forms of occurrence (Vinogradov 1959; Wedepohl 1969).

It is known that Se is enriched in sedimentary Fe ores and it is true that Se is retained in Fe-rich soils, e.g. Hawaiian soils. Byers *et al*. (1938) have shown that selenite is bound more strongly than selenate by Fe oxide.

From the data available it seems likely that a portion of soil Se occurs as selenite adsorbed by Fe oxides. Some may be in the organic fraction.

BROMINE

Little is known of the way in which Br occurs in igneous rocks, but to some extent it may substitute for hydroxyls. The increased content of soils over rocks is due to additions to the soil by dust and rain, the Br being of oceanic origin. It is likely that a large proportion of Br in soils is organically combined (Vinogradov 1959).

MOLYBDENUM

In igneous rocks, Mo occurs as the sulphide, and in weathering it is oxidized to the molybdate ion. Goldschmidt (1958) notes that Mo is often concentrated in sedimentary Fe, and particularly Mn, oxide deposits. Both Vinogradov (1959) and Mitchell (1964) in reviewing the subject suggest that molybdate ions in the soil are bound by sesquioxides and clay minerals. In extracting the Mn oxides from soils, Taylor & McKenzie (1966) found no correlation between the fraction of Mo and the fraction of Mn extracted, and in most cases less than 10% of the total Mo was dissolved.

Against this rather negative result for Mn oxide, there is a consistency in investigations that show Mo is concentrated with the Fe oxides (Stanfield 1935; Oertel & Prescott 1944; Wells 1956; Turton *et al*. 1962; Taylor & Giles 1970), and Trobisch & Schilling (1963) found a correlation between the amounts of Mo and Fe dissolved by extracting agents. Jones (1957) showed that Fe oxides adsorb Mo more strongly than Al oxides and clay minerals. Figure 7 shows microprobe analyses of a weathered shale in which the Mo and Fe concentrations have a correlation coefficient of 0.92. Table 4, sample 2, gives an analysis of this goethite. In view of the many adsorption studies demonstrating fixation by Al oxides, a result of Wells (1956) is worth noting. He found that ironstone nodules from soils contained up to 10 ppm of Mo, whereas aluminous nodules contained < 1 ppm.

All the above results indicate that, in the inorganic form, Mo is associated with Fe oxides in soils, probably as an adsorbed phase. Although it has been suggested that the Mo and Fe form a distinct insoluble crystalline phase, there is little evidence to support such an hypothesis.

IODINE

This is the most dispersed of all elements, most rocks and minerals averaging around 0.3 ppm (Goldschmidt 1958). Because of its importance in nutrition, large numbers of I determinations have been made on soils and its geographic distribution is reasonably understood. Soils contain about 20 times as much as rocks. There is very considerable evidence that I is added to the soil from rain, the I entering the atmosphere from sea spray (Goldschmidt 1958; Vinogradov 1959).

The way in which I occurs in many rocks and minerals is not understood, but it can proxy for hydroxyls in micas, and a small amount of I may be retained in soils in the structures of such minerals.

Soils of high clay content and high organic matter content retain I but little is known of how it is held. Whitehead (1973*a,b*) has obtained significant correlations between organ-

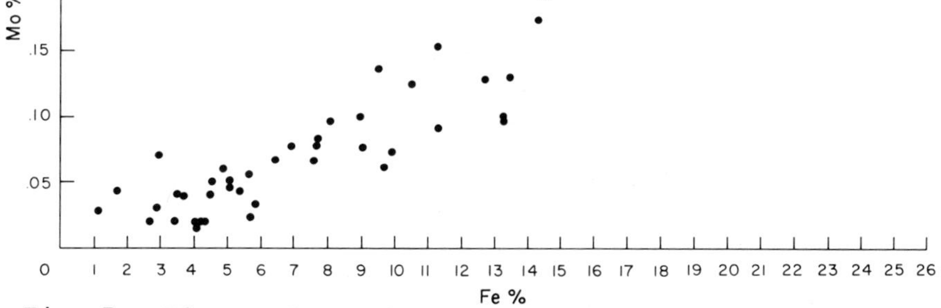

Fig. 7. Microprobe analyses showing the association of Mo with Fe in a weathered shale.

ic matter content and I content, and I adsorbed by soils. He also obtained high and significant correlations between total I and extractable Al and Fe oxides, and suggests that the free oxides in soils may retain some I in the same way that they fix phosphate and sulphate.

BARIUM

In the minerals of rocks and soils, Ba can replace K so that Ba can be high in micas and potash feldspars. Barite, $BaSO_4$, exists in Ba-rich soils and it is soluble enough to cause plants growing on these soils to be toxic to animals. In this laboratory, barite has been identified as nodules in particular Australian soils. Tables 2 and 3 show clearly that Ba is concentrated in Mn oxides in the soil and this is in accord with the usually high Ba content of sedimentary Mn deposits. Norrish (1968) and others (Adams *et al.* 1973) have shown that gorceixite, barium aluminium phosphate, exists in some soils (Fig. 8) but, except for particular soils, the Ba of this mineral would be a small part of the total soil Ba.

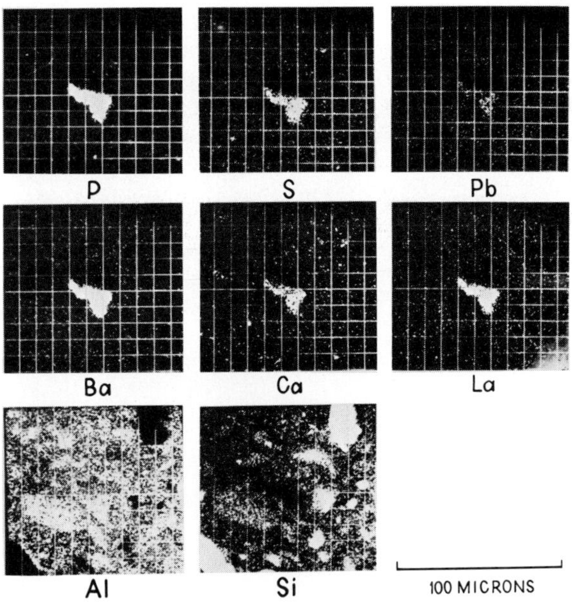

Fig. 8. Microprobe displays showing Ba, Pb and other elements in a complex phosphate mineral, in a laterite from Western Australia.

LEAD

In rocks, Pb commonly occurs as the sulphide, but its radius is such that it often replaces K, Ba, Sr and even Ca in minerals.

In extracting free Fe (and Mn) oxides from soils, Le Riche & Weir (1963) removed about 60% of the total Pb. Vinogradov (1959) points out that some soils, particularly the Gray Forest Soils, show an accumulation of Pb in the A_1 horizon. He also shows that much of the Pb of soils is removed by nitric acid and by acetic acid (60% and 30% respectively). Taylor *et al.* (1964) did not determine the Pb content of their soil Mn oxides, but noted that Pb can be high in such minerals. Taylor & McKenzie (1966) report Pb contents of several thousand ppm for two of their Mn oxides, but most were below the detection limit of 100 ppm. Against this, Table 3 shows that in some surface soils the Mn oxides have very high Pb contents, sufficient to account for most of the Pb in the soil.

Norrish (1968) has reported 1% Pb in a phosphate concentrate, gorceixite, from a laterite. In more recent studies the author has identified hinsdalite, in a Pb-rich lateritic podzolic soil from near Adelaide. The electron microprobe photos (Figs. 8 and 9) show the distribution of elements in

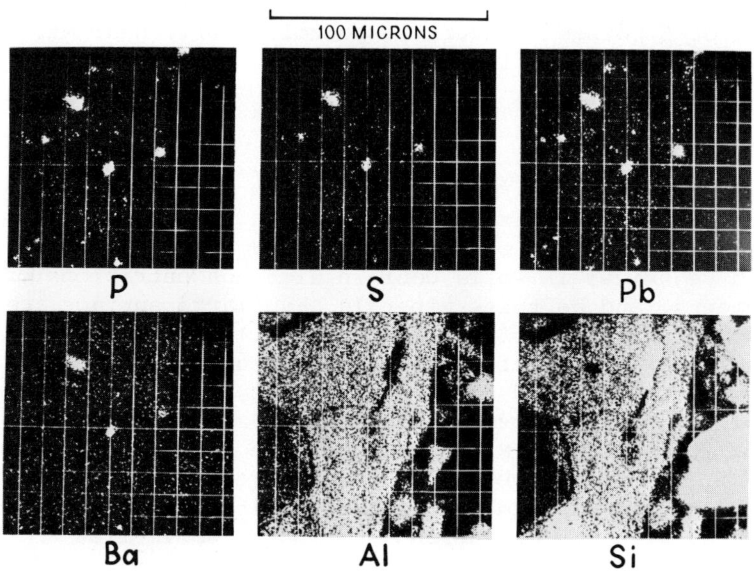

Fig. 9. Microprobe displays showing Pb in very high concentration associated with Al, P and S, in a lateritic podzolic soil, Mt. Torrens, South Australia.

these soils, and Table 7 gives microprobe analyses of the minerals. The Pb of these phosphate minerals would be much less available than that associated with the Mn oxides.

TABLE 7

Electron microprobe analyses of Pb-rich phosphate minerals in soils

	Sample 1 (%)	Sample 2 (%)
P_2O_5	23.8	14.64
SO_3	2.9	7.78
Al_2O_3	29.5	21.5
Fe_2O_3	3.2	10.1
PbO	1.3	35.2
BaO	11.9	2.9
La_2O_3	4.0	-
CeO_2	3.8	-
CaO	1.2	-
SrO	2.7	0.3
SiO_2	2.6	0.8

1 Average of 6 points. Laterite, Western Australia. Profile 31 of Norrish (1968).

2 Average of 5 points. Lateritic podzolic soil, near Mt. Torrens, South Australia.

CONCLUSIONS

That fraction of the trace elements held in the structure of resistant minerals is of little interest agriculturally. The Mn and Fe oxides in soils contain high concentrations of trace elements, and as these minerals have large surface areas and are comparatively mobile compared to silicates, it is likely they play an important role in retaining and supplying trace elements. In the case of Co at least, its availability is controlled by the Mn oxides in the soil. On the data available, it seems that Mn oxides can have high concentrations of Fe, Co, Ni, Zn, Ba and Pb, while Fe oxides retain V, Mn, Ni, Cu, Zn, As, Se and Mo. Note that it is Fe oxide that is associated with the oxy-anions and to the above list should be added sulphate, phosphate and silicate.

Laboratory adsorption experiments show that Al hydroxides can adsorb a variety of trace elements, and some workers, interpreting extraction experiments, consider that in soils the role of Al is more important than Fe in retaining some nutrient elements. There is little direct evidence to support this view. Likewise, at many times the fixation of elements has been attributed to reactions that result in the formation of distinct crystalline phases, and these have been demonstrated in experiments using concentrated solutions. There is no convincing evidence that such phases are formed by adding fertilizers to soils.

With the exception of a few elements, our knowledge of the mineralogy of trace elements is largely indirect and based on results the interpretation of which is not always unambiguous. In recent years there has been a considerable advance in methods and instrumentation, in chemistry and mineralogy, and the application of these techniques to soils should answer, with certainty, many of the long-standing questions concerning the nature of trace elements in soils.

REFERENCES

ADAMS, J.A., HOWARTH, D.J. & CAMPBELL, A.S. (1973) - *J. Soil Sci. 24* : 224

ADAMS, S.N., HONEYSETT, J.L., TILLER, K.G. & NORRISH, K. (1969) - *Aust. J. Soil Res. 7* : 29

BREWER, R., PROTZ, R. & McKEAGUE, J.A. (1973) - *Can. J. Soil Sci. 53* : 349

BROWN, G. & STEPHEN, I. (1959) - *Am. Miner. 44* : 251

BYERS, H.G., MILLER, J.T., WILLIAMS, K.J. & LAKIN, H.W. (1938) - *U.S. Dept. Agr. Techn. Bull. 601* : 1

COUCH, E.L. & GRIM, R.E. (1968) - *Clays Clay Miner. 16* : 249

DAY, F.H. (1963) - "The Chemical Elements in Nature". George G. Harrap & Co. : London

DEGENS, E.T. (1965) - "Geochemistry of Sediments". Prentice-Hall Inc. : New Jersey

FORDHAM, A.W. & NORRISH, K. (1974) - *Aust. J. Soil Res. 12* : 165

GREENLAND, D.J., OADES, J.M. & SHERWIN, T.W. (1968) - *J. Soil Sci. 19* : 123

GRIMME, H. VON (1968) - *Z. Pfl-Ernähr. Bodenk. 121* : 58

GOLDSCHMIDT, V.M. (1958) - "Geochemistry" (A. Muir, ed.). Clarendon Press : Oxford

GONZALEZ GARCIA, F. & GARCIA GOMEZ, A.M. (1964) - *An. Edafol. Agrobiol. 23* : 563

HATHAWAY, J.C. (1959) - *U.S. Geol. Surv. Profess. Papers No. 320* : 133

HEMWALL, J.B. (1957) - *Adv. Agron. 9* : 95

HINGSTON, F.J., POSNER, A.M. & QUIRK, J.P. (1972) - *J. Soil*

Sci. 23 : 177
HODGSON, J.F. (1963) - *Adv. Agron. 15* : 119
HOFMANN, J., WEISS, A., KOCH, G., MEHLER, A. & SCHOLZ, A. (1956) - *Proc. 4th Natl Conf. Clays Clay Miner.*, p. 273
HUBNER, M. (1969) - *Ber. dt. Ges. geol. wiss. B. Miner. Lagerslallenforsch. 14* : 5
JACOBS, L.W., SYERS, J.K. & KEENEY, D.R. (1970) - *Proc. Soil Sci. Soc. Am. 34* : 750
JASMUND, K. & LINDNER, B. (1973) - *Proc. Int. Clay Conf., Madrid, 1972*, p. 399
JENNE, E.A. (1968) - *Adv. Chem. Series 73* : 337
JONES, L.H.P. (1957) - *J. Soil Sci. 8* : 313
KARDOS, L.T. (1964) - *In* "Chemistry of the Soil" (F.E. Bear, ed.), 2nd Edn. Reinhold Publishing Corp. : New York
KOJIMA, M. (1964) - *Nippon Dofo-Hiryogaku Zasshi 35* : 171
LE RICHE, H.H. & WEIR, A. (1963) - *J. Soil Sci. 14* : 223
MASON, B. (1952) - "Principles of Geochemistry". John Wiley & Sons Inc. : New York
MATSUSAKA, Y. & SHERMAN, G.D. (1961) - *Soil Sci. 91* : 239
MACKENZIE, R.C. (1949) - *Nature, Lond. 164* : 244
McKENZIE, R.M. (1967) - *Aust. J. Soil Res. 5* : 235
McKENZIE, R.M. (1970) - *Aust. J. Soil Res. 8* : 97
McKENZIE, R.M. (1971) - *Pfl-Ernähr. Bodenk. 131* : 221
MITCHELL, R.L. (1964) - *In* "Chemistry of the Soil" (F.E. Bear, ed.), 2nd Edn. Reinhold Publishing Corp. : New York
MULJADI, D., POSNER, A.M. & QUIRK, J.P. (1966) - *J. Soil Sci. 17* : 212
MURRAY, D.J., HEALY, J.W. & FUERSTENSU, D.W. (1968) - *Adv. Chem. Series 79* : 74
NORRISH, K. (1968) - *Trans. 9th Int. Congr. Soil Sci., Adelaide II* : 713
NORRISH, K. & TAYLOR, R.M. (1961) - *J. Soil Sci. 12* : 294
OADES, J.M. & TOWNSEND, W.N. (1963) - *J. Soil Sci. 14* : 179
OERTEL, A.C. & GILES, J.B. (1963) - *Aust. J. Soil Res. 1* : 215
OERTEL, A.C. & GILES, J.B. (1964) - *Aust. J. Soil Res. 2* : 162
OERTEL, A.C. & PRESCOTT, J.A. (1944) - *Trans. Roy. Soc. S. Aust. 68* : 173
RANKAMA, K. & SAHAMA, T.G. (1950) - "Geochemistry". Univ. of Chicago Press : Chicago
SILANPÄÄ, M. (1972) - FAO Soils Bull. 17
SIMS, J.R. & BINGHAM, F.T. (1967) - *Proc. Soil Sci. Soc. Am. 31* : 728
SIMS, J.R. & BINGHAM, F.T. (1968) - *Proc. Soil Sci. Soc. Am. 32* : 364
STANFIELD, K.H. (1935) - *Ind. Engng Chem. analyt. Edn 7* : 273
SWAINE, D.J. (1951) - Thesis, University of Aberdeen.
SWAINE, D.J. (1955) - Commonw. Bur. Soil Sci. Tech. Commun. No. 48

SWAINE, D.J. & MITCHELL, R.L. (1960) - *J. Soil Sci. 11* : 345
TAYLOR, R.M. (1959) - *J. Soil Sci. 10* : 309
TAYLOR, R.M. & GILES, J.B. (1970) - *J. Soil Sci. 21* : 203
TAYLOR, R.M. & McKENZIE, R.M. (1966) - *Aust. J. Soil Res. 4* : 29
TAYLOR, R.M., McKENZIE, R.M. & NORRISH, K. (1964) - *Aust. J. Soil Res. 2* : 235
TILLER, K.G. (1963) - *Aust. J. Soil Res. 1* : 74
TILLER, K.G. & HODGSON, J.F. (1962) - *Proc. 9th Natl Conf. Clays Clay Miner.*, p. 393. Pergamon Press : New York
TROBISCH, S. & SCHILLING, G. (1963) - *Chemie Erde 23* : 91
TURTON, A.G., MARSH, N.L., McKENZIE, R.M. & MULCAHY, M.J. (1962) - CSIRO Aust. Div. Soils, Soil Publ. No. 20
VINOGRADOV, A.P. (1959) - "The Geochemistry of Rare and Dispersed Chemical Elements in Soils", 2nd Edn (English translation by Consultants Bureau Inc.). Chapman and Hall : London
WEDEPOHL, K.H. (ed.) (1969) - "Handbook of Geochemistry". Springer-Verlag : Berlin
WELLS, N. (1956) - *N.Z. Jl Sci. Technol. 37* : 482
WHITE, M.L. (1957) - *Econ. Geol. 52* : 645
WHITEHEAD, D.C. (1973*a*) - *J. Soil Sci. 24* : 260
WHITEHEAD, D.C. (1973*b*) - *J. Sci. Fd Agric. 24* : 547
WILD, A. (1950) - *J. Soil Sci. 1* : 221

THE MINERALOGY AND CHEMISTRY OF SOIL COBALT

R.M. McKenzie

Division of Soils, CSIRO
Glen Osmond, South Australia 5064

INTRODUCTION

An adequate supply of Co in pastures is essential to the health of grazing animals. Ruminants in many countries have been afflicted with a disease known variously as Bush Sickness, Denmark Wasting Disease, Coasty Disease and Pining, due to a deficiency of Co. The forms in which Co occurs in soils are of considerable interest from the points of view of the availability of naturally occurring Co and of the fate of Co applied as fertilizer.

Various authors have claimed that Co accumulates in the Fe oxides in soils. Le Riche (1973) extracted soils with hydrogen peroxide and ammonium oxalate under ultra-violet light, and found that almost all of the Co was extracted. He concluded that the Co was associated with the sesquioxides. Kubota (1965) extracted soils with dithionite and concluded that Co was associated with the Fe oxide. Gonzalez Garcia & Garcia Gomez (1964) found correlations between extractable Fe and Co in some soils, and between total Mn and Co in others. Grimme (1968) has shown that Co is adsorbed by Fe oxides, and Tiller *et al.* (1962,1963) have studied the adsorption of Co by layer silicates and soil clays. However, in a study of the lateritic soils of Western Australia, Turton *et al.* (1961) found no evidence to suggest that Co was significantly enriched in the ironstone nodules relative to the soil matrix. More recent work leaves little doubt that the most important factor in the chemistry of soil Co is its association with the Mn oxide minerals.

MATERIALS

The Mn oxides used in the studies described here were the following.

Naturally Occurring Oxides

These consisted of Mn nodules hand-picked from soils, containing the Mn minerals birnessite, hollandite, todorokite and lithiophorite. These minerals occur mixed with Fe oxides and other soil materials. They are non-stoichiometric varieties of Mn dioxide and contain a number of other elements, as well as the structural Ba in hollandite and Al and Li in lithiophorite.

Synthetic Oxides

These included the non-stoichiometric minerals birnessite, cryptomelane (a K-bearing mineral) and ρ-MnO_2, and the stoichiometric minerals pyrolusite (MnO_2), partridgeite (α-Mn_2O_3), manganite (γ-MnOOH) and hausmannite (Mn_3O_4). Partridgeite and manganite contain trivalent Mn, and hausmannite contains both di- and tri-valent Mn. Synthesis of these minerals is described in detail by McKenzie (1971).

ANALYTICAL DATA

Rankama & Sahama (1950) mentioned that Co accumulates with Mn in the oxidate sediments, and various references to Co accumulation in Mn minerals in soils have been made. Tiller (1963) reported the occurrence of lithiophorite containing 1% of Co in a krasnozem soil in Tasmania, and Giles (1962) found an accumulation of Co in a Mn nodule in a Queensland soil. Giles (1962) and Oertel & Giles (1964) noted a correlation between Co and Mn in soil profiles in Queensland, and suggested that Co was retained by the deposited Mn. However, the extent of the association and its importance from an agricultural point of view have only become evident since a comprehensive study of the Mn oxides in soils was undertaken in the laboratories of the Division of Soils, CSIRO.

Taylor & McKenzie (1966) analysed soils and Mn nodules, and results for the whole soils and nodules hand-picked from them are shown in Table 1. These figures show that Co accumulates in the nodules to a much greater extent than Ni.

Mineralized Mn was extracted from a number of soils and nodules to determine the amounts of other elements extracted with the Mn (Taylor & McKenzie 1966). These samples were

TABLE 1

Analyses of whole soils and hand-picked nodules (ppm)

Soil description	whole soil			nodules		
	Mn	Co	Ni	Mn	Co	Ni
Light brown sandy soil from Chewco, Qld.	92	4.6	22	3600	82	67
Sandy red earth, Esk district, Qld.	2500	20	8.3	72000	380	41
Red-yellow podzolic, Esk district, Qld.	1300	5.6	14	40000	110	39

Data from Aust. J. Soil Res. 4 : 33 (1966)

ground to pass a 0.1 mm sieve, shaken for 1 min with 0.1 M nitric acid, centrifuged, and then allowed to stand for 1 hr in 0.001 M nitric acid. These acid washings were designed to remove soluble and exchangeable ions, and the washings were discarded. The residues were then extracted with 10% hydrogen peroxide in 0.001 M nitric acid, small amounts of peroxide and nitric acid being added as required to maintain the reaction and pH. The extracts were finally recovered by centrifugation, evaporated to dryness and analysed. Typical results for Mn, Co and Ni in soils and extracts are shown in Table 2. Although some of these soils did not contain visible nodules or Mn stains, the results in Table 2 again show that Co is closely associated with the extractable Mn, while Ni is less closely associated.

Figure 1 shows plots of Co and Ni *versus* Mn extracted from these and other soils and nodules, expressed as the ratio of

TABLE 2

Analyses of soils and peroxide extracts (ppm)

Soil			Peroxide extract		
Mn	Co	Ni	Mn	Co	Ni
310	8.1	18	55000	860	210
110	8.9	20	8100	360	440
880	29	140	69000	1800	260
160	15	130	4700	200	320
590	28	46	64000	1500	390

Data from Aust. J. Soil Res. 4 : 33 (1966)

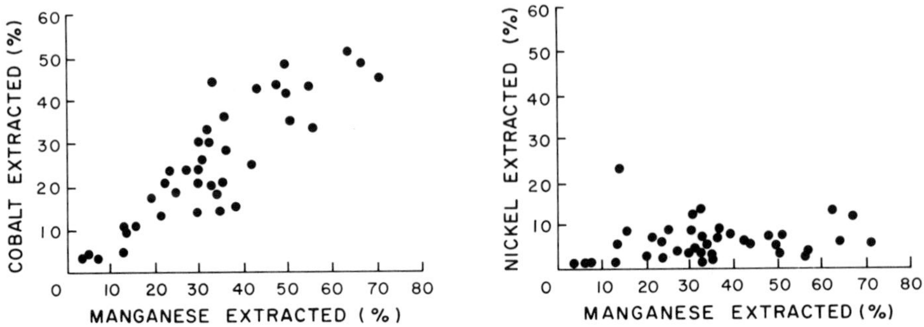

Fig. 1. Cobalt and nickel versus *manganese extracted from soils and nodules by H_2O_2, expressed as the ratio of amount extracted to amount present. [Adapted from Aust. J. Soil Res. 4 : 35 (1966)].*

amount extracted to total amount present. These show a marked linear relationship between Mn and Co extracted, with a correlation coefficient of 0.87, while there is no such relationship between Mn and Ni ($r = 0.11$). Similar plots for Mo, V, Cr and Zn showed that there was no relationship between these elements and Mn. The slope of the regression line for the Co-Mn relationship suggests that an average of about 80% of the total Co is associated with the Mn oxide minerals. While Ni is clearly also concentrated in the Mn minerals, it is evident that a much smaller proportion of the total amount is in this form.

ADSORPTION STUDIES

Adsorption studies were carried out on ground Mn nodules (McKenzie 1967) by shaking 0.1 g of the ground nodules in 20 ml of 0.1 M calcium chloride solution containing 5 mg of Cu^{2+}, Co^{2+} or Ni^{2+} per gram of sample. pH was adjusted by adding HCl or NaOH, and the amount sorbed was determined by measuring the amount left in solution after shaking for 4 days. The amount sorbed, expressed as a percentage of the amount added, is plotted against pH in Figure 2(a). Figure 2(b) is a similar plot for the same elements sorbed by goethite, from solutions containing 1.2 mg of adsorbing element per gram of goethite. The shapes of these two groups of curves are different due to the smaller amounts added to goethite. However, they show that, while Co is sorbed to a slightly greater extent, relative to Ni, by the Mn oxide, the amounts of Cu sorbed in both cases are far greater than those of Co. These curves, then, do not suggest any reason for the accumulation of Co in the Mn oxides in soils.

When the sorbed elements were removed by extraction with 2.5% acetic acid, it was found that, in the case of goethite, a relatively constant proportion of each element was removed, so that the amounts of non-extractable elements remaining were still in the order Cu > Co > Ni. A much smaller proportion of the Co sorbed on the Mn oxide was removed, so that in this case the relative amounts remaining after extraction were in the order Co > Cu > Ni. To study the effect of aging on the extraction of sorbed Co, a number of samples of ground nodules were prepared for each element, containing 0.99 mg/g of sorbed Cu, 0.90 of Co and 0.71 of Ni. These samples were washed once with water, and the amounts of Cu, Co or Ni extractable by 2.5% acetic acid were determined on one sample for each element. The remainder were stored wet for periods of up to 7 months before extraction with acetic acid. As shown in Figure 3, the amounts of non-extractable Co increased sharply with

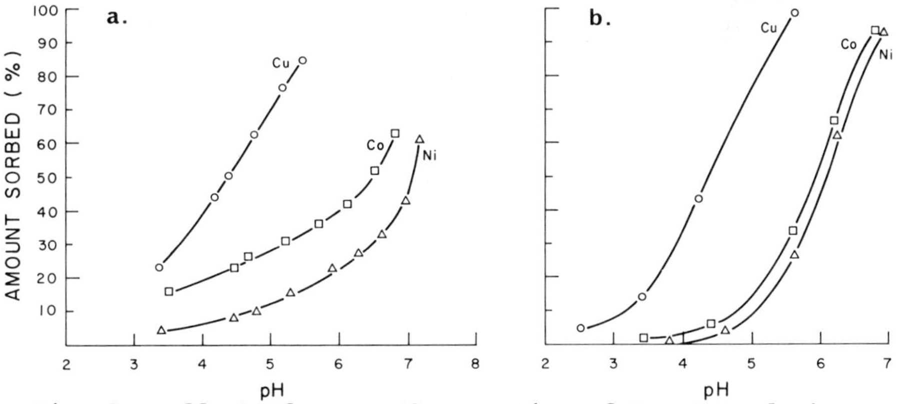

Fig. 2. Effect of pH on the sorption of Cu, Co and Ni. (a) ground Mn nodules; (b) goethite. [Reproduced from Aust J. Soil Res. 5 : 238 (1967)]

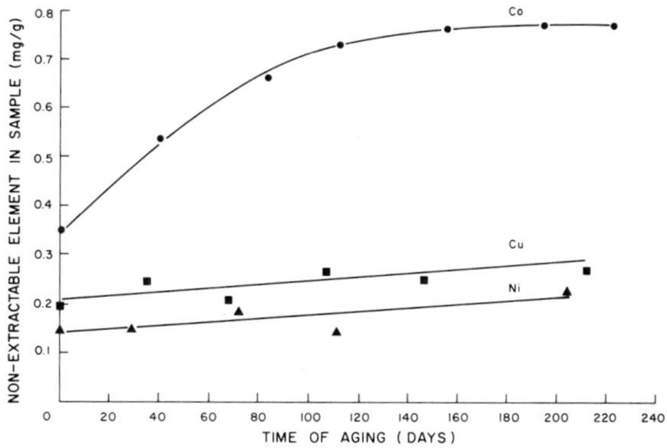

Fig. 3. Effect of aging on the non-extractable fraction of sorbed Co, Cu and Ni on ground Mn nodules. [Reproduced from Aust. J. Soil Res. 5 : 244 (1967)]

time of aging until, after 180 days, 86% of the Co originally sorbed had become non-extractable. This was more than double the amount found immediately after sorption, while results for Cu and Ni showed only a slight increase in the non-extractable fraction.

Further sorption studies were made on synthetic Mn oxide minerals (McKenzie 1970), because the naturally occurring oxides in soil nodules could not be separated from the large amounts of other minerals present. The minerals used were two birnessites, prepared by different methods, two cryptomelanes, ρ-MnO_2 and pyrolusite. Two series of experiments were carried out on these samples. In the first, samples were saturated with Co, Cu or Ni by standing, with occasional shaking, in 0.3 M solutions of $CoCl_2$, $CuCl_2$ or $NiCl_2$. The solutions were renewed at intervals of 2 weeks, and at the end of 8 weeks the samples were washed and dried for analysis. In the second series of experiments, 0.1 g samples were shaken for periods varying from 1 to 42 days with 20 ml portions of 0.1 M $CoCl_2$, $CuCl_2$ or $NiCl_2$, after which the samples were analysed for Co, Cu or Ni, and the supernatant for Mn, K (where K was present in the original material) and pH. Analytical data for the first series of experiments, in which samples were saturated with Co, Cu or Ni, are presented for two samples in Table 3. The values for MnO represent Mn of valence less than 4, and the fact that they are reported as MnO is not intended to imply that this fraction is actually present as Mn^{2+} rather than Mn^{3+}. Table 3 shows that Cu is the most effective in replacing exchangeable K, but Co is most effective in replacing lower-valence Mn. Similar results were obtained for the other sample of birnessite and the two cryptomelanes.

Results for the second series of experiments (sorption *versus* time) are plotted for two samples in Figure 4, in which the sorption of Co, Cu or Ni, and the release into solution of Mn, K and H ion are plotted as meq/g of sample against time

TABLE 3

Analyses of samples saturated with Co, Cu or Ni (%, material dried at 105°C)

Mineral	Treatment	MnO_2	MnO	K_2O	CoO	CuO	NiO	H_2O	Total
Birnessite	Untreated	69.2	13.4	10.5	0	0	0	7.3	100.4
	Co	65.4	4.8	1.6	17.2	0	0	10.4	99.4
	Cu	62.8	8.4	0.6	0	16.7	0	10.8	99.3
	Ni	65.0	12.4	1.8	0	0	8.3	11.5	99.0
ρ-MnO_2	Untreated	88.7	4.4	0	0	0	0	5.9	99.0
	Co	80.2	0.4	0	11.2	0	0	ND	91.8
	Cu	82.7	2.6	0	0	7.4	0	6.7	99.4
	Ni	84.5	4.0	0	0	0	3.5	ND	92.0

Data from Aust. J. Soil Res. 8 : 99 (1970)

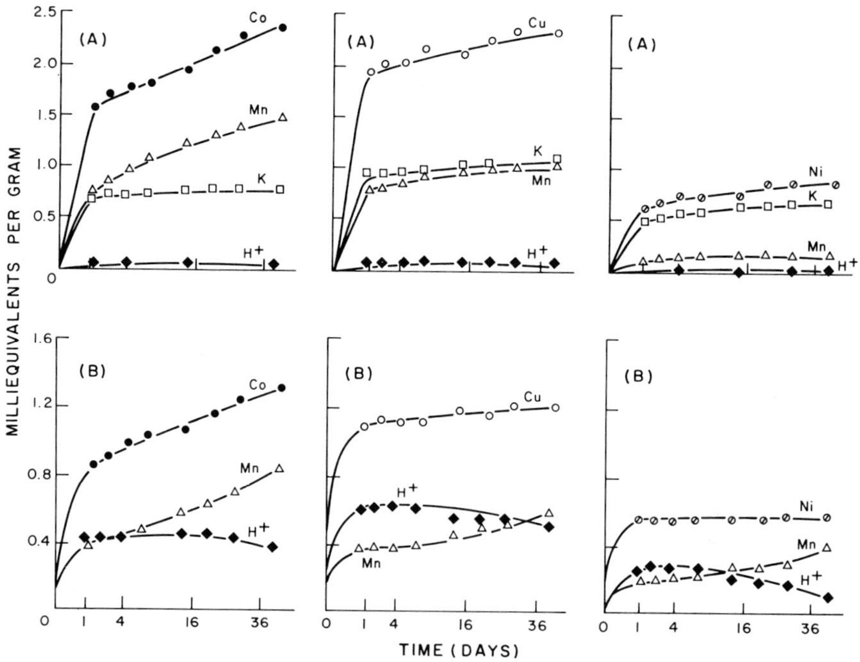

Fig. 4. Variation in uptake of Co, Cu and Ni, and release of Mn, K and H^+ *with time. (a) birnessite; (b)* ρ*-*MnO_2*. [Adapted from Aust. J. Soil Res. 8 : 100 (1970)]*

in days. These curves, and others not given, show that, at any time, the amount of metal ion sorbed was balanced by the sum of Mn^{2+}, K^+ and H^+ released into solution. There was a large initial uptake of metal ions on the first day, replacing rapidly exchangeable Mn^{2+}, K^+ and H^+. After the initial reaction there was a slow uptake accompanied by a further release of Mn^{2+}, K^+ and H^+, or in some cases by the disappearance from solution of K^+ or H^+. For all samples except pyrolusite, Co replaced less K^+ and H^+ but more Mn^{2+} than did Cu, and, although the initial uptake of Co was smaller than that of Cu, the amount slowly sorbed between days 1 and 42 was greater. Furthermore, where the slow uptake of Cu was significant, it was accounted for by the slow release of H^+ or K^+, and where there was a significant slow release of Mn during the reaction with Cu, this was at least partly accounted for by the disappearance of H^+ from solution, and not by the uptake of Cu. That part of the slow uptake of these ions occurring between 1 and 42 days, which was accompanied by release of Mn, is shown in Table 4. These figures were obtained from the amount of uptake during this period, or the amount of Mn released

TABLE 4

Cobalt, copper or nickel taken up by replacement of manganese during the slow phase of the reactions (meq/g of sample)

Sample	Co	Cu	Ni
Birnessite	0.71	0.27	0.086
Birnessite	0.51	0.044	0
Cryptomelane	0.30	0.02	0
Cryptomelane	0.34	0.18	0.052
ρ-MnO_2	0.44	0.11	0.027
Pyrolusite	0.01	0.029	0.003

Data from Aust. J. Soil Res. 8 : 102 (1970)

during this period, whichever was the smaller. The slow replacement of Mn by Co was much greater than that by Cu or Ni, for all samples except pyrolusite. This shows the specific nature of the reaction of Co with Mn minerals, and corresponds to the slow fixation of Co previously found. While Cu and Ni can replace some Mn during both the rapid and the slow reactions, there is a large fraction of the slowly replaceable Mn which is replaceable by Co, but not by Cu or Ni. It is probable that this extra Mn came from the crystal lattice, since, if it were merely exchangeable Mn, one would expect it to be replaced more readily by Cu than by Co.

The substitution of lower-valence Mn for Mn^{4+} is a common feature of the higher oxides of Mn, and all of the oxides in which Co accumulates contain appreciable amounts of Mn in this form. Of the oxides studied, only pyrolusite contains only Mn^{4+}, and this was the only oxide in which there was no slow replacement of Mn by Co. It is therefore proposed that Co replaces the lower-valence Mn in these oxides, where lower-valence Mn substitutes for Mn^{4+}.

Similar studies (McKenzie 1972) on the lower oxides partridgeite, manganite and hausmannite show that Co is slowly sorbed to a marked degree by partridgeite, but not by manganite or hausmannite. Plots of Co, Cu, Zn and Ni adsorption against time are shown for partridgeite in Figure 5. The broken line shows Mn appearing in solution as Co is sorbed. This supports the proposition that Co replaces Mn of valence less than four, and suggests that the replaced species is Mn^{3+}.

It has been proposed that the driving force of this reaction is the high crystal field stabilization energy of the low-spin Co^{3+} ion (McKenzie 1970,1972). Of the elements Cu, Co, Ni and Zn, only Co could be oxidized by Mn^{3+}, and only

Fig. 5. Variation in sorption of Co, Cu, Zn and Ni with time by partridgeite. Broken line shows Mn displaced by Co. [Adapted from Geoderma 8 : 31 (1972)]

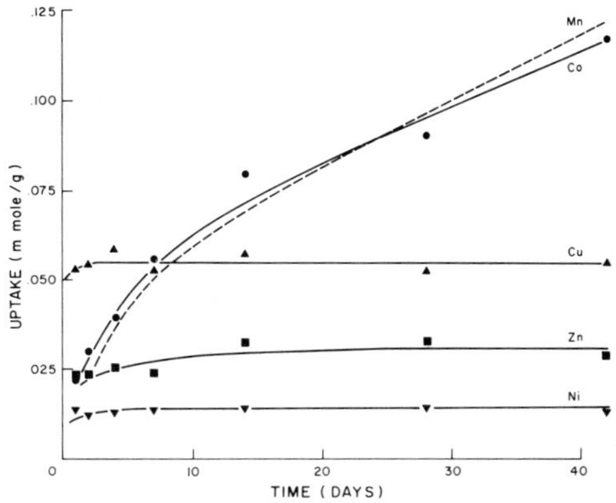

the low-spin Co^{3+} ion has a higher crystal field stabilization energy than Mn^{3+}. The proposed mechanism is that Co^{2+} is at first sorbed in a manner similar to the other heavy metals, then slowly oxidized to Co^{3+}, and becomes incorporated into the surface layers of the crystal lattice; the resulting Mn^{2+} ion is expelled into solution. The reason why this does not occur in manganite and hausmannite may be the greater Mn-O distance in these two minerals. The mean Mn-O distance in the Mn dioxide minerals, where these are known, are cryptomelane 1.89 Å, ramsdellite 1.893 Å, psilomelane 1.93 Å, lithiophorite 1.94 Å and hollandite 1.98 Å. In the tri-valent oxide partridgeite, the mean Mn-O distance in the Type 1 octahedra is 2.00 Å. With the exception of ramsdellite and psilomelane, which have not been tested, all of these oxides have been shown by experiment or by observation in soils, to accumulate Co. The mean Mn-O distance in manganite is 2.04 Å, and in the octahedral sites in hausmannite it is 2.05 Å. The greater Mn-O distance in these two minerals may reduce the crystal field strength sufficiently to prevent the transition of Co^{3+} to the low-spin state. This would make the oxidation of Co energetically unfavourable, and prevent the replacement of Mn^{3+} by Co.

ELECTRON MICROPROBE STUDIES

A number of undisturbed soil aggregates were vacuum impregnated with an epoxy resin, and polished sections were prepared for analysis by the electron probe microanalyser. By this technique, analyses are made on areas about 1 μ in diameter at many points on the sample. These results showed that Co was correlated with Mn in all samples. A plot of Co *versus* Mn for one of the soils is shown in Figure 6. Although Ni concentrations were often higher in the high Mn spots than

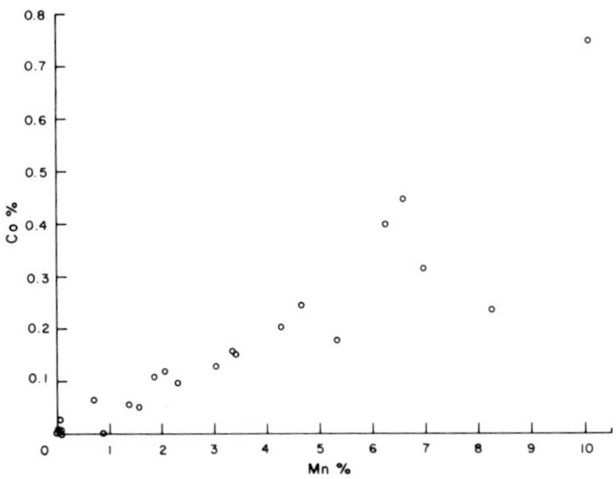

Fig. 6. Relationship be-between Co and Mn concentrations in Urrbrae loam.

elsewhere, Ni was correlated with Mn in only half of the samples. Calculations based on the Co to Mn ratios in the whole soil and in the high Mn spots showed that 30-50% of the total Co in the soils was associated with the Mn minerals, while similar calculations showed that only 0.3-19% of the Ni occurred in this way.

PRACTICAL IMPLICATIONS

The association of Co with the Mn dioxide minerals in soils has important implications for agriculture. Tiller *et al.* (1969) found that the Co sorption capacity of all the soils they studied was highly correlated with Co content and surface area, and to a lesser extent with Mn and clay content and pH. When soils of high montmorillonite content were excluded, highly significant correlations were obtained only with Mn and Co contents and surface area.

Adams *et al.* (1969) studied the availability of applied Co to plants in pot and field experiments. They found that the uptake of applied Co by clover was inversely related to the total Mn content of the soil, and that the soils could be broadly separated as to the practicability of using added Co salts as a means of alleviating Co deficiency in pasture. On the moderately acid soils with less than about 100 ppm Mn, the basal dressing of 1.1 kg/hectare of $CoSO_4 \cdot 7H_2O$ appeared to be adequate. On the second class of soils, containing from about 100 to 1000 ppm Mn, higher dressings were required, and for certain soils with more than about 1000 ppm Mn, the application of Co salts to the soil was not effective. On these soils the authors suggested the use of foliar sprays or Co oxide pellets for stock. These experiments showed that the fixation of Co by the Mn dioxides was rapid and that, once fixed, the Co was unavailable to plants.

REFERENCES

ADAMS, S.N., HONEYSETT, J.L., TILLER, K.G. & NORRISH, K. (1969) - *Aust. J. Soil Res. 7* : 29

GILES, J.B. (1962) - *CSIRO Div. of Soils, Div. Rep. No. 12/62*

GONZALEZ GARCIA, F. & GARCIA GOMEZ, A.M. (1964) - *An. Edafol. Agrobiol. 23* : 563

GRIMME, H. (1968) - *Z. Pfl-Ernähr. Bodenk. 121* : 58

KUBOTA, J. (1965) - *Soil Sci. 99* : 166

LE RICHE, H.H. (1973) - *Geoderma 9* : 43

McKENZIE, R.M. (1967) - *Aust. J. Soil Res. 5* : 235

McKENZIE, R.M. (1970) - *Aust. J. Soil Res. 8* : 97

McKENZIE, R.M. (1971) - *Mineralog. Mag. 38* : 493

McKENZIE, R.M. (1972) - *Geoderma 8* : 29

OERTEL, A.C. & GILES, J.B. (1964) - *Aust. J. Soil Res. 2* : 162

RANKAMA, K. & SAHAMA, T.G. (1950) - "Geochemistry". University of Chicago Press : Chicago

TAYLOR, R.M. & McKENZIE, R.M. (1966) - *Aust. J. Soil Res. 4* : 29

TILLER, K.G. (1963) - *Aust. J. Soil Res. 1* : 74

TILLER, K.G. & HODGSON, J.F. (1962) - *Clays Clay Miner. 9* : 393

TILLER, K.G., HODGSON, J.F. & PEECH, M. (1963) - *Soil Sci. 95* : 392

TILLER, K.G., HONEYSETT, J.L. & HALLSWORTH, E.G. (1969) - *Aust. J. Soil Res. 7* : 43

TURTON, A.C., MARSH, N.L. & McKENZIE, R.M. (1961) - *CSIRO Div. of Soils, Div. Rep. No. 8/61*

TRACE ELEMENT ADSORPTION BY SOIL MINERALS

J.P. Quirk* and A.M. Posner

Department of Soil Science & Plant Nutrition
Institute of Agriculture
University of Western Australia
Nedlands, Western Australia 6009

INTRODUCTION

The simplest process governing the concentration of ionic species in solution is the 'solubility product principle'. However this is not often obeyed and then only under special conditions rarely found in soils. In many situations in the soil, pure phases are not formed or the concentrations of the ions are not high enough to reach the solubility product. Under these circumstances, adsorption equilibria govern the ionic concentrations. This is likely to be the case for trace elements such as Co, Zn, Mo, Cu and B.

Earliest attempts to formulate a distribution between surfaces and the equilibrium solution used an analogy with adsorption from the gas phase, where the surface and the molecule are treated as uncharged. Of recent times, soil physical chemists have been trying to model the distribution of ions between the surfaces of soil colloids and the soil solution.

Bolt (1955) and Shainberg & Kemper (1967) have considered the distribution of ions between a charged clay mineral surface and the soil solution. Here the charge is due to isomorphous substitution and its magnitude is essentially unaffected by pH, although the diffuse layer charge will be affected by the proportion of ions strongly bound to the clay surface in the Stern and inner Helmholtz layer. This has the effect of changing the selectivity of an ion as its proportion changes.

---

*Present address: Waite Agricultural Research Institute, Private Bag No. 1, Glen Osmond, South Australia 5064

Jenne (1968) has suggested that in very many situations the level of ions in natural waters including the soil solution is controlled by adsorption on the surface of oxides, particularly those of Fe and Mn. This would be likely to be true for many soils of Australia.

One dramatic example of the effect of adsorption on oxide surfaces on availability of Co has been reported by Adams *et al.* (1969). These authors conclude that for soils containing greater than 1000 ppm of Mn the use of Co salts as a soil amendment does not appear to be a practical solution for relieving Co deficiency. This result suggests that Co has a very strong affinity for the surface of manganese dioxides and such a suggestion is reinforced by the studies of McKenzie (1967,1970), which demonstrate that Co^{3+} replaces Mn^{3+} in a manganese dioxide crystal lattice.

In another paper, McKenzie (1971) concludes that the reactivity (as measured by the rate of reduction of manganese dioxide by oxalic acid) of Mn minerals is reduced when 20% or more of the surface octahedra are occupied by Co. From this result and the relative levels of Mn and Co in soils, he concludes that the amount of Co in soils could affect the availability of Mn itself.

Thus work has recently aimed at trying to formulate a model for the pH-dependent charge due to oxide surfaces and the binding of ions, both anion and cation, to these surfaces. It should be borne in mind that the edge surfaces of clay minerals also have an oxide character (Schofield & Samson 1953).

ADSORPTION MODEL

Adsorption of Potential Determining Ions

For oxides and similar surfaces the charge at the interface arises from the adsorption of potential determining ions (PDI's), which in the simplest case of oxide surfaces are H^+ and OH^-. These ions are in equilibrium with ionic species of the oxide lattice and hence have a special relationship to it. The adsorption and desorption may be represented by the reactions

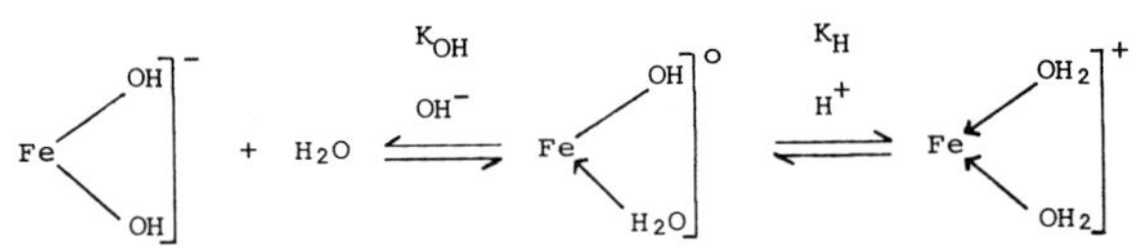

Thus, at any pH value, there will be present on the surface three types of site: positive, negative and neutral. There will be a pH value at which the number of positive sites

equals the number of negative sites when the surface as a whole will be electrically neutral. This pH value is termed the isoelectric point—IEP (Parks 1967). In the absence of specifically adsorbing ions (see below) the excess charge on the surface on either side of the IEP will be balanced by ions of any indifferent electrolyte such as KNO_3 that is present. Since the level of indifferent electrolyte affects the potential distribution at the charged interface, the adsorption of potential determining ions increases with ionic strength (Atkinson *et al.* 1967).

The PDI's may be regarded as being in the Stern or Inner Helmholtz layer (Grahame 1947) of the surface, while the counter-ions to their charge will be in the diffuse and outer Helmholtz layer.

Levine & Smith (1971) and Wright & Hunter (1973) have developed mathematical models for the above scheme, and these have been applied by Bowden *et al.* (1973) to the charging curves of goethite (αFeO·OH) surfaces.

The equation describing the adsorption of H^+ and OH^- is

$$\sigma_s = \frac{N_S\left[K_H\,(a_H)\exp(-F\psi_s/RT) - K_{OH}\,(a_{OH})\exp(F\psi_s/RT)\right]}{1 + K_H\,(a_H)\exp(-F\psi_s/RT) + K_{OH}\,(a_{OH})\exp(F\psi_s/RT)} \quad \ldots (1)$$

where σ_S is the net charge on the surface,
K_H and K_{OH} are the binding constants for H and OH at activities of a_H and a_{OH} respectively,
ψ_S is the potential on the surface, and
N_S is the maximum site density adsorption.

It takes into account the fact that the ions charge up the surface, rendering the addition of an ion of the same sign more difficult and easier for an ion of the opposite sign. This equation is based on a Langmuir model isotherm of competition between H^+ and OH^- species.

If K_H and K_{OH} are very large and the charge is not too high, this equation becomes the Nernst equation.

$$\psi_s = \frac{RT}{F}\,\ln a_H/a_{H_O} \quad \ldots (2)$$

where a_{H_O} is the activity of the protons in the solution at the IEP.

The counter-ions to the charge will be in a diffuse layer. There will be a minimum distance (β) of closest approach of these ions to the surface. The potential drops linearly from ψ_S to ψ_d over this distance and thereafter in an approximately exponential manner. The effect of this model is to relate ψ_S, ψ_d, and σ_d through the following equations for a symmetrical electrolyte of valency Z:

$$\sigma_d = -1.22 \times 10^{-10}\ C^{\frac{1}{2}} \sinh\ (0.0195/z/\psi_d) \qquad \ldots (3)$$

$$\sigma_s + \sigma_d = 0 \qquad \ldots (4)$$

$$(\psi_s - \psi_d) = (4\pi\beta/\varepsilon_\beta)\ \sigma_s = \sigma_s/G \qquad \ldots (5)$$

where ε_β is the relative permittivity of the surface region and G its electrical capacity,
ψ_s and ψ_d are in mV,
C is the ionic strength of the medium in moles/ℓ.
Solution of the above equations simultaneously enables the relationship between charge and pH to be determined.

Adsorption of Ions with a Specific Affinity for the Surface

The above description is that of a situation where the counter-ions have no specific affinity for the surface. However, many ions including those of the trace elements have a specific affinity for the surface, the effect of which is to allow adsorption of ions in excess of the surface charge and, as an extreme example, gives rise to the adsorption of cations and anions on positively and negatively charged surfaces respectively. Such effects are seen in the adsorption of the transition metals such as Cu, Co, Zn (Forbes 1973) and anions such as selenite and molybdate on goethite (Hingston *et al.* 1972).

The adsorption of these specifically adsorbing ions is modelled (Bowden *et al.* 1973,1974) by placing them in another layer some distance from the potential determining ions layer which for simplicity has been put equal to β where the potential is ψ_d. The equation governing their adsorption is of similar form to (1) and takes into account all specifically adsorbing species present in solution other than PDI's.

$$\sigma_i = \frac{N_T\ z_i K_i C_i \exp\ (-z_i F\psi_d/RT)}{1 + K_i C_i \exp\ (-z_i F\psi_d/RT)} \qquad \ldots (6)$$

$$\sigma_s + \sigma_i + \sigma_d = 0$$

where N_T is the maximum possible adsorption in the second layer in moles/cm^2,
z_i is the valency (including sign) of ionic species (i),
K_i is the binding constant of ionic species (i).
Many cations undergo hydrolysis reactions, e.g.

$$Zn^{2+} + OH^- \rightleftharpoons Zn(OH)^+ + OH^- \rightleftharpoons Zn(OH)_2\ ,$$

so that, in the use of equation (6), K_i's are assigned to all these species and any others known to adsorb. Bowden *et al.*

have accounted for the adsorption of Zn(II) on goethite and found it necessary to postulate the adsorption of Zn^{2+} and $Zn(OH)^{+}$. The values of K_i taken were 1×10^6 and 2.5×10^7 respectively.

SPECIFIC ADSORPTION OF METAL IONS

Figure 1 shows adsorption isotherms for Zn(II) at constant pH. It can be noted that the model discussed here gives a better fit to the results than a simple Langmuir plot, which is frequently used. Unlike the case where the simple Langmuir equation is used, it is possible to achieve good fits to the data with only one value of N_T for all pH values. The present model also allows for a change in binding energy as the charge on the surface varies either through a change in pH or adsorption; thus it may not be necessary to postulate

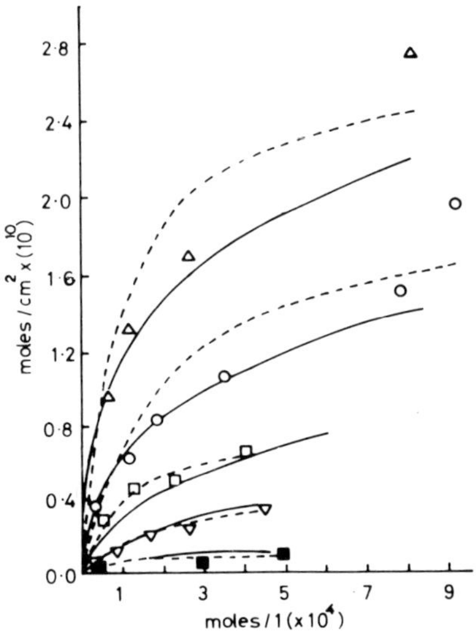

FIG. 1

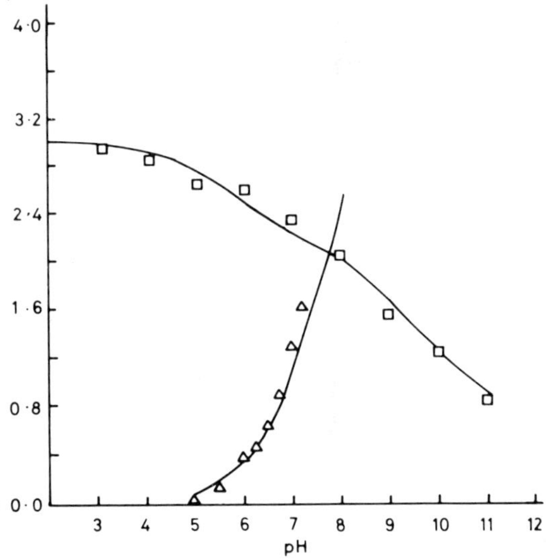

FIG. 2

Fig. 1. Adsorption isotherms of Zn(II) on goethite in 0.1 M NaCl. ----, Langmuir isotherm; ——, Bowden et al. *model. Points are experimental data. [Reproduced by courtesy of Nature. See Bowden, J.W., Bolland, M.D.A., Posner, A.M. & Quirk, J.P. (1973) - Physical Science 245 : 81.]*

Fig. 2. Adsorption on goethite as a function of pH. Δ, Zn 1×10^{-4} M (0.1 M NaCl); □, phosphate (0.01 M NaCl); ——, Bowden et al. *model. Points are experimental data. [Reproduced by courtesy of Nature. See Bowden, J.W., Bolland, M.D.A., Posner, A.M. & Quirk, J.P. (1973) - Physical Science 245 : 81.]*

different types of sites as has been done in the past even on pure mineral surfaces. Increase in charge of the surface occasioned by adsorption at constant pH renders subsequent adsorption more difficult, hence the flattening of the isotherms as in Figure 1 long before all possible sites become filled. The simple Langmuir model does not allow for these changes in charge. The data for a phosphate adsorption envelope (Hingston *et al.* 1972) in Figure 2 also show a good fit to the model results, and are included because of the similarities of the chemistry of phosphate, and molybdate and selenite (Fig. 3).

It may be noted from Figure 2 that the amount of Zn adsorbed increased markedly with pH in the region pH 5-7 and the model gives a good fit to experimental data. Similarly shaped curves have been obtained by Grimme (1968) for Cu, Zn, Co and Mn. The position of the curves relative to pH varies in relation to the pK of the first hydrolysis product (Table 1).

K_i is related to the standard free energy of adsorption (ΔG) through

$$\Delta G = -RT\ell nK_i \qquad \ldots (7)$$

This has been divided by James & Healy (1972) into a change in solvation energy term (ΔG_{solv}) and an intrinsic chemical interaction term ${}_{Int}\Delta G_{chem}$. ΔG_{solv} is related to the size of the ion and the fact that it has to be partially desolvated before it can be adsorbed. Because this term is smaller the lower the charge on the ion, adsorption is favoured the greater the degree of hydrolysis of a cation. This effect

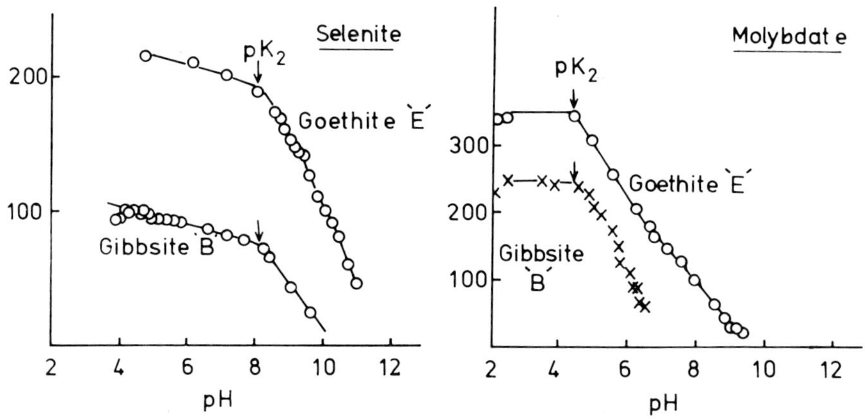

Fig. 3. Adsorption of molybdate and selenite as a function of pH on goethite (81 m^2/g) and gibbsite (58 m^2/g). [Reproduced by courtesy of the Journal of Soil Science. See Hingston, F.J., Posner, A.M. & Quirk, J.P. (1972)-Vol. 23 : 177.]

TABLE 1

Adsorption of heavy metal ions (M) as a function of anion and pH (results expressed as % of initial amount of M in solution)

M	Anion	pH							
		4.7	5.2	5.5	5.9	6.4	7.2	7.5	8.0
Cu	NO_3	9	44	76	89	-	-	-	-
	Cl	17	55	75	90	-	-	-	-
Pb	NO_3	-	29	38	63	-	-	-	-
	Cl	-	43	56	75	-	-	-	-
Zn	NO_3	-	-	-	11	19	61	-	-
	Cl	-	-	-	13	22	68	-	-
Cd	NO_3	-	-	-	-	11	27	41	-
	Cl	-	-	-	-	23	44	53	-
Co	NO_3	-	-	-	-	-	39	52	78
	Cl	-	-	-	-	-	39	54	78

Approximate pK_a values are: Cu = 8.0; Pb = 8.0; Zn = 9.0; Cd = 9.5; Co = 9.5. Initial concentration of M = 3.2×10^{-5} M : 20 mg goethite/40 ml. Equilibrium time = 25 min. Adsorption carried out at 20 ± 1°C in 0.075 M $NaNO_3$ or NaCl.

outweighs the lessening coulombic attraction with a negative surface and is in the same direction as the lessening repulsion with a positive surface.

The factors contributing to the magnitude of the ${}_{Int}\Delta G_{chem}$ have yet to be determined, although it is to be expected that it would parallel the solution behaviour of the ions because of the correlation of adsorption with the pK_a values both for cations and anions.

Two other features of the reaction scheme are noteworthy. These are that Zn(II) and Cu(II) can adsorb at pH values as low as 4 to 5, i.e. onto a positive surface since goethite has an IEP of approximately 8. Thus adsorption takes place despite an unfavourable electrical potential term due to the large chemical affinity (K_i). The other feature is that adsorption of Zn(II) and Cu(II) takes place well below the pK_a values of 9 and 8 respectively and involves proton release; thus the adsorption can be viewed as an hydrolysis promoted by the surface. An equivalent way of regarding this process is that the surface has a strong affinity for an hydrolysis product (e.g. $ZnOH^+$) and, even though such a species would be at exceedingly low concentrations at pH 4-5, the removal of the hydrolysis reaction products by the surface causes displacement of the solution equilibrium and appreciable amounts of the metal ion in the hydrolysed form are adsorbed.

The adsorption of Zn(II) at an oxide surface may be represented as a bridging ligand between two neutral sites thus:

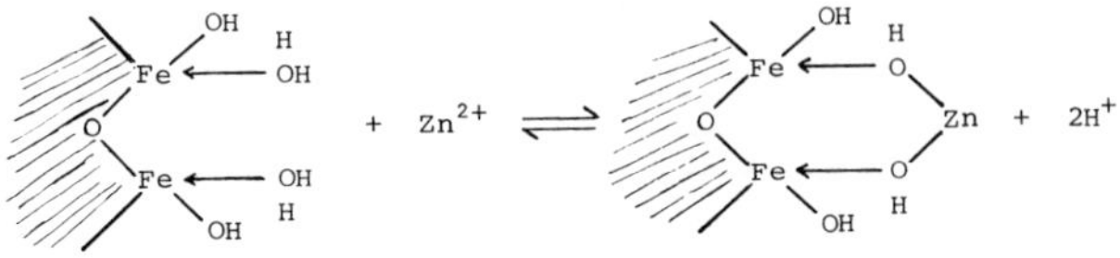

In this scheme an olation bridge and ring structure is postulated to form. Such adsorption can be viewed as a growth or an extension of the surface and its properties would accord with the lack of complete reversibility of such an adsorption (Padmanabham, private communication).

An extremely significant feature is the concomitant desorption of protons when an ion such as Zn(II) or Cu(II) is sorbed. Figure 4 shows data for Cu(II) adsorption on goethite (Forbes 1973) and indicates that, for each Cu(II) species adsorbed, on average 1.78 protons are released. The fact that this value is < 2 over the pH range 4.3-5.2 may indicate some monodentate binding.

Recently Forbes *et al.* (1974) have investigated further the role of the OH in the adsorption of heavy metal cations on goethite. They showed that if the M-OH link were too co-

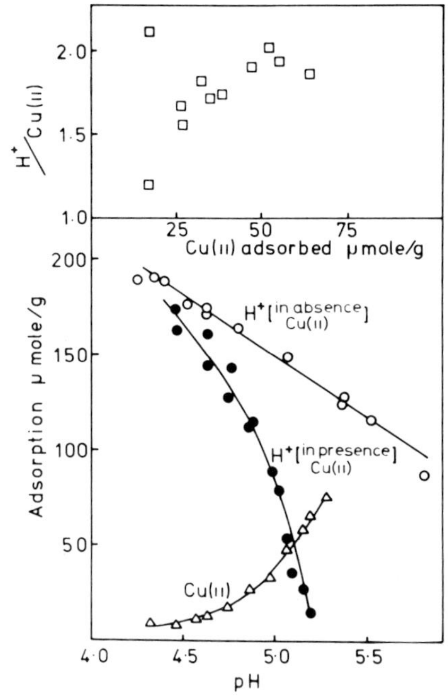

Fig. 4. Lower curve: *Adsorption of Cu(II) and H^+ ions on goethite (0.075 M KNO_3; initial [Cu], 6.5×10^{-5} M; 10 mg goethite in 20 ml; 20°C).* Upper curve: *Mole ratio, H^+ displaced/Cu(II) adsorbed as a function of adsorption.*

valent, so that no sharing of the OH with the surface was possible, then adsorption was weak, as in the case of $Hg(OH)_2^{\circ}$. Introduction of Cl in place of OH rendered it even weaker because of an increase in covalency of the bond. The same results were observed in the absorption of Co^{3+} complexes where there was no specific adsorption of the complex ions due to the large degree of covalency of the Co-OH link. However, where the M-OH link was not too strong, then complex formation with Cl tended to enhance adsorption because of an increase in the ΔG_{solv} term due to an effective reduction of the cation charge.

The model described above assumes that all the sites on the surface have the same intrinsic reactivity. However, in the soil situation there are likely to be many different types of sites, each with a somewhat different K_i. It has been shown that such a situation can lead, in uncharged systems, to the system obeying a Freundlich isotherm (Equation 8) if the adsorption energies are distributed in an exponential manner (Young & Crowell 1962):

$$\log A = k \log C + Y \qquad \ldots (8)$$

While no formal treatment has yet been given to the adsorption of charged species in a charged heterogeneous system, it is intuitively expected that this will also obey a Freundlich isotherm, especially since the effect of charge is to spread out the concentration range over which adsorption occurs.

The Freundlich isotherm has been found to describe the adsorption properties of soils (e.g. for phosphate, Barrow 1973), although no detailed study of the effect of conditions of adsorption on the constants k and Y has been published.

INFLUENCE OF PROTONS ON Zn(II) DIFFUSION

It follows from the above that the amount of Zn adsorbed may be appreciably influenced by pH, and since Zn adsorption releases protons and desorption causes them to be taken up, then pH changes are to be expected as a result of adsorption-desorption reactions during the diffusion of Zn in soils.

Bar Yosef *et al.* (1974) combined the above model with diffusion theory to account for the diffusion of Zn(II) in goethite pastes. They used the following equation (9) to determine the diffusion coefficient ($\bar{D}$) from the experimentally determined relationship between Zn(II) (Q) and distance across a diffusion profile produced after two goethite paste plugs, one with Zn(II) and one without, had been stood in contact for a given time.

$$\frac{\delta Q}{\delta t} = \frac{\partial}{\partial x}\left[\bar{D}\,\frac{\delta Q}{\delta x}\right] \qquad \ldots (9)$$

$$\bar{D} = D^{\ell}\left[\frac{1}{b+\theta} - \frac{1}{\theta}\cdot\left(\frac{\partial C}{\partial H}\right)\cdot\frac{\delta H/\delta x}{\delta Z/\delta x}\right] \qquad \ldots (10)$$

where D^{ℓ} is the diffusion coefficient of Zn(II) in water and includes the effect of tortuosity,

b is the slope of the adsorption isotherm, and

θ is the volumetric water content of the paste.

The slope of the isotherm (b) was obtained from the Bowden *et al.* (1973,1974) model.

It should be especially noted that, as well as the term involving the slope of the adsorption isotherm (Crank 1956; Nye 1966; Olsen & Kemper 1968), a term involving $\delta H/\delta x$ has to be included in Equation 10. This arises because increases in H ion concentration are generated in the solution when Zn^{2+} and $Zn(OH)^{+}$ are adsorbed, and the reverse process occurs on desorption. The schematic effect of such a process is shown in Figure 5 (inset). Because $\delta H/\delta x$ could not be measured, a comparison is made only between the experimental and calculated values of D when Zn(II) adsorption is relatively small, that is when $\delta H/\delta x$ is insignificant.

The form of the curve in Figure 5 shows the usual increase in $\bar{D}$ as a function of an amount adsorbed, followed by a marked decrease, which is caused by the increase in pH near the boundary of high and low concentration halves of the cell. This is due to desorption of Zn(II) on the high concentration side. On the low concentration side, there is a subsequent fall in pH due to the release of protons arising from Zn(II) adsorption. Where the amount of Zn(II) adsorbed is small, the value for $\bar{D}$ is the same as that on the high concentration side of the cell, where the amount desorbed is not large (Fig. 5).

ROOT SOIL INTERFACE

Wilkinson *et al.* (1968) by means of an autoradiographic technique using ^{65}Zn were able to show that depletion zones 2-4 mm wide developed in the vicinity of wheat roots (Plate 1). The depletion zones were not affected by the rate of transpiration but broadened with time, indicating diffusion to be the dominant mechanism. The significant aspect of this result was that little Zn(II) was removed from the soil on leaching with 0.01 M $CaCl_2$ and the above result therefore suggests that the pH in the immediate vicinity of the wheat root was less than that of the soil. The source of nitrogen

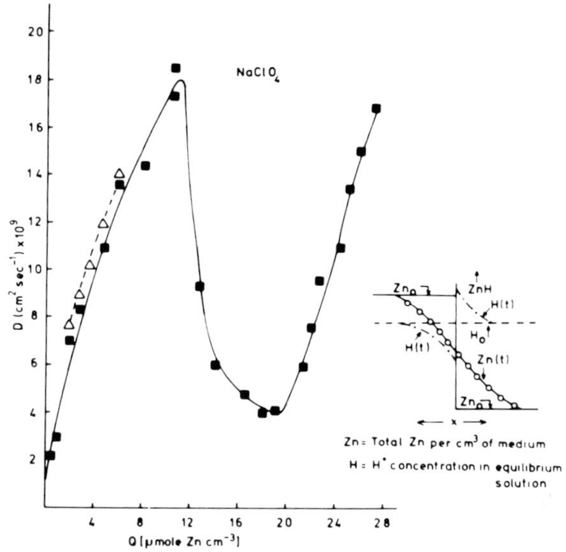

Fig. 5. Inset: *Schematic profiles of diffusible Zn and H$^+$ concentrations in a goethite paste before (t = 0) and after a certain diffusion period (t).* Main diagram: *The diffusion coefficient of Zn in a goethite paste as a function of diffusible Zn. The open and filled symbols were obtained by solving equations (9) and (10) respectively. [Reproduced by courtesy of the Journal of Soil Science. See Bar-Yosef, B., Posner, A.M. & Quirk, J.P. (1974).]*

was NH_4NO_3, a salt that is associated with a fall in pH around an absorbing root surface. As with many anions on oxide surfaces (Hingston *et al.* 1974), some hysteresis also exists for Cu(II), Zn(II) and other heavy metal cations. If, however, the pH is lowered, reversibility is more readily achieved (Padmanabham, unpublished), and this is in accordance with the observations of Wilkinson *et al.* (1968).

ACKNOWLEDGEMENTS

We gratefully acknowledge our students and co-workers, whose work (see references) forms the basis for the major part of this paper.

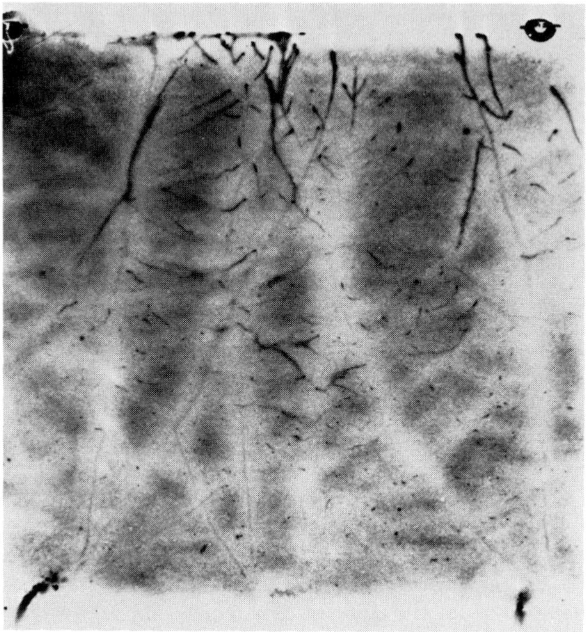

Plate 1. Depletion of ^{65}Zn by wheat roots in Wongan loamy sand after 14 days. Depletion zones were similar for high and low transpiration rates. [See Wilkinson, H.F., Loneragan, J.F. & Quirk, J.P. (1968) - Proceedings of the Soil Science Society of America 32 : 831.]

REFERENCES

ADAMS, S.N., HONEYSETT, J.L., TILLER, K.G. & NORRISH, K. (1969) - *Aust. J. Soil Res. 7* : 29

ATKINSON, R.J., POSNER, A.M. & QUIRK, J.P. (1967) - *J. phys. Chem. 71* : 550

BARROW, N.J. (1973) - *Aust. J. Soil Res. 11* : 57

BAR-YOSEF, B., POSNER, A.M. & QUIRK, J.P. (1974) - *J. Soil Sci. (in press)*

BOLT, G.H. (1955) - *Soil Sci. 79* : 267

BOWDEN, J.W., BOLLAND, M.D.A., POSNER, A.M. & QUIRK, J.P. (1973) - *Nature, Phys. Sci. 245* : 81

BOWDEN, J.W., POSNER, A.M. & QUIRK, J.P. (1974) - *10th Int. Congr. Soil Sci., Moscow II* : 29

CRANK, J. (1956) - *In* "Mathematics of Diffusion". Oxford

University Press : Oxford
FORBES, E.A. (1973) - Ph.D. Thesis, University of Western Australia
FORBES, E.A., POSNER, A.M. & QUIRK, J.P. (1974) - *J. Colloid Interface Sci.* (in press)
GRAHAME, D.C. (1947) - *Chem. Rev. 41* : 441
GRIMME, H. (1968) - *Z. Pfl-Ernähr. Bodenk. 121* : 58
HINGSTON, F.J., POSNER, A.M. & QUIRK, J.P. (1972) - *J. Soil Sci. 23* : 177
HINGSTON, F.J., POSNER, A.M. & QUIRK, J.P. (1974) - *J. Soil Sci. 25* : 16
JAMES, R.O. & HEALY, T.W. (1972) - *J. Colloid Interface Sci. 40* : 65
JENNE, E.A. (1968) - *Adv. Chem. Ser. 73* : 337
LEVINE, S. & SMITH, A.L. (1971) - *Discuss. Faraday Soc. 52* : 290
McKENZIE, R.M. (1967) - *Aust. J. Soil Res. 5* : 235
McKENZIE, R.M. (1970) - *Aust. J. Soil Res. 8* : 97
McKENZIE, R.M. (1971) - *Aust. J. Soil Res. 9* : 55
NYE, P.H. (1966) - Transactions Commissions II and IV International Society of Soil Science, Aberdeen, p. 317
OLSEN, S.R. & KEMPER, W.D. (1968) - *Adv. Agron. 20* : 91
PARKS, G.A. (1967) - *Adv. Chem. Ser. 67* : 121
SCHOFIELD, R.K. & SAMSON, H. (1953) - *Clay Miner. Bull. 2* : 45
SHAINBERG, I. & KEMPER, W.D. (1967) - *Soil Sci. 103* : 4
WILKINSON, H.F., LONERAGAN, J.F. & QUIRK, J.P. (1968) - *Proc. Soil Sci. Soc. Am. 32* : 831
WRIGHT, H.J.L. & HUNTER, R.J. (1973) - *Aust. J. Chem. 26* : 1191
YOUNG, D.M. & CROWELL, A.D. (1962) - *In* "Physical Adsorption of Gases". Butterworths : London

THE AVAILABILITY AND ABSORPTION OF TRACE ELEMENTS IN SOIL-PLANT SYSTEMS AND THEIR RELATION TO MOVEMENT AND CONCENTRATIONS OF TRACE ELEMENTS IN PLANTS

J.F. Loneragan

School of Environmental and Life Sciences
Murdoch University
Murdoch, Western Australia 6153

INTRODUCTION

This paper reviews the present knowledge of the processes governing the absorption of trace elements from soils and of their relationship to the concentrations of trace elements in plants and to plant growth. It also reviews present knowledge of the movement of trace elements within plants since this, together with the supply and the absorption of trace elements, governs the relationship between trace element concentrations in plants and plant growth.

ABSORPTION FROM SOILS

The concept that only part of the total amount of a plant nutrient in a soil is available for absorption by plants was put forward as early as 1845 (Daubeny cited by Russell 1973). In trace element nutrition the concept appears in the early classical studies of Samuel and Piper, who clearly recognized that their field problem of Mn deficiency in oats involved the availability of Mn in the soil to the plant rather than a deficiency of total Mn within the soil; indeed they were able to overcome Mn deficiency by acidification or heat sterilization of the soil (Samuel & Piper 1928,1929; Piper 1931). The concept of availability has featured prominently in explanations of the behaviour of other trace elements subsequently studied.

Later workers have attempted to measure the quantities of

available nutrients in soils by chemical extraction procedures. Beckwith (1963) pointed out that all such procedures suffer from the lack of a suitable standard. Plants are the final arbiter of the amounts of nutrients which are available and plants vary greatly in the quantity of nutrients which they absorb. The quantity varies with the environmental conditions and with species, density of sowing, and length and stage of growth of the plant. Moreover, following nutrient removal by plants, a slow redistribution of soil nutrients may occur, resulting in the conversion of some less available forms to more available ones. For these reasons Beckwith (1963,1973) has preferred to restrict the term "available" to the amount of a nutrient absorbed by a plant under specified conditions. Like Steenbjerg (1954), he favoured using the quantity of nutrient absorbed rather than plant growth or its concentration in the plant.

In the present state of our understanding of soil-plant relationships, Beckwith's use of the term "available" is readily appreciated. Yet it has the serious disadvantage that it no longer permits us to distinguish the contribution of soil factors in supplying nutrients to the plant root from that of plant factors in absorbing nutrients from the soil. Such a distinction is essential if we are to understand the absorption of trace elements from soils. For this reason and because the term "available" has been used in so many different ways, I will avoid this term and will discuss the processes involved in the absorption of trace elements by plants in soils under three headings:-

- those controlling the supply of trace elements from the solid phase to the soil solution;
- those governing the absorption of trace elements by plants from solution; and
- those involving interaction between soil and plant.

Supply to the Soil Solution

Concentrations and forms in solution. The concentrations and forms of trace elements in soil solutions have been studied in relatively few soils. The most extensive work has been that of Hodgson and his colleagues on a range of New York and Colorado soils. The concentrations of Co, Cu and Zn in solutions displaced from the surface horizons of these soils were generally less than 1 μM (Table 1). Values for Co were one-tenth those of Cu and Zn in the New York soils (Hodgson *et al.* 1965), while values for Zn were particularly low in the calcareous Colorado soils (Hodgson *et al.* 1966). The concentrations of Mn in solutions displaced from the New York soils ranged very widely, from 0.02 to 68 μM; moreover, in

TABLE 1

Concentrations and degree of complexing of some trace elements in solutions displaced from the surface horizons of some New York and Colorado soils (Hodgson et al. *1965; Hodgson* et al. *1966; Geering* et al. *1969)*

Element	Total element in soil solution (μM)	Degree of complexing (%)
Cobalt	< 0.007-0.2	8-50
Zinc	< 0.03-3	28-99
Manganese	< 0.02-68	84-99
Copper	< 0.01-0.6	89-99.8

some soils the values fluctuated by at least 100-fold during the course of a year and were very sensitive to conditions of storage and of preparation (Geering *et al.* 1969).

In acid soils the dominant inorganic ions of Co, Cu, Mn and Zn are the divalent cations M^{2+}. In neutral and alkaline soils the monovalent hydroxy cation $M(OH)^{+}$ would also be important (Lindsay 1972). However, in solutions extracted from the surface horizons of all of the soils studied, appreciable quantities of these trace elements were present as complexes. In general, about 25% of the Co, 50% of the Zn, 90% of the Mn and more than 90% of the Cu were complexed in the solutions extracted from most of the soils (Table 1 : Hodgson *et al.* 1965; Hodgson *et al.* 1966; Geering *et al.* 1969). In many of the soils more than 99% of the Cu was present as a complex.

The complexes of Cu and presumably of other transition metals in soil solutions are anionic (Hodgson *et al.* 1966; Beckwith 1973) but their chemical nature has not been established. They are believed to be organic compounds since the degree of complexing of Cu and Zn correlated with organic matter content of soil solutions (Hodgson *et al.* 1965). Detailed studies on solution displaced from one soil have also established that at least three groups of organic compounds with weakly acidic properties were associated with Cu and Zn : a dialyzable fraction had acidic characteristics attributable to aliphatic and amino acid residues (Geering & Hodgson 1969).

Much less is known of the concentrations and forms of Fe, B and Mo in soil solutions. Concentrations of Fe in soil solutions have been reported from 0.4 to 5 μM (Bowen & Cawse 1964; Oliver & Barber 1966; Bowen 1966) : most of the Fe could be expected to be strongly complexed. Saturation extracts of 11 Indiana soils ranged in Mo concentration from 0.02 to 0.08 μM (Lavy & Barber 1964). Concentrations of B in soil solutions have been reported to range from 3 to 1000 μM (Bowen 1968), but analytical techniques for low concentrations of B have been unsatisfactory and most interest has centred on soils where B is toxic.

The dominant form of inorganic B in soil solutions is believed to be the undissociated acid, H_3BO_3. Since pKa = 9.2 for its first dissociation constant, appreciable quantities of borate ion, $H_2BO_3^-$, would only appear with increasing alkalinity above pH 7. The predominant inorganic form of Mo in soil solutions is believed to be the divalent molybdate anion, MoO_4^{2-}. Since pKa = 4.3 for the second dissociation constant of molybdic acid, the monovalent anion, $HMoO_4^-$, would be negligible in neutral soils but would approach the concentration of the divalent ion in extremely acid soils (Lindsay 1972). Nothing is known of organic complexes of B or Mo in the soil solution although organic complexes of B have been reported in sea water (Noakes & Hood 1961). Compounds such as sugars and phenols containing *cis*-diol groups form complexes with B and may be important in soil solutions : mannitol in $CaCl_2$ has been reported to desorb B more readily from clays than did $CaCl_2$ alone (Hingston 1964).

The concentrations of trace elements in the soil solution are generally so low that the total quantity present at any one time would sustain relatively little plant growth. To provide sufficient for growth, the trace elements would need to be rapidly replenished from the solid phase as they were depleted.

Equilibria with the solid phase. The solid phase of most soils contains relatively large quantities of trace elements in a variety of forms. Hodgson (1963) lists five possible forms of combination as follows: (1) precipitation with other soil components to form a new solid phase; (2) occlusion during formation of new solid phases in which the nutrient is not a principal constituent; (3) inclusion in soil minerals — by incorporation during crystallization of the mineral, or by reaching their locations later by solid state diffusion; (4) association with soil surfaces, either organic or inorganic; and (5) incorporation as insoluble compounds in biological materials.

In soils, each trace element probably occurs to some extent in several or all of these combined forms. All could release trace elements into the soil solution. They would do so at different rates which would vary with soil conditions. Some, such as (3), would generally release trace elements extremely slowly and others, such as (4), would do so very rapidly.

Chemical tests of availability generally measure only the fast reactions. However, in some soils, the slow reactions could contribute significantly to the supply of trace elements to plants.

Examples of several forms of combination of trace elements with soil solids have already been given in this Symposium. Brief mention will be made here of one important precipitation reaction (the formation of MnO_2) and of the reaction of trace elements with organic matter in soils. The incorporation of trace elements into biological materials has been too little studied for comment, other than to suggest that changes in the growth of soil microorganisms could be expected to affect trace elements in the soil solution in essentially the same way as they do N (e.g. Russell 1973).

Precipitation of Mn as MnO_2 appears to be an important reaction governing the concentration of Mn^{2+} in soil solutions. The relationship of Mn^{2+} to MnO_2 is complicated by factors influencing the kinetics of the reaction, which appear to depend more on current microbiological activity than on physico-chemical factors alone. This may partly explain the very large and rapid fluctuations which occur in the Mn concentrations of soil solutions, as well as their sensitivity to soil pH, aeration, and sterilization (Leeper 1947,1970; Mulder & Gerretsen 1952; Geering *et al.* 1969).

The reactions of trace elements with soil organic matter have been reviewed recently by Stevenson & Ardakani (1971) and by Schnitzer & Khan (1972). The reactions include ion exchange, chelation and complex formation. Because of the relatively high concentrations of Ca^{2+} in soil solutions, ion exchange is unlikely to be important. On the other hand, trace elements react strongly with organic matter even in the presence of high concentrations of alkaline earth and alkali cations. Other evidence supports the formation of complexes and chelates between metal ions and organic matter, probably through their COOH, phenolic OH, and possibly C=O and NH_2 functional groups. However, the order of bonding of divalent metal ions varies with organic matter in different soils. Of the divalent metal ions, Cu^{2+} is generally the most strongly bound. The binding of Zn^{2+} and Mn^{2+} is usually appreciably weaker. Both Fe^{3+} and Al^{3+} are more strongly bound than Cu^{2+} at acid pH values. With increasing pH, the stabilities of the metal-organic complexes increase up to a point where the complexes break up and hydroxides form. Schnitzer & Skinner (1963) found that equimolar complexes of fulvic acid with Al^{3+} ions broke up at pH 8 and with Fe^{3+} at pH 9, but those with Cu^{2+} did not break up until pH 10.

Some workers have attempted to estimate stability constants

for metal-organic matter complexes. Such constants are difficult to determine accurately. In addition, great care must be taken in extrapolating them to soils for a number of reasons — the procedure for extraction of the organic matter may have changed its properties; the values will vary with the pH and the ionic strength of the soil solution; and, perhaps most significantly, the values obtained present an average behaviour for sites of varying bonding strengths. The average behaviour has generally been determined at relatively high concentrations of trace elements: at the very low concentrations which exist in the soil solution, specific sites with very much higher bonding strengths than the average may assume the dominant role.

Organic matter in soils appears able to compete strongly with oxides for several metals. Fulvic acid added to goethite, gibbsite, or soil has been shown to complex up to equimolar quantities of Fe or Al (Schnitzer & Skinner 1963). A mixture of humic and fulvic acids also complexed Mn from Mn^{3+} and Mn^{4+} oxides and occluded Mn^{2+} hydroxides (Rosell & Babcock 1968). Such reactions may be very important in keeping these metals in soluble forms in alkaline soils.

Present evidence suggests that the behaviour of B and Mo in soil solutions is largely controlled by the reaction of their oxy-anions with Fe oxides. These reactions are very sensitive to pH, being strongest at pH values close to the pKa values for the dissociation of the acids: thus the adsorption of B is strongest around pH 9 and decreases with decreasing pH, while the adsorption of Mo is strongest around pH 4 and decreases with increasing pH (Jones 1957*a*,*b*; Hingston 1964; Hingston *et al.* 1972; Quirk & Posner, this Symposium).

The evidence also suggests that the behaviour of the metals of the first transition series is dominated by the reaction of the hydrated ions of the metals and their hydrolysis products with various soil components. The behaviour of Zn in many soils may be controlled by Fe oxides (Quirk & Posner, this Symposium): the effect of phosphate in increasing the bonding of Zn to Fe oxides may be particularly important in some soils (Stanton & Burger 1967). The behaviour of Co and Mn in many soils is controlled by the reaction and formation of MnO_2 (McKenzie, this Symposium), while that of Cu and Fe in most soils and of Zn, Mn and Co in others may be dominated by reactions with both soluble and insoluble forms of organic matter.

Absorption from Solution

The plant processes operative in the absorption of trace elements have been little studied. Moreover, of those studies which have been made, few have been done under conditions which have any relevance to soils. Nevertheless, studies of absorption of trace elements from solution have revealed characteristics of the process which are likely to be important in soils, viz.:

(1) it operates effectively at extremely low concentrations in solution;
(2) at low concentrations, the rate of absorption of trace elements varies almost directly with concentration;
(3) it is active, being sensitive to temperature and probably also to aeration;
(4) it reacts strongly to H^+ and other ions;
(5) its intensity varies with plant species;
(6) it sometimes involves reductive activity at the root surface.

In most cases the evidence in support of these generalizations rests upon relatively little evidence for one or two of the trace elements only.

Absorption from low concentrations. When the very low concentration of 0.01 μM Zn^{2+} was maintained in a recirculating nutrient solution, eight plant species were observed to absorb Zn at average rates over 46 days of 2-4 ng atoms/g fresh roots/day. Such rates would provide sufficient Zn for relatively fast rates of plant growth although they were not sufficient to prevent Zn deficiency from developing at the very fast rates of growth of all species in the experiments described. Concentrations of 0.1 to 2.5 μM Zn^{2+} eliminated Zn deficiency and gave maximal yields of plants with rates of Zn absorption between 10 and 30 ng atoms of Zn/g fresh roots/day (Fig. 1; Carroll & Loneragan 1969).

In preliminary experiments we have also observed that wheat plants can obtain their Cu requirements from solutions

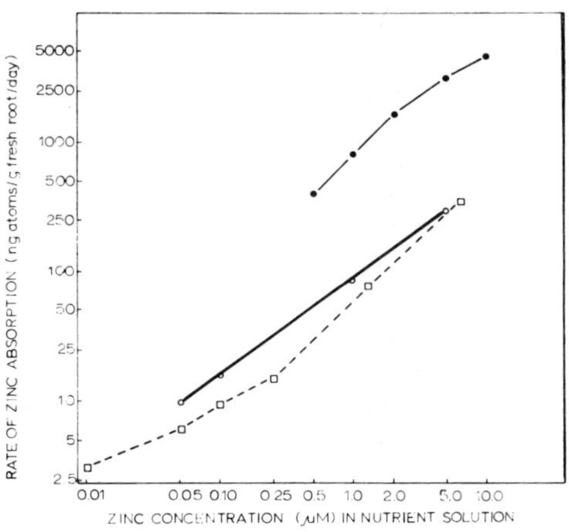

Fig. 1. Rates of Zn absorption in (●) short-term studies by excised barley roots in 500 μM $CaCl_2$ (Schmid et al. *1965); (○) long-term studies by whole wheat plants in flowing, complete nutrient solutions (Carroll & Loneragan 1969); (□) short-term studies by young wheat seedlings in complete nutrient solution. [Reproduced with permission from Chaudhry & Loneragan (1972a)]*

with less than 0.01 μM Cu^{2+}. Only at concentrations approaching 0.001 μM did wheat plants show any signs of Cu deficiency when the concentration of Cu^{2+} in solution was replenished daily at the nominated value (J.F. Loneragan & K. Snowball, unpublished data). Naturally, in culture systems where trace element concentrations are not replenished the nominal threshold concentration for deficiency is very much higher, depending more upon the total supply of element than upon its initial concentration (see e.g. Carroll & Loneragan 1969).

The data available indicate that most plants can probably obtain their requirements for the transition metals from extremely low concentrations when the metals are present as simple cations. There are no suitable data for Mo and B. Nor are there suitable data for the absorption of trace elements when present in organic complexes. Some data obtained at relatively high concentrations of trace elements showed a 5- to 10-fold reduction in the rate of absorption of Zn, Cu or Co by plant roots when the synthetic chelate EDTA (ethylenediaminetetraacetate) was present (De Kock & Mitchell 1957) and of Fe by isolated leaf cells when EDDHA [ethylenediamine di(*o*-hydroxyphenylacetate)] was present (Kannan 1969).

The relation between concentration and absorption. As Zn^{2+} concentrations in flowing nutrient solutions increased from 0.01 to 6.25 μM, rates of Zn absorption by eight plant species increased almost linearly from 2 to 400 ng atoms Zn/g fresh roots/day (Fig. 1; Carroll & Loneragan 1969). This linear response to increasing Zn concentrations contrasts sharply with the markedly hyperbolic response which characterizes the relationship between concentration and absorption at higher concentrations of ions, both of macronutrients (see e.g. Loneragan 1968*a*) and of micronutrients. Thus the absorption of B (Bowen 1968) and of Cu, Zn and Mn (Bowen 1969) by sugarcane leaf discs and of Mn by excised barley roots (Maas *et al.* 1968) responded strongly to increasing concentrations up to but not beyond 100 μM. The absorption of Fe by isolated tobacco leaf cells showed a similar strong response up to but not beyond 1000 μM Fe^{2+} (Kannan 1969). The fact that below 100 μM the absorption of all trace elements responded strongly to the concentration of their ions in solution suggests that at the yet lower concentrations present in soil solutions their behaviour would be similar to that shown for dilute concentrations of Zn^{2+}.

The nature of the absorption process. Suggestions have been made that Zn absorption is under the control of non-metabolic physical processes (Gutknecht 1961,1963; Findenegg & Broda 1965; Pickering & Puia 1969; Rathore *et al.* 1970).

However, Schmid *et al.* (1965) have shown that a high proportion of Zn^{2+} taken into plant roots in short-term experiments is adsorbed on cell surfaces and, unless removed, can lead to serious errors in estimates of Zn absorption into the interior of cells. When adsorption was properly taken into account, absorption of Zn was shown to proceed *via* a metabolically mediated mechanism, sensitive to metabolic poisons and temperature. The temperature response for Zn absorption shows a Q_{10} of nearly 2 with increasing temperatures from 10 to 30°C and a Q_{10} of 6 with decreasing temperatures from 10 to 2°C (Chaudhry & Loneragan 1972*a*).

The absorption of Zn, Cu and B by leaf discs (Bowen 1968, 1969) and of Fe by intact pea and bean plants (Branton & Jacobson 1962; Wallace *et al.* 1963) were also inhibited by low temperatures and by metabolic poisons, indicating that even at relatively high concentrations metabolic processes were involved. The data support the suggestion that, at the concentrations generally present in soil solutions, the absorption of trace elements by plant roots is controlled by metabolic processes within the root.

Interaction with other ions. In general, increasing pH strongly depresses the rates of absorption of B (Bowen 1968) and of Mo (Stout *et al.* 1951) but promotes those of the metallic cations when they are present as inorganic ions. Thus, the rate of Mn absorption from recirculating solution cultures containing 0.3 μM Mn^{2+} increased up to 20-fold in legumes as solution pH increased from 5.4 to 7.0 (Robson & Loneragan 1970). Similarly the rate of Zn absorption from solutions containing 1 μM Zn^{2+} increased with increasing pH (Chaudhry & Loneragan 1972*b*). By contrast, when chelating agents are present, increasing pH may decrease Zn absorption as, for example, in bush-bean seedlings from solutions containing 100 μM ^{65}Zn DTPA (diethylenetriaminepentaacetate) (Wallace 1962). Apparently the effect of increasing pH in increasing the bonding of Zn^{2+} to DTPA more than offset any effect of pH in increasing the rate of absorption of Zn^{2+} from solution.

The concentrations of other macronutrient and micronutrient ions in solution may also strongly influence trace element absorption. The effects are not simple but vary greatly with the concentration of the absorbing and interfering ions. The complexity of these interactions has not generally been realized and there are relatively few data at those concentrations of trace elements and other ions which commonly occur in soil solutions. The data available indicate that the absorption of the transition metal cations from concentrations of the order of 1 μM is not affected by anions but is strongly inhibited by the alkali and alkaline earth cations and by other transition metal cations. Thus, the absorption of Zn from 1 μM Zn^{2+} was strongly inhibited by increasing concentrations of nitrate salts of the alkali and alkaline earths, but replacing NO_3^- with Cl^-, SO_4^{2-} or $H_2PO_4^-$ had no effect (Chaudhry & Loneragan 1972*a*,*c*) : the inhibitory effect of these cations appeared to be mutually competitive but, at least under the conditions likely to be encountered in soils, to be additive to those of H^+ (Chaudhry & Loneragan 1972*a*,*b*). Differences in the concentrations of cations could largely account for the large differences between the results of two groups of workers in the relationship between concentrations of Zn^{2+} in solution and rates of Zn absorption by plants (Fig. 1).

Additive effects of Ca^{2+} salts and of H^+ ions in inhibiting absorption of Mn from nutrient solutions containing 0.3 μM Mn^{2+} have also been reported in four legumes (Robson & Loneragan 1970).

By contrast with the absorption of the metallic cations, the absorption of B from the lowest concentrations (2-100 μM) studied was highly specific, showing no effect of Ca^{2+}, K^+, Na^+, SO_4^{2-}, Cl^- or $H_2PO_4^-$ up to concentrations of 50 mM (Bowen 1969). The absorption of Mo was also inhibited by SO_4^{2-} but greatly enhanced by $H_2PO_4^-$ in short-term experiments (Stout & Meagher 1948; Stout *et al.* 1951).

Interactions among the trace elements may also be important. The absorption of Zn from Zn^{2+} was strongly inhibited by Cu^{2+}, weakly by Co^{2+}, and not at all by Mn^{2+} or Fe^{2+} (Hawf & Schmid 1967; Bowen 1969; Chaudhry & Loneragan 1972*b*). At low concentrations of Zn^{2+} and Cu^{2+} and in the presence of Ca^{2+} at concentrations commonly present in soil solutions, the effect of Cu^{2+} was competitive, suggesting that it would be additive to those of Ca^{2+} and of H^+. Both Zn^{2+} and Mn^{2+} ions inhibited Fe absorption from Fe^{2+} by tomato roots (Tiffin 1967) and by tobacco leaf cells (Kannan 1969).

Plant species. The limited data available suggest that the general physiological characteristics of trace element absorption by plants are probably similar in most species. But while the broad characteristics are similar, plant species may differ considerably in the extent to which they respond to varying factors in the environment of the absorbing solution. For example, while all species responded strongly to increasing concentrations of Zn^{2+}, subterranean clover absorbed Zn almost twice as fast per g fresh roots as did wheat and oats grown in the same recirculating nutrient solutions containing 0.25 μM Zn^{2+} or less (Carroll & Loneragan 1969). Similarly, two medic species absorbed Mn from complete nutrient solutions containing 0.33 μM Mn^{2+} between 2 and 5 times as fast per g fresh root as did two subterranean clover cultivars. Changing the concentrations of H^+ and Ca^{2+} had marked and proportionately similar effects on Mn absorption in all four species and cultivars. As a result, the rates of Mn absorption and the Mn concentrations in the tops of the plants grown in the same solutions for 10 days were strikingly different (Robson & Loneragan 1970).

Reductive activity of roots. Plant roots have been shown to reduce Fe^{3+} to Fe^{2+} (Brown *et al.* 1961; Ambler *et al.* 1970). This reduction appears to be an essential prerequisite to the absorption of Fe from solution. It also appears to provide an effective mechanism for releasing Fe^{3+} from chelated forms, which bind it much more strongly than Fe^{2+}. Both the reduction and the release of Fe^{3+} from chelates appears to proceed outside the cell since addition of the Fe^{2+}-specific chelate BPDS (bathophenanthroline disulfonate) to nutrient solutions containing Fe^{3+} chelates inhibited the absorption of Fe and resulted in the formation of Fe^{2+} BPDS (Chaney *et al.* 1972).

Soil-Plant Interactions

The study of soil-plant interactions is frequently hampered by the difficulty of sampling roots. As a result, conclusions on absorption are often reached on the basis of analyses of plant tops alone. Such conclusions should be regarded as tentative, since plant roots frequently retain a high proportion of the absorbed trace element and this proportion frequently varies with treatment and with species (e.g. Robson & Loneragan 1970).

The proximity, extent, and pattern of contact between soil and root are important factors in the absorption of ions from soils and must be especially important for the absorption of those trace elements which are tightly bonded to the soil colloids. Jenny (1961) has proposed that plants absorb these ions by direct 'contact exchange' between roots and soil particles. However, present evidence suggests that there is no reason to postulate specific processes of 'contact exchange'. Nevertheless, plant roots and soil do interact in ways which modify the properties and behaviour both of root systems and of adjacent soil.

Root systems in soil. Physical and chemical properties may influence the form and distribution of plant roots in soil in ways which affect the absorption of trace elements. For example, lucerne roots developed poorly in soil regions of low B content, and this retarded root development and nutrient uptake from those regions (Simpson & Lipsett 1973).

Biological activity in the soil may also influence trace element absorption by affecting root form and function. Damage by nematodes or by root-rot fungi could be expected to impair the absorption of trace elements by roots. Soil fungi may also form intimate associations with plant roots resulting in mycorrhizal associations with a greater extent of contact between plant and soil than the uninfected root (Harley 1969; Sanders & Tinker 1971). In addition to their larger size, mycorrhizal roots may also be more active than non-mycorrhizal roots. Thus Bowen and his colleagues (1974) showed that ectomycorrhizas of *Pinus radiata* and endomycorrhizas of *Araucaria cunninghamii* took up more Zn from solution culture as a result partly of their larger size and partly of their greater activity. In a Zn-deficient soil, peach seedlings with endomycorrhizal roots took up more Zn and did not suffer from Zn deficiency when compared with uninfected seedlings (Gilmore 1971).

The interaction of plant and soil may also change the composition of the soil solution at the root surface in ways which influence the absorption of trace elements from the solution. The processes involved in these changes will be discussed in the next section.

Influence of plant on soil supply. The plant root modifies the soil environment in its immediate vicinity both by excreting chemically active substances and by absorbing water and ions.

Mucilaginous material, possibly pectin, coats the surface of young roots and extends some distance into the soil (Scott *et al.* 1958; Jenny & Grossenbacher 1963). This and other excretions, such as organic acids, amino acids, HCO_3^-, and H^+ ions, may affect the release of trace elements from soil colloids. For example, an organic compound obtained by washing oat roots increased the release of Mn from insoluble oxides (Bromfield 1958*a*,*b*). Wheat roots also mobilized Fe in their vicinity in agar (Weavind & Hodgson 1971) while solutions of dilute acid, chelates, and organic compounds circulating through a thin, porous tube increased the movement of Zn from surrounding soil (Elgawhary *et al.* 1970; Wilkinson 1970). Compounds with reductive activity have been found in solutions in which plants had previously grown (Chaney *et al.* 1972). Excretions may also stimulate the activity of micro-organisms, which may in turn affect the supply of trace elements both by competing with the plant root for their absorption and by affecting their release from soil colloids. For example, organisms isolated from soil rhizospheres have been shown to excrete ketogluconate and to accelerate the release of Zn from insoluble silicates (Webley *et al.* 1960).

The transpiration of water by plants moves trace elements to the root surface by 'mass flow' in the soil solution (Barber 1962). At the root surface the concentration of trace elements will depend upon the balance between their rate of arrival and their rate of absorption by the root. Where absorption proceeds faster than arrival, the concentration of trace elements will fall. The resulting concentration gradient will cause trace elements to move from the bulk soil solution to the root by diffusion. From the concentration of trace elements in the soil solution and the loss of water by transpiration, the amount of trace element moved to the root by mass flow may be calculated. Such calculations suggest that in some soils Mo moves primarily by mass flow while in others with lower concentrations of Mo in solution, relatively little Mo moves by mass flow and diffusion appears important (Lavy & Barber 1964). Diffusion also appears important in the movement of Cu, Fe, Mn and Zn since, where the movement of these elements has been studied, mass flow could account for relatively little movement (Oliver & Barber 1966). Autoradiographs have also demonstrated depletion of ^{99}Mo and ^{65}Zn from soil near the surface of roots, suggesting that in these soils diffusion could be involved in the movement of Mo and Zn (Lavy & Barber 1964; Wilkinson *et al.* 1968).

Whatever the mechanisms involved, the autoradiographs of ^{99}Mo and ^{65}Zn indicate that, in the soils studied, plants obtained most of these elements from a narrow zone of soil immediately adjacent to the root. Clearly then, the rate and extent of the exploration of soil by roots and the nature of the interaction between root and soil are major factors in the absorption of most trace elements from soils.

Plant species. The properties and behaviour of individual species or cultivars are extremely important in determining the quantities of trace elements which plants obtain from soils. An example of the variation among plants is given in Figure 2, where the concentrations of Cu, Zn, Fe and Mn in the tops of 24 species and cultivars grown on the same soil are compared. While concentrations in plant tops respond to processes other than absorption (see later), the data of Figure 2 clearly indicate that plant species differ considerably in their abilities to absorb trace elements. Concentrations of Mn showed a particularly wide variation, ranging from an average of 30 ppm in *Medicago truncatula* to nearly 500 ppm in *Lupinus albus*. The same two species showed relatively little difference in their concentrations of Fe and of Cu.

Many other examples of differences among species in their

trace element contents are known (e.g. Fig. 4; Piper & Beckwith 1949; Peterson 1971). They are not always consistent and may vary with changing conditions, as is seen for example in Figure 4, where the Mo content of different species varies with changing pH. Some of the differences may have important consequences resulting in deficiencies or toxicities of trace elements to plants or to animals grazing them. The ability of *Astragalus* to absorb and accumulate Se to toxic levels in their tops is well known (e.g. Rosenfeld & Beath 1964). It is possible too that differences in the concentrations of trace elements in plant species available to grazing animals may be important in the development of deficiencies of Se, I, Cu and Co in animals (Johnson & Butler 1957; Ehlig *et al.* 1968; Kubota 1968; Gladstones *et al.* 1974). Examples of differences among plant species in sensitivity to deficiency due to differences in absorption of trace elements are also known, as in the sensitivity to Mn deficiency of oats compared with vetch (Jones 1957*c*). The reasons for their different behaviour is not understood.

Response to changing soil conditions. Changing soil pH frequently induces large changes in the absorption of trace

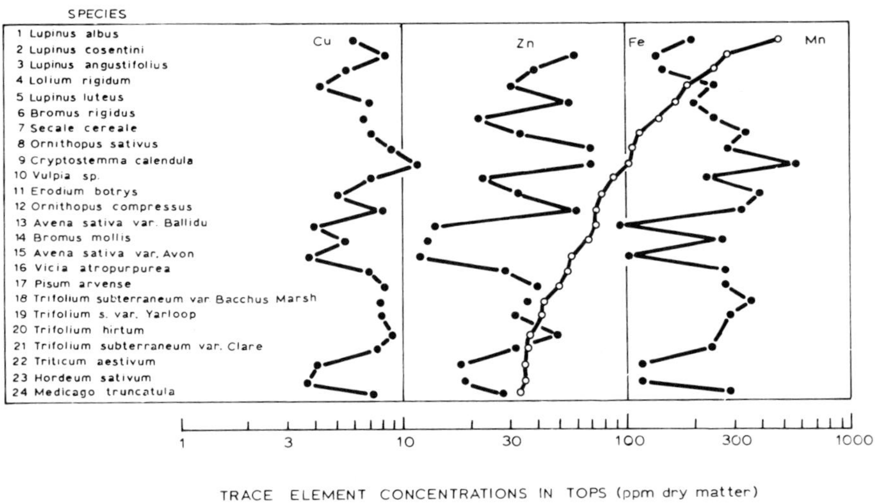

Fig. 2. Geometric mean concentrations of different trace elements in tops of 24 annual crop and pasture species grown at Gidgegannup, Western Australia. Data for Mn, Cu and Zn are means for 3 harvests and 3 rates of trace element application. Those for Fe are at the intermediate rate of trace element application only. [*Reproduced with permission from data of Gladstones & Loneragan (1970)*]

elements by plants. In these cases the dominant processes involved appear to be those of adsorption of trace elements to soil colloids. For example, the Zn content of sorghum tops decreased progressively with increasing pH of Norfolk sandy loam to which calcium carbonate, sodium carbonate or calcium sulphate was added (Fig. 3; Wear 1956). Since equivalent quantities of $CaSO_4$, which acidified the soil, and of $CaCO_3$, which made it more alkaline, had opposite effects on the Zn content of the sorghum, the effects of the fertilizers must be attributed to pH rather than to cations.

The results correspond with the effect of pH in increasing adsorption of Zn by soil colloids (Quirk & Posner, this Symposium) but are opposite to those observed for the absorption of Zn from solution cultures. Apparently the effects of the fertilizers on the adsorption of Zn by soil colloids dominated any physiological effect of Ca^{2+} and H^+ on the absorption of Zn by the plants.

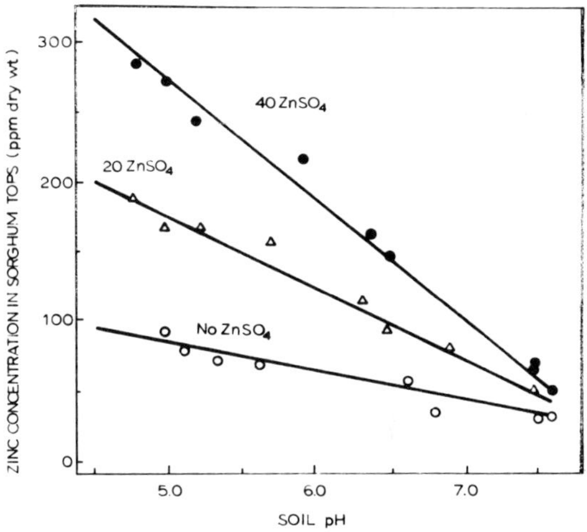

Fig. 3. Relationship between Zn concentration in sorghum tops and pH of Norfolk sandy loam. [Reproduced with permission from Soil Science, after Wear (1956)]

The effect of changing soil pH on the absorption of Mo by plants (Fig. 4; Piper & Beckwith 1949) also parallels its effect on the adsorption of Mo by colloids (Jones 1957*d*; Quirk & Posner, this Symposium) and proceeds in the opposite direction to the effects of pH on the absorption of Mo by plants described earlier.

By contrast with Mo and Zn, changing soil pH does not affect the absorption of Cu by plants in the way predicted from studies of the adsorption of Cu^{2+} ions on organic (Peech 1941) and inorganic (Quirk & Posner, this Symposium) colloids. Indeed, the absorption of Cu by plants may change remarkably little with treatments which change soil pH (Fig. 4; Piper & Beckwith 1949). In these cases it is likely that the behaviour of Cu is dominated by the formation of soluble complexes of Cu with organic compounds in the soil solution, as Hodgson *et al.* (1966) have suggested to explain the insensitivity of plants to Cu deficiency on calcareous soils. The behaviour of Fe in soil solutions is also largely dominated by its chelation in soluble organic compounds. However, many plants are more susceptible to Fe deficiency than to Cu deficiency on calcareous soils and this may be explained by the lower stability at alkaline pH of Fe chelates compared with Cu chelates.

The processes controlling absorption of Mn by plants in many soils seem to be largely those controlling the equilibrium between divalent Mn^{2+} and MnO_2 in the soil. The extensive literature has been reviewed by Leeper (1970) and only two examples are given here.

On most soils, liming depresses Mn absorption and on many it induces severe Mn deficiency, as demonstrated long ago by Samuel & Piper (1928). However, liming an organic soil from

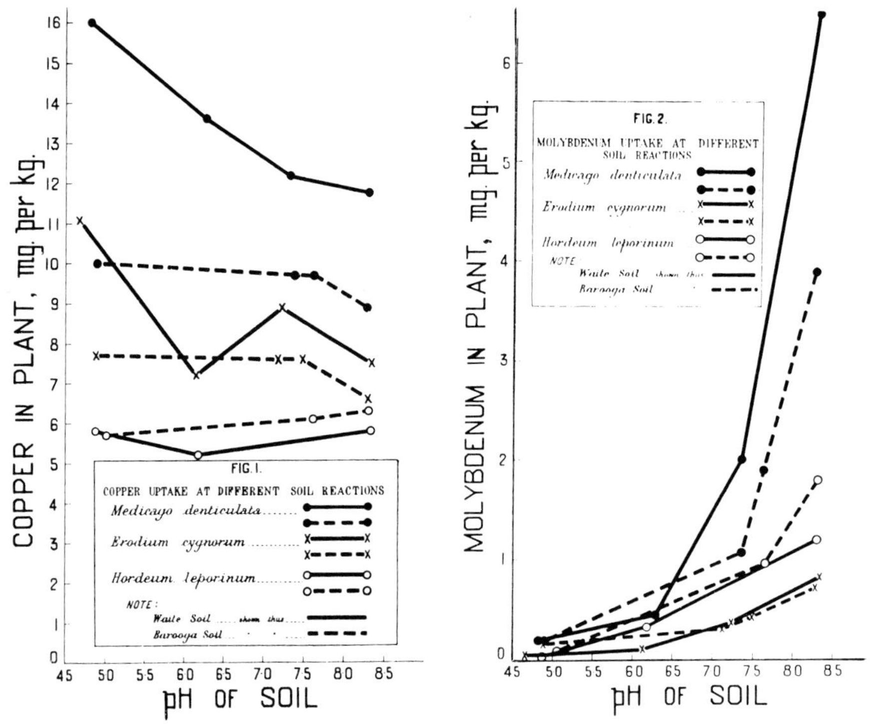

Fig. 4. Relationship between pH of soil and concentration of Cu (Fig. 1) and of Mo (Fig. 2) in several plant species. [Reproduced with permission from Piper & Beckwith (1949)]

Penola, South Australia, generally increased the absorption of Mn by oats and alleviated Mn deficiency. But when the same soil was sterilized prior to sowing, liming depressed Mn absorption in the usual way (Jones 1957*a*). This complex behaviour of Mn absorption by plants paralleled changes in the quantities of Mn^{2+} which were extracted from moist soils incubated with MnO_2 (Jones 1957*b*). Concentrations of Mn in the soil solution and in plants also fluctuate widely and rapidly with changing soil aeration as, for example, in the development of Mn toxicity in waterlogged soils, which is believed to result from reduction and solution of MnO_2 (Grasmanis & Leeper 1966).

The formation of MnO_2 and other Mn oxides also appears to be very important in controlling the level of Co in solution in many soils (McKenzie, this Symposium) and, through these reactions, in controlling the absorption of Co by plants. For example, the relationship between increasing levels of Co fertilizer and Co concentrations in the foliage of subterranean clover plants varied widely for seven Australian soils. But when the level of Co fertilizer added was divided by the Mn concentration of the soil, the same data fell on a single line for six of the soils suggesting that, in them, reactions of Co with Mn controlled the availability of Co fertilizer to the plant. Only on a humus podzol did the relationship differ, indicating that, in it, factors other than Mn content may have controlled Co absorption by plants (Adams *et al.* 1969).

Two other interesting reactions involving absorption of trace elements will be mentioned briefly. The first involves an effect of Zn in decreasing the absorption of Cu by plants. The second involves an effect of phosphate in depressing the absorption of Zn.

It has long been known that application of Zn fertilizers to soils low in Cu and Zn can induce disastrous Cu deficiency in wheat crops (Fig. 5; Dunne 1956). Recent work has shown that Zn may depress Cu absorption by wheat plants in soils (Chaudhry & Loneragan 1970). The cause of this depression is unlikely to involve soil reactions since Zn^{2+} displaces Cu^{2+} from specific adsorption sites on colloids (Tiller & Hodgson 1962). It seems more likely that Zn^{2+} competes in the physiological absorption of Cu in the same way as Cu^{2+} has been shown to inhibit absorption of Zn^{2+} as discussed earlier. However, there is as yet no suitable data.

Phosphate-induced Zn deficiency has produced a voluminous literature. In many cases the application of phosphate stimulated plant growth to such an extent that it diluted the supply of Zn and so induced Zn deficiency (Boawn *et al.* 1954).

In a relatively few cases, application of phosphatic fertilizers actually depressed the amount of Zn absorbed (Rogers & Wu 1948; Loneragan 1951). Two processes may have contributed to depressing Zn absorption — (a) phosphate has been shown to enhance the adsorption of Zn by Fe oxides (Stanton & Burger 1967) and (b) Ca^{2+} added with phosphatic fertilizers may have inhibited the absorption of Zn^{2+} by plants, as discussed earlier. Grove (1969) has recently obtained evidence that the first mechanism may be important in soils : dressings of calcium phosphate inhibited absorption of Zn by subterranean clover from a soil high in Fe oxides but had no effect whatever on the absorption of Zn from a soil low in Fe oxides.

Trace element applications vary considerably in their effectiveness when applied to soil. However, time does not permit the development of this topic other than to note that applications of B, Cu, Mo and Zn fertilizers to soils generally overcome deficiencies and increase the concentrations of these trace elements in plants. However, applications of Co, Fe, Mn or Se fertilizers are often ineffective and at best have only transitory effects.

ABSORPTION IN RELATION TO PLANT CONCENTRATIONS AND GROWTH

General Relationships

The total quantity of any nutrient absorbed by a plant may be primarily determined either by the total quantity of nutrient present in the growth medium or by the time and rate of nutrient absorption from the medium. When the total quantity of nutrient in the medium is the prime determinant, the concentration of nutrient in a plant varies directly and inversely with the *amount* of growth. When the quantity of nutrient absorbed by a plant is determined primarily by the time and rate of its absorption, nutrient concentrations vary inversely with the *rate* of plant growth (Williams 1948; Loneragan 1968*a*).

$$I_m = \frac{RGR \, . \, [M]}{W_R}$$

where I_m = rate of absorption of mineral nutrient M per unit root weight,

RGR = relative growth rate of the whole plant,

M = the average concentration of nutrient M in the whole plant,

W_R = ratio of weights of root to whole plant.

Clearly, anything which promotes plant growth without concomitantly increasing the rate of absorption or the proportion of root will decrease mineral concentrations in the plant.

Thus, unless the rate of absorption increases, increasing plant production dilutes nutrient concentrations in plants. As a result, we can expect more widespread deficiencies of trace elements on Australian soils as crop and pasture production respond to the progressive improvement in soil fertility resulting from the continuing use of legumes and phosphatic fertilizers. The effect of phosphatic fertilizers in inducing Zn deficiency by promoting growth has already been mentioned. An example of the effect of nitrogenous fertilizer in inducing Cu and Zn deficiencies on a virgin loamy sand from Badgingarra is given in Figure 5. Grain yield of wheat showed no response to Cu or Zn on this soil in the absence of N fertilizer. Addition of NH_4NO_3 promoted growth, depressed Cu and Zn concentrations, and induced substantial responses to both Zn and Cu. The effect of NH_4NO_3 was particularly striking in the presence of Zn fertilizer, which depressed Cu absorption : the combined fertilizers induced such severe Cu deficiency that the grain yield dropped substantially below that of the N-deficient control plants.

Fig. 5. Effects of N, Cu and Zn fertilizers on the grain yield of wheat grown on a loamy sand from Badgingarra, Western Australia. [*Reproduced with permission from Australian Journal of Agricultural Research 21 : 865*]

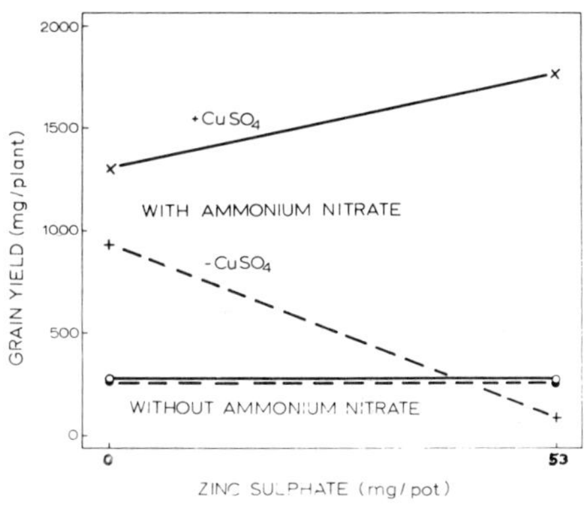

Movement Within the Plant

So far we have related nutrient absorption to the *average* concentration of nutrients in the whole plant. The concentration of nutrients in different parts of the plant may vary greatly depending upon the rate of nutrient absorption and the movement of nutrients among organs. Some nutrients such as N, P and K move freely and rapidly among organs. These nutrients occur in reasonably high concentrations in young developing organs regardless of the current rate of absorption or nutrient status of the plant : by contrast, their concentrations in mature and older organs vary greatly in response to the general nutrient status of the plant for which they are much better indicators than young organs (Ulrich 1952; Biddulph 1959). Other nutrients such as Ca move freely with the transpiration stream in the xylem but appear unable to move out of leaves in which they accumulate. These nutrients vary widely in their concentrations in young tissues, reflecting their current rates of absorption by plant roots : by contrast, their concentrations in mature and older leaves depend on the rate of their absorption at the time of development of those leaves and are largely independent of current rates. These nutrients become deficient as soon as their rate of supply from the soil falls below values which can sustain their absorption at the rates needed to maintain their concentrations at the levels required by developing tissues and organs (Loneragan 1968*b*; Loneragan & Snowball 1969*a*,*b*).

For those nutrients which are freely mobile within the plant, the concentration of nutrients in whole plant tops is a useful guide to the nutrient status of plants, but it may be misleading for those nutrients which are not mobilized from old leaves. For example, plants transferred from a high to a low supply of Ca became Ca deficient even though they had concentrations of Ca in their tops 5 times those of healthy plants given a moderately low and continuously replenished supply of Ca (Loneragan 1968*b*; Loneragan & Snowball 1969*a*).

The trace elements B and Fe appear to behave like Ca, becoming deficient in the developing tissues of plants as soon as their supply to the roots is exhausted (Haynes & Robbins 1948). The B concentrations of young leaves has also been observed to drop to very low values while old leaves remained high when sugar beets were transferred from a high to a low B supply (Brandenburg 1939). By contrast with B and Fe, Co and Mo appear to move readily out of older leaves into new leaves and roots (Meagher *et al.* 1952; Gustafson 1956).

The trace elements Mn, Zn and Cu appear to have a variable and limited mobility from old leaves. In extensive studies with subterranean clover, Riceman & Jones (1958,1960) observed that up to 25% of the Zn in leaves and petioles moved into developing fruits. However, the movement of Zn depended on the Zn status of plants, those with luxury levels of Zn in their leaves moving a large amount and those with marginal or deficient levels of Zn in their leaves moving little or none at all. Oats appeared to behave in a similar way (Wood & Sibly 1950; Williams & Moore 1952). There are fewer data for Mn and Cu. Leaves of onions lost 40% of their Mn and Cu contents to bulbs over a period of 52 days (Carolus 1962). Leaves of sugar cane lost most of their Mn and part of their Cu to the stem during senescence (Mukherjee 1969), while leaves of oats lost about 25% of their Cu to developing grain (Williams & Moore 1952). In other studies with oats, leaves lost no Cu (Wood & Womersley 1946) and no Mn (Williams & Moore 1952) during grain development. Clearly, the movement of Mn, Zn and Cu is restricted so that attempts to diagnose their deficiencies by analyses of whole tops could, like diagnoses for Fe and B, suffer the same difficulties inherent in the analysis of the whole tops for Ca.

Significance of Interrelations for Deficiency-Diagnosis — An Example

We have recently attempted to clarify the relationships between grain yield and the supply, movement and concentrations of Cu in organs during growth of wheat plants. The results indicate that Cu can move very rapidly from old leaves but that its movement is strongly dependent on the Cu status of the plants. Leaves of plants given a luxury supply of Cu lost more than 70% of their Cu during grain development. By contrast, leaves of Cu-deficient plants lost less than 20%. Apparently Cu, like Zn (Riceman & Jones 1960), is relatively immobile from leaves when it is deficient.

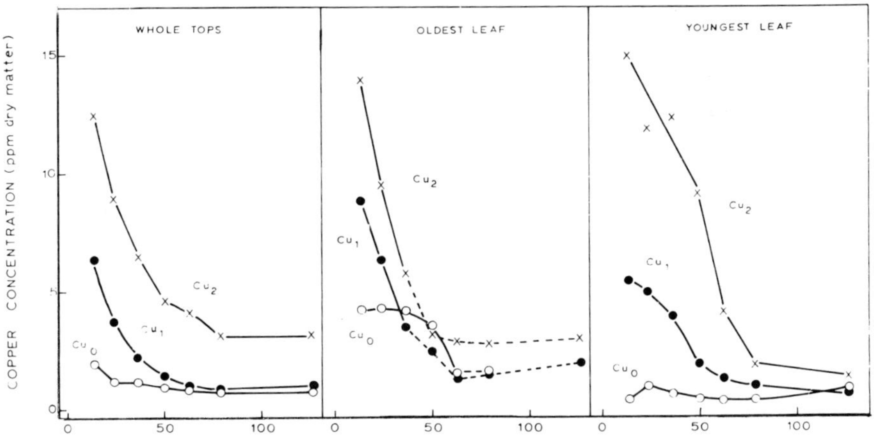

Fig. 6. Effect of supply on the concentration of Cu in whole tops and in oldest and youngest leaves of wheat plants grown on a sand from Lancelin, Western Australia. [Reproduced with permission from Loneragan, Snowball & Robson (in preparation)]

The low mobility of Cu in Cu-deficient plants may largely account for the characteristic development of symptoms of Cu deficiency in newly developing tissues. It also produces some peculiar anomalies in the relationship between Cu concentrations in plants and yield. For example, Figure 6 shows that from midway through the growing season the Cu concentrations in the tops of Cu_1 plants, which produced maximal grain yield, were almost the same as those for Cu_0 plants, which produced no grain at all. The Cu concentrations of the oldest leaves gave an even poorer indication of the Cu status of plants. By contrast, the young developing leaves reflected the Cu status of the plants quite well until they became mature and lost their Cu to other developing tissues.

When the data for the mid-growth harvests of the oldest leaves in Figure 6 are plotted as Cu concentrations against yield, we obtain a response curve with the lowest Cu treatment having higher Cu concentrations in tissues than plants receiving higher Cu applications. These peculiar "Piper-Steenbjerg" response curves (Ulrich 1952) have puzzled workers in plant nutrition ever since they were reported for Cu in oats by Piper (1942) and in barley by Steenbjerg (1950, 1951). Similar relationships appear in the data of even earlier investigations of Cu deficiency in South Australia (Riceman *et al.* 1940). They have also been reported for Mn in oats (Jones 1957*c*) and for Zn in sugar beets (Rosell & Ulrich 1964). The present results suggest that the complex interrelationships between supply and movement of trace elements in plants may explain these phenomena. I hope so, because it would give me a particular pleasure to help advance our understanding of one small portion of the pioneering and monumental work on trace elements which has been done here at the Waite Institute and to which I have referred so frequently in this review.

REFERENCES

ADAMS, S.N., HONEYSETT, J.L., TILLER, K.G. & NORRISH, K. (1969) - *Aust. J. Soil Res. 7* : 29

AMBLER, J.E., BROWN, J.C. & GAUCH, H.G. (1970) - *Pl. Physiol. 46* : 320

BARBER, S.A. (1962) - *Soil Sci. 93* : 39

BECKWITH, R.S. (1963) - *Agrochimica 7* : 296

BECKWITH, R.S. (1973) - Ph.D. Thesis, University of Western Australia

BIDDULPH, O. (1959) - *In* "Plant Physiology" (F.C. Steward, ed.), Vol. 2. Academic Press : New York

BOAWN, L.C., VIETS, F.G. & CRAWFORD, C.L. (1954) - *Soil Sci. 78* : 1

BOWEN, G.D., SKINNER, M.F. & BEVEGE, D.I. (1974) - *Soil Biol. Biochem. 6* : 141

*BOWEN, H.J.M. & CAWSE, P.A. (1964) - *Soil Sci. 98* : 358

BOWEN, J.E. (1968) - *Pl. Cell Physiol. 9* : 467

BOWEN, J.E. (1969) - *Pl. Physiol. 44* : 255

BRANDENBURG, E. (1939) - *Phytopath. Z. 12* : 1

BRANTON, D. & JACOBSON, L. (1962) - *Pl. Physiol. 37* : 539

BROMFIELD, S.M. (1958*a*) - *Pl. Soil 9* : 325

BROMFIELD, S.M. (1958*b*) - *Pl. Soil 10* : 147

BROWN, J.C., HOLMES, R.S. & TIFFIN, L.O. (1961) - *Soil Sci. 91* : 127

CAROLUS, R.L. (1962) - *Int. Hort. Congr. Rep., 16th (Brussels, Belgium) 2* : 202

CARROLL, M.D. & LONERAGAN, J.F. (1969) - *Aust. J. agric. Res. 20* : 457

CHANEY, R.L., BROWN, J.C. & TIFFIN, L.O. (1972) - *Pl. Physiol. 50* : 208

CHAUDHRY, F.M. & LONERAGAN, J.F. (1970) - *Aust. J. agric. Res. 21* : 865

CHAUDHRY, F.M. & LONERAGAN, J.F. (1972*a*) - *Proc. Soil Sci. Soc. Am. 36* : 323

CHAUDHRY, F.M. & LONERAGAN, J.F. (1972*b*) - *Proc. Soil Sci. Soc. Am. 36* : 328

CHAUDHRY, F.M. & LONERAGAN, J.F. (1972*c*) - *J. exp. Bot. 23* : 552

DE KOCK, P.C. & MITCHELL, R.C. (1957) - *Soil Sci. 84* : 55

DUNNE, T.C. (1956) - *Proc. Aust. Conf. Pl. Nutr., Melbourne 1* : 278 (CSIRO Aust. : Melbourne)

EHLIG, C.F., ALLAWAY, W.H., CARY, E.E. & KUBOTA, J. (1968) - *Agron. J. 60* : 43

ELGAWHARY, S.M., LINDSAY, W.L. & KEMPER, W.D. (1970) - *Proc. Soil Sci. Soc. Am. 34* : 211

FINDENEGG, G. & BRODA, E. (1965) - *Nature, Lond. 208* : 196

GEERING, H.R. & HODGSON, J.F. (1969) - *Proc. Soil Sci. Soc. Am. 33* : 54

GEERING, H.R., HODGSON, J.F. & SDANO, C. (1969) - *Proc. Soil Sci. Soc. Am. 33* : 81

GILMORE, A.E. (1971) - *J. Am. Soc. hort. Sci. 96* : 35

GLADSTONES, J.S. & LONERAGAN, J.F. (1970) - *Proc. XI Int. Grassld Congr., Brisbane*, p. 350

GLADSTONES, J.S., LONERAGAN, J.F. & SIMMONS, W.J. (1974) - *Aust. J. agric. Res.* (in press)

GRASMANIS, V.O. & LEEPER, G.W. (1966) - *Pl. Soil 25* : 41

GROVE, T.S. (1969) - B.Sc. (Hons.) Thesis, University of Western Australia
GUSTAFSON, F.G. (1956) - *Am. J. Bot. 43* : 157
GUTKNECHT, J. (1961) - *Limnol. Oceanogr. 6* : 426
GUTKNECHT, J. (1963) - *Limnol. Oceanogr. 8* : 31
HARLEY, J.L. (1969) - "The Biology of Mycorrhiza", 2nd Edn. Leonard Hill Ltd. : London
HAWF, L.R. & SCHMID, W.E. (1967) - *Pl. Soil 27* : 249
HAYNES, J.L. & ROBBINS, W.R. (1948) - *J.Am.Soc.Agron. 40* : 795
HINGSTON, F.J. (1964) - *Aust. J. Soil Res. 2* : 83
HINGSTON, F.J., POSNER, A.M. & QUIRK, J.P. (1972) - *J. Soil Sci. 23* : 177
HODGSON, J.F. (1963) - *Adv. Agron. 15* : 119
HODGSON, J.F., GEERING, H.R. & NORVELL, W.A. (1965) - *Proc. Soil Sci. Soc. Am. 29* : 665
HODGSON, J.F., LINDSAY, W.L. & TRIERWEILER, J.F. (1966) - *Proc. Soil Sci. Soc. Am. 30* : 723
JENNY, H. (1961) - *Agrochimica 5* : 281
JENNY, H. & GROSSENBACHER, K. (1963) - *Proc. Soil Sci. Soc. Am. 27* : 273
JOHNSON, J.M. & BUTLER, G.W. (1957) - *Physiologia Pl. 10* : 100
JONES, L.H.P. (1957*a*) - *Pl. Soil 8* : 301
JONES, L.H.P. (1957*b*) - *Pl. Soil 8* : 315
JONES, L.H.P. (1957*c*) - *Pl. Soil 8* : 328
JONES, L.H.P. (1957*d*) - *J. Soil Sci. 8* : 313
KANNAN, S. (1969) - *Pl. Physiol. 44* : 1457
KUBOTA, J. (1968) - *Soil Sci. 106* : 122
LAVY, T.L. & BARBER, S.A. (1964) - *Proc. Soil Sci. Soc. Am. 28* : 93
LEEPER, G.W. (1947) - *Soil Sci. 63* : 79
LEEPER, G.W. (1970) - "Six Trace Elements in Soils. Their Chemistry as Micronutrients". Melbourne University Press : Melbourne
LINDSAY, W.L. (1972) - *In* "Micronutrients in Agriculture" (J.J. Mortvedt, P.M. Giordano & W.L. Lindsay, eds.), p. 41. Soil Science Society of America Inc. : Wisconsin
LONERAGAN, J.F. (1951) - *Aust. J. scient. Res. B 4* : 108
LONERAGAN, J.F. (1968*a*) - *Trans. 9th Int. Congr. Soil Sci., Adelaide, II* : 173
LONERAGAN, J.F. (1968*b*) - *Nature, Lond. 188* : 1307
LONERAGAN, J.F. & SNOWBALL, K. (1969*a*) - *Aust. J. agric. Res. 20* : 465
LONERAGAN, J.F. & SNOWBALL, K. (1969*b*) - *Aust. J. agric. Res. 20* : 479
MAAS, E.V., MOORE, D.P. & MASON, B.J. (1968) - *Pl. Physiol. 43* : 527

MEAGHER, W.R., JOHNSON, C.M. & STOUT, P.R. (1952) - *Pl. Physiol.* *27* : 223
MUKHERJEE, K.L. (1969) - *J. Indian bot. Soc.* *47* : 180
MULDER, E.G. & GERRETSEN, F.C. (1952) - *Adv. Agron.* *4* : 222
NOAKES, J.E. & HOOD, D.W. (1961) - *Deep Sea Res.* *8* : 121
OLIVER, S. & BARBER, S.A. (1966) - *Proc. Soil Sci. Soc. Am.* *30* : 468
PEECH, M. (1941) - *Soil Sci.* *51* : 473
PETERSON, P.J. (1971) - *Sci. Prog., Oxf.* *59* : 505
PICKERING, D.C. & PUIA, I.L. (1969) - *Physiologia Pl.* *22* : 653
PIPER, C.S. (1931) - *J. agric. Sci., Camb.* *21* : 762
PIPER, C.S. (1942) - *J. agric. Sci., Camb.* *32* : 143
PIPER, C.S. & BECKWITH, R.S. (1949) - Brit. Commonwealth Spec. Conf. Agric., Australia
RATHORE, V.S., WITTWER, S.H., JYUNG, BAJAJ, Y.P.S. & ADAMS, M.W. (1970) - *Physiologia Pl.* *23* : 908
RICEMAN, D.S., DONALD, C.M. & EVANS, S.T. (1940) - CSIRO Aust. Pamphlet No. 96
RICEMAN, D.S. & JONES, G.B. (1958) - *Aust. J. agric. Res.* *9* : 730
RICEMAN, D.S. & JONES, G.B. (1960) - *Aust. J. agric. Res.* *11* : 162
ROBSON, A.D. & LONERAGAN, J.F. (1970) - *Aust. J. agric. Res.* *21* : 223
ROGERS, L.H. & WU, C. (1948) - *J. Am. Soc. Agron.* *40* : 563
ROSELL, R.A. & BABCOCK, K.L. (1968) - *In* "Isotopes and Radiation in Soil Organic Studies". International Atomic Energy Agency : Vienna
ROSELL, R.A. & ULRICH, A. (1964) - *Soil Sci.* *97* : 152
ROSENFELD, I. & BEATH, O.A. (1964) - "Selenium". Academic Press : New York
RUSSELL, E.W. (1973) - "Soil Conditions and Plant Growth", 10th Edn. Longmans : London
SAMUEL, G. & PIPER, C.S. (1928) - *J. Dep. Agric. S. Aust.* *32* : 696, 789
SAMUEL, G. & PIPER, C.S. (1929) - *Ann. appl. Biol.* *16* : 493
SANDERS, F.E.T. & TINKER, P.B. (1971) - *Nature, Lond.* *233* : 278
SCHMID, W.E., HAAG, H.P. & EPSTEIN, E. (1965) - *Physiologia Pl.* *18* : 860
SCHNITZER, M. & KHAN, S.U. (1972) - "Humic Substances in the Environment". Dekker : New York
SCHNITZER, M. & SKINNER, S.I.M. (1963) - *Soil Sci.* *96* : 181
SCOTT, F.M., HAMNER, K.C., BAKER, E. & BOWLER, E. (1958) - *Am. J. Bot.* *45* : 449
SIMPSON, J.R. & LIPSETT, J. (1973) - *Aust. J. agric. Res.* *24* : 199

STANTON, D.A. & BURGER, R. duT. (1967) - *Geoderma 1* : 13
STEENBJERG, F. (1950) - *Trans. 4th Int. Congr. Soil Sci., Amsterdam, 1* : 309
STEENBJERG, F. (1951) - *Pl. Soil 3* : 97
STEENBJERG, F. (1954) - *Pl. Soil 5* : 226
STEVENSON, F.J. & ARDAKANI, M.S. (1971) - *In* "Micronutrients in Agriculture" (J.J. Mortvedt, P.M. Giordano & W.L. Lindsay, eds.), p. 79. Soil Science Society of America Inc. : Wisconsin
STOUT, P.R. & MEAGHER, W.R. (1948) - *Science 108* : 471
STOUT, P.R., MEAGHER, W.R., PEARSON, G.A. & JOHNSON, C.M. (1951) - *Pl. Soil 3* : 51
TIFFIN, L.O. (1967) - *Pl. Physiol. 42* : 1427
TILLER, K.G. & HODGSON, J.F. (1962) - *Proc. 9th Natl Conf. Clays Clay Miner. 9* : 393 (Pergamon Press : New York)
ULRICH, A. (1952) - *A. Rev. Pl. Physiol. 3* : 207
WALLACE, A. (1962) - "A Decade of Synthetic Chelating Agents in Inorganic Plant Nutrition". Arthur Wallace, UCLA : Los Angeles
WALLACE, A., CARMACK, C. & WALLACE, G.A. (1963) - *In* "Solute Uptake by Intact Plants" (A. Wallace, ed.), p. 147. Arthur Wallace, UCLA : Los Angeles
WEAR, J.I. (1956) - *Soil Sci. 81* : 311
WEAVIND, T.E.F. & HODGSON, J.F. (1971) - *Pl. Soil 34* : 697
WEBLEY, D.M., DUFF, R.B. & MITCHELL, W.A. (1960) - *Nature, Lond. 188* : 766
WILKINSON, H.F. (1970) - Ph.D. Thesis, University of Western Australia
WILKINSON, H.F., LONERAGAN, J.F. & QUIRK, J.P. (1968) - *Proc. Soil Sci. Soc. Am. 32* : 831
WILLIAMS, C.H. & MOORE, C.W.E. (1952) - *Aust. J. agric. Res. 3* : 343
WILLIAMS, R.F. (1948) - *Aust. J. scient. Res. B 1* : 333
WOOD, J.G. & SIBLY, PAMELA M. (1950) - *Aust. J. scient. Res. B 3* : 14
WOOD, J.G. & WOMERSLEY, H.B.S. (1946) - *Aust. J. exp. Biol. Med. Sci. 24* : 79

**Added in proof:*
BOWEN, H.J.M. (1966) - "Trace Elements in Biochemistry". Academic Press : London & New York

THE EFFECTS OF FLOODING ON THE AVAILABILITY OF TRACE METALS TO RICE IN SOILS OF DIFFERING ORGANIC MATTER STATUS

R.S. Beckwith, K.G. Tiller & E. Suwadji

Division of Soils, CSIRO
Glen Osmond, South Australia 5064

INTRODUCTION

Studies of the effects of flooding on the chemistry of trace metals in soils and sediments are important because of their implications for plant health, especially for rice, and in relation to the cycling of metals in the environment.

The chemical changes which occur when a soil is submerged are complex and have been well reviewed recently by Ponnamperuma (1972). On current understanding, the changes affect redox potential, pH and the breakdown of organic matter with the production of CO_2 and new organic compounds. A different chemical environment is created, resulting in major shifts in ionic composition of the soil solution and in chemical equilibria related to sorption/desorption and stability of solid phases. In particular, levels of Fe and Mn in the soil solution are affected by the changed acidity, redox potential and complexing organic ligands.

The chemistry of other metals, such as Cu and Zn, in flooded soils has been generally less studied than that of Fe and Mn and knowledge is limited (Ponnamperuma 1972). These metals do not undergo valence change in flooded soils but their equilibria are sometimes dependent on the stability of other solid phases which are sensitive to these conditions. Thus Tiller (1958) ascribed the losses of Cu and Zn from seasonally waterlogged soil horizons to the breakdown of host Fe oxide minerals. The release of Co in waterlogged soils is undoubtedly attributable to instability of Mn minerals, which often contain other metals, especially Co. As a consequence, plant uptake of Co is increased by waterlogging soils.

In so far as the solid-solution equilibria of trace metals are controlled by sorption reactions, even minor dissolution of the adsorbing surfaces of Fe and Mn oxides can significantly affect trace metal distributions. When such changes are accompanied by concurrent changes in acidity, in nature and amounts of ligands, and in their distributions between soil surfaces and solution, the final effects on trace metals may become very considerable. It is the complexity of these possible interactions which has limited progress in understanding the effects of submergence of soils on such trace metals as Cu and Zn.

Nutritional problems of field-grown rice involving trace metals have mainly concerned Fe, Mn and Zn deficiencies and Fe toxicity (Tanaka & Yoshida 1970). Rice yields have occasionally been increased by applications of copper sulphate in India (Joshi & Joshi 1957; Rahaja *et al.* 1959; Rajan & Rao 1964), in Portugal (Candido *et al.* 1968) and in Japan (Yamasaki 1965). Ostrovskaya (1956) stimulated seedlings growth of rice grain containing only 2 μg Cu per gram by presoaking seeds in $CuSO_4$ solution.

Several experimental studies of effects of flooding on levels of trace metals in soil solutions have been reported. Clark & Resnicky (1956) included addition of straw as a soil treatment and measured changes in nutrient concentrations with time of submergence. Most nutrients, including Mn, Fe and the majors, tended to reach a peak concentration after about a month, with decreases subsequently. Copper in solution did not increase with time of flooding and was apparently unaffected by added organic matter. Ng & Bloomfield (1962) incubated soils anaerobically with and without added dry plant material and studied amounts of extractable metals. Contrasting effects on extractable Cu, in the presence and absence of added grass, were ascribed to sorption on grass fibres and to acid production respectively. Zinc and manganese were mobilized in water-soluble forms by flooding in the presence of grass and the effect did not decrease with time. But in the absence of grass, no such mobilization was detected.

Detailed studies at the International Rice Research Institute (Philippines) described the changes in water-soluble Fe and Mn in paddy soils as affected by soil type, organic matter and time of flooding (IRRI 1964; IRRI 1970). Rapid decrease in soluble Mn, after the initial increase, was attributed to the formation of $MnCO_3$. Soluble Fe behaved similarly and the secondary decrease was here interpreted as a pH-controlled formation of Fe hydroxides. Ponnamperuma (1972) noted that soil solutions were commonly supersaturated with respect to siderite ($FeCO_3$). Analogous decreases in water-soluble Zn, after several weeks' flooding of soils, are even less well understood (Ponnamperuma 1972) but may be controlled by zinc sulphide solubility in some soils (IRRI 1972).

This paper reports effects of organic matter on availability of metals to rice on some Australian soils. The cultural practice in the rice-growing lands of the Murrumbidgee Irrigation Area (MIA) is to include rice in a rotation based on a long period (usually 5 years) of leguminous pasture. Hence soil organic matter was varied both by selection of pairs of near-adjacent soils at different stages in this rotation and by laboratory additions of ground rice straw. Chemical analyses, and isotope dilution procedures for Zn, were used to assess effects of treatments and of previous land use on the size of the total pool of available metals. With the flooded soils, solution pH and metal concentrations were monitored throughout growth of the rice to assist in the interpretation of results.

MATERIALS AND METHODS

Some characteristics of the soils used in the investigation are given in Table 1. These four paired soils were collected from adjacent fields of contrasting management, either in pasture or fallowed after cereal cropping. The soils E447, E448, E450 and E451 have each grown 5 or 6 rice crops during the last 30 years approximately.

Surface soil (0-10 cm) was sampled at several locations within a radius of 50 metres at each site and bulked for glasshouse and laboratory work.

Glasshouse Methods

Basal fertilizer equivalent to 200 kg N, 40 kg P, 44 kg K and 18 kg S per hectare was supplied as analytical grade urea, NaH_2PO_4 and K_2SO_4, and mixed throughout pots containing about 2.5 kg air-dried soil. When required, ground rice straw was also mixed with the soils at the rate of 3.7 g/pot (about 3 tonnes/hectare). Carrier-free ^{65}Zn was mixed well in the dampened soil (approximately 30 µCi/2.5 kg soil) as described by Tiller *et al.* (1972).

During growth of the rice plants, non-flooded soils were maintained at 60% of water holding capacity by weighing and a free water layer of 2.5 cm depth was maintained over flooded soils using de-ionized, distilled water. Pre-germinated seeds of the variety "Calrose" were sown in non-puddled soils and thinned to six plants per pot after establishment. Day length was extended artifically to 14 hr for the first two

TABLE 1

Some characteristics of the soils used

Soil No.	Location	Land use prior to sampling	Chemical data				
			pH	Clay %	C %	N %	CEC Meq%
E444[1]	Urrbrae, S.A.	Pasture	5.8	19	1.52	0.16	15
E445	" "	Wheat	5.3	18	0.92	0.10	14
E447[2]	Yanco, NSW	Pasture	7.5	49	1.29	0.15	30
E446	" "	Rice	7.3	42	1.17	0.14	27
E448[3]	Deniliquin, NSW	Pasture	6.6	20	1.24	0.16	15
E449	" "	Rice	6.3	39	0.72	0.11	23
E451[4]	Leeton, NSW	Pasture	5.4	33	1.20	0.15	22
E450	" "	Cereal	5.7	38	0.94	0.12	25

[1] Soils E444 and 445 are from the permanent rotation plots, Waite Agricultural Research Institute, University of Adelaide.

[2] Soils E446 and 447 belong to the Gogeldrie series (van Dijk 1961).

[3] Soils E448 and 449 belong to the Riverina Clay series (Johnston 1953).

[4] Soils E450 and 451 belong to the Willbriggie series (van Dijk 1961).

months after seeding. Three representative plants were harvested from each pot after 8 weeks; the remainder were grown to maturity. Straw and panicles were separated at harvest. Plant material was dried, ground and digested in $HNO_3/HClO_4$ in Kjeldahl flasks before analysis by atomic absorption.

Extraction Procedures

Trace metal status of soils was assessed by three extraction procedures.

EDTA. Air-dried soil (10 g) was shaken for 7 days with 25 ml 0.1 M EDTA at pH 6.0. The equilibrium solution was analysed after filtration and centrifugation.

DTPA. The procedure of Lindsay & Norvell (1969) was followed.

Water. Soil solution from the flooded pots with growing rice plants was removed periodically using a filter cartridge under gentle suction.

RESULTS

Yields of Experimental Plants

Dry matter. The effects of cropping history, addition of ground rice straw and flooding are illustrated for three pairs of soils in Figures 1 and 2, presenting results for the two Waite Institute soils and for two of the three pairs of soils from the MIA. Statistically, the yields of dry matter (and of metals) are dependent on site, cropping history and flooding and, at harvest 1, yields of dry matter, Mn and Zn are strongly affected by straw additions (Table 2). Because of the great diversity of sites, effects of other significant factors, although strong, show as interactions rather than directly.

In Figures 1 and 2, mean yields at harvest 1 (8 weeks) and subsequent mean yield increments, between 8 weeks and maturity, are plotted. This partition of yields shows that there were some relatively large differences during the first 8

Figure 1

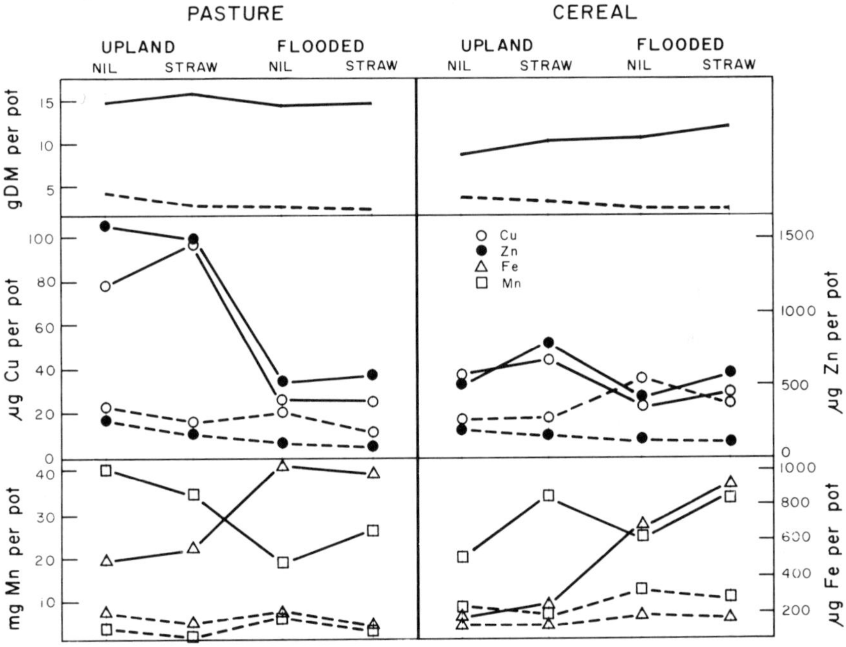

Fig. 1. Yields of dry matter (g DM/pot) and of metals (µg/pot) for rice at two harvests from the Waite Institute soils. Dotted lines show values at 8 weeks and continuous lines the increments produced between 8 weeks and maturity.

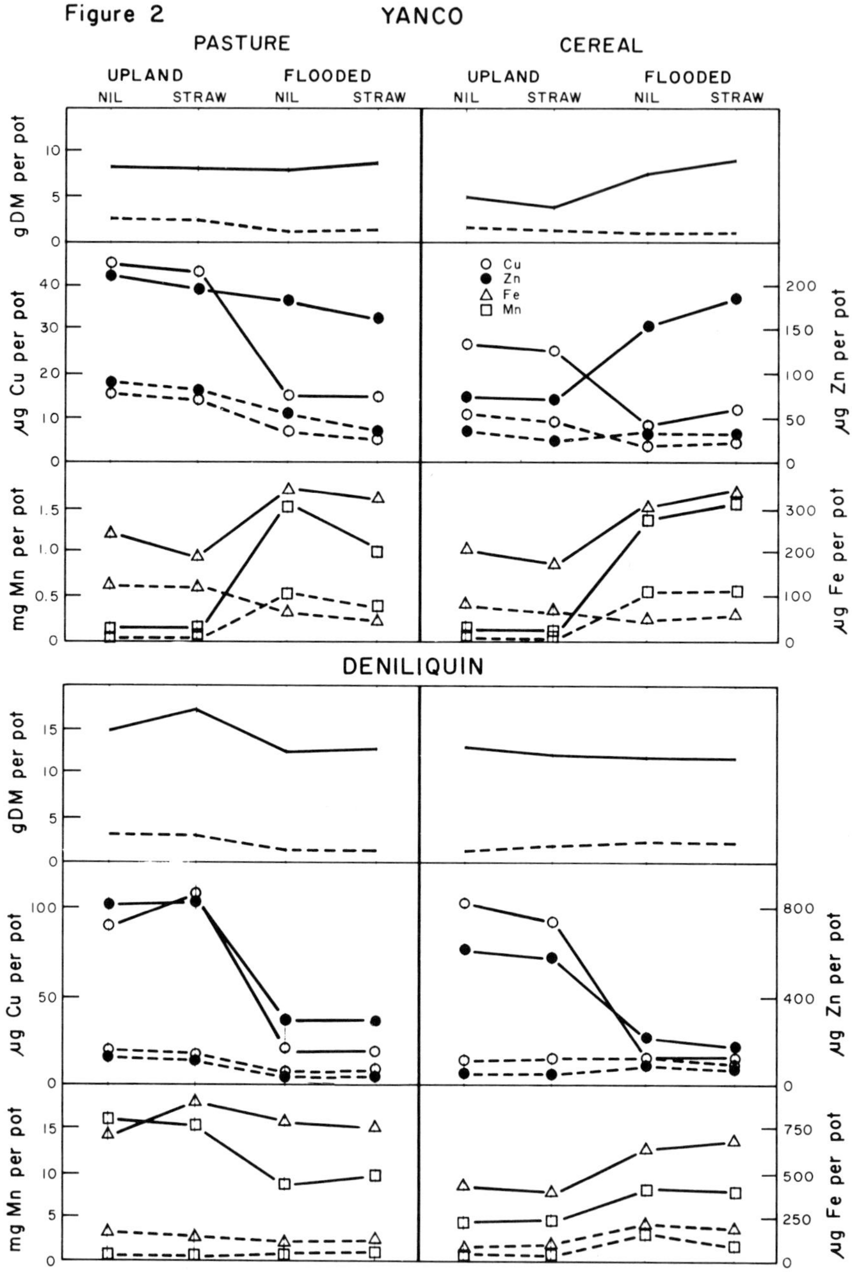

Fig. 2. Yields of dry matter (g DM/pot) and of metals (µg/pot) for rice at two harvests from two pairs of MIA soils plotted as in Figure 1.

TABLE 2

The statistical significance of the treatments and their interactions

Variables	Yield of DM		Cu uptake		Zn uptake		Fe and Mn uptakes	
	At H1	H2-H1	At H1	H2-H1	At H1	H2-H1	At H1	H2-H1
Sites (S)	xxx	xxx	xxx	xxx	xxx	xxx	xxx	xxx
History (H)	NS	x	x	x	NS	NS	NS	NS
Flooding (F)	NS	NS	NS	x	NS	NS	NS	NS
Straw (C)	NS	NS	NS	NS	NS	NS	NS	NS
	INTERACTIONS							
S × H	xxx	xx	xxx	xxx	xxx	xxx	xxx	xxx
S × F	xxx	xxx	xxx	xxx	xxx	xxx	xxx	xxx
S × C	xxx	x	x	NS	xxx	x	xxx*	NS
H × F	NS	NS	NS	xxx	NS	NS	NS	NS
S × H × F	xxx	x	xxx	xxx	xxx	xxx	xxx	xxx

* significant only for Mn, not for Fe

weeks of growth but that these early effects, associated with the period when soil conditions were undergoing most rapid change immediately after waterlogging (IRRI 1964), did not usually persist during subsequent growth when most dry matter was produced and soil conditions were more stable. The most notable exception is the cereal soil from Yanco (Fig. 2, top right).

For the four pairs of soils, the pasture soil was usually more productive than the near-adjacent cereal soil, especially under upland conditions. With the exception of the Yanco cereal soil (discussed later), only small differences were found in final yields of tops under upland and flooded conditions. But with all three of these soils from which roots were recovered (cereal soils from Waite Institute, Deniliquin and Leeton respectively), flooding appeared to promote root growth. Panicle weight was often larger for upland than for flooded rice in this experiment, but the least productive soil (Yanco cereal) was again exceptional.

Iron. Flooding consistently increased Fe uptake by rice shoots after the first 8 weeks except with the Deniliquin pasture soil, on which it had no clear effect (Fig. 2, lower left). In some instances, such as the Yanco pasture (Fig. 2, top left), flooding decreased Fe uptake at harvest 1 together with dry matter, but subsequently its effect was reversed. Pasture soils usually supplied more Fe to rice than corresponding cereal soils but effects of added rice straw on Fe uptake were small.

Manganese. Flooding did not always increase Mn uptake by rice shoots; with the acid soils, from which upland rice absorbed most Mn, flooding increased soil pH and sometimes decreased Mn uptake. The contrasting effects are illustrated by the two Deniliquin soils (Fig. 2). With the Leeton pasture soil (results not included here), flooding failed to increase Mn in rice shoots either at harvest 1 or maturity.

For the upland Deniliquin after 8 weeks, and for flooded and upland Leeton soils at both harvests, rice absorbed markedly more Mn from the pasture than from the cereal soil, but in other instances differences were quite small. Added rice straw usually decreased Mn uptake from the Waite Institute and Yanco soils at harvest 1; higher organic content in these soils apparently does not increase availability of Mn in the short term although, with flooded Waite Institute soils, Mn uptake after 8 weeks appeared to be increased.

Copper. On average, flooding decreased final uptake of Cu by rice shoots to less than half that of upland plants; only with the Waite Institute cereal soil was the decrease, during growth from 8 weeks to maturity, less than this. This strong decrease with flooding did not always apply to Cu uptake at harvest 1.

As with Fe, Cu uptake after the first harvest was usually larger from pasture soil than from the corresponding cereal soil, particularly for upland rice. Again like Fe, effects of added rice straw on Cu uptake were small compared with the above effects. Under flooded conditions, the amount of Cu absorbed during the first 8 weeks' growth was often disproportionately large compared with the dry matter yield of the small plants (e.g. Fig. 2, lower right).

The levels of Cu in rice on the flooded MIA soils were generally low (1.4-2.3 ppm in mature straw) and especially so for the Yanco cereal soil. These values compare with the recommended 3-4 ppm for straw of mature, high-yielding, field-grown rice (IRRI 1970) and with the claimed "critical level" of 6 ppm in straw of rice grown in solution culture (Karim & Vlamis 1962).

Zinc. Flooding decreased Zn uptake by rice shoots for all soils except the Yanco cereal soil (Fig. 2, top right) and often did so before harvest 1. Generally, this effect seems attributable to increase of soil pH on flooding but was also marked for Yanco pasture soil of pH 7.5. The slightly alkaline Yanco cereal soil was unique in showing *increased* Zn uptake under flooded conditions after the first 8 weeks, but this effect can be partly attributed to increased yield of dry matter (Fig. 2, top right). The Zn concentration in plants on this soil was low (16-18 ppm in mature straw).

After the first 8 weeks, pasture soil released more Zn to rice than cereal soil under upland conditions, and, with Deniliquin and Leeton soils, under flooded conditions also. However, added rice straw commonly decreased Zn uptake somewhat at harvest 1.

Other nutrients. Despite the application of basal N, P, K and S fertilizers to all pots, analysis of the mature straw of selected plants revealed S concentrations of only 0.042-0.066%. Rice straw from plants in the Waite Institute soils was only marginally richer in S (0.064-0.086%). These levels appear sub-optimal compared with published recommended levels in mature straw of 0.10% S (Tanaka & Yoshida 1970) and 0.06% (IRRI 1971).

Analysis of Soil Solutions from Flooded Pots

The changes with time of pH and of concentrations of Fe, Mn and Zn in solutions from flooded pots were similar to those reported by others and are not shown here (IRRI 1964; IRRI

1970). The Mn levels resembled the concentrations reported for rice soils of low Mn status, i.e. soils of high pH or low in organic matter and/or active Mn (IRRI 1964). Levels of Fe and Zn in solution in our soils were very low by comparison with most previous values. Thus soluble Fe in flooded MIA soils did not exceed 10 ppm and usually remained below 1 ppm after the first 10 weeks of flooding. Soluble Zn decreased to 0.001-0.02 ppm after the first few weeks' submergence of soils compared with previously reported "low" values of 0.04-0.05 ppm (IRRI 1970) including some solutions from Zn-responsive soils. Our soluble Zn levels resemble the very low values reported for aerobic, alkaline soils (Hodgson *et al.* 1966; Tiller *et al.* 1972).

Levels of soluble Cu were too small for regular monitoring by direct analysis. After 14 weeks' flooding, all soil solutions contained only about 0.002 ppm Cu. These concentrations also are similar to the lowest reported for solutions from drained soils (Kubota *et al.* 1964; Hodgson *et al.* 1965) and to those in 1/2.5 water extracts of two Cu-deficient calcareous soils (Beckwith 1973).

Analysis of Pot Soils After Rice Harvest

As shown in Table 3, *more* Cu, Zn and Fe were extracted by EDTA from flooded than from corresponding upland soils. The differences for Cu and Zn were relatively small but the flooded MIA soils released about twice as much Fe as corresponding upland soils, and for non-rice soils, including those from Waite Institute, the difference was larger (unpublished data). Under our test conditions, extractable Mn was not affected by flooding. DTPA, in place of EDTA, extracted smaller amounts of metals, but these amounts also failed to reflect the larger availabilities of Cu and Zn in upland than in flooded soils and in pasture soils compared with cereal soils.

Estimates of Total Available Zinc in Pot Soils (Zn_L)

Values for Zn_L, derived from isotope dilution in the plant tops, were near constant at the two harvests but, unlike Zn uptake by rice, tended to be larger for flooded than for upland soils (Table 3). Hence it is the ease with which plants absorb Zn from a near-constant "pool" which differs under upland and flooded conditions rather than the size of the "pool" of total available soil Zn. In the case of the Yanco pasture soil, at least, this difference is not confounded by any yield or pH differential.

TABLE 3

EDTA-extractable metals in pot soils after experiment (ppm)

Soil No.	Zinc U	Zinc F	Copper U	Copper F	Manganese U	Manganese F	Iron U	Iron F
444	4.5 (6.6)	6.6 (8.3)	5.3	7.6	470	420	540	2500
445	3.2 (6.1)	5.3 (6.8)	5.6	7.5	460	410	410	2300
447	6.1 (6.4)	6.1 (7.7)	9.2	11.2	180	160	900	1700
446	3.1 (3.0)	3.4 (3.6)	9.4	10.7	180	180	1300	1600
448	4.2 (5.7)	5.0 (5.8)	4.9	6.6	220	210	890	1800
449	2.1 (2.6)	2.6 (3.0)	8.0	10.7	470	390	940	2200
451	1.5 (3.7)	2.7 (4.1)	7.0	9.9	140	150	2200	4500
450	2.7 (5.8)	3.7 (5.8)	8.5	10.8	210	220	2400	4200

Means of quadruplicates
"L" values for Zn shown in bracket, based on specific activity of Zn uptake in mature plant

DISCUSSION

With the single exception of the Yanco cereal soil (discussed later), there were only relatively small differences, on any one soil, in the yields of dry matter produced after harvest 1. Hence the amounts of metals absorbed per pot should reliably reflect the effects of treatments on the abilities of the soils to supply nutrients to rice. In at least some instances, flooding increased root yields relative to corresponding upland pots, in contrast to yields of shoots. In these circumstances, flooding possibly decreased availabilities of some metals by rather more than our data indicate.

Some of the differences in metal uptakes between matched pasture and cereal soils appear to be partly attributable to yield differences. But the consistency of the effects and their usually large size relative to yield differences are evidence of a possible soil influence as well. On a similar basis, rice straw sometimes appeared to decrease availabilities of Zn and Mn to rice for several weeks after its addition. Such a decrease agrees with reports of similar effects of added cellulose on availability of Zn to rice (IRRI 1970). Indeed, from those and other studies, it was concluded that "a high pH, high content of organic matter and a high content of available P depress the availability of Zn in flooded soils as they do in upland soils" (IRRI 1970). Our agreement with this statement must be qualified by the observation that, after the first 8 weeks, Deniliquin and Leeton pasture soils supplied more Zn than the corresponding cereal soils, under both flooded and upland conditions, and no pasture soil appeared to supply *less* Zn than its paired cereal soil. Assuming that organic content is the major difference between our soil pairs, it seems that the *kind* as well as the *amount* of organic matter is important to its effects on Zn availability to rice under both flooded and upland conditions.

It has been reported that organic matter indirectly inhibits root uptake of Zn through build-up of CO_2 and of organic acids in solutions in flooded soils (IRRI 1970). The importance of its effects on dissolution of sorbing soil mineral surfaces, creation of sorbing organic residues and metal solubilization by complex formation still remains unclear. However, from the present results, total available Zn is little affected by flooding and it seems probable that similar concentrations of Zn occur in solution in *Zn-deficient* drained and flooded soils, at least after the first few weeks of submergence. Further, as quoted above, the same factors generally aggravate Zn deficiency in paddy rice and in upland crops. Hence it seems rather unlikely that different, specific Zn compounds (IRRI 1970) are responsible for limiting Zn uptake by rice from neutral and alkaline soils.

There is a further complication in so far as "soluble" Zn does not always appear readily available to rice. Uptake of metals by rice often appears strangely low during the first weeks after submergence when, it is generally agreed, solution concentrations of Mn, Zn and Fe reach their peaks. Indeed, it has been observed that pre-submergence of Zn-deficient soils for 12 weeks before transplanting rice seedlings sometimes *prevents* occurrence of Zn deficiency in the crop (IRRI 1971) even though Zn in solution after 12 weeks would be expected to be much lower than during the first 3 weeks. Again, in the present work, water-soluble Fe, Mn and Zn all attained their maximal levels in the flooded soils before the first rice harvest (unpublished data), yet there was little tendency for uptakes of these metals at harvest 1, relative to the final uptake, to be larger than dry matter production relative to the final harvest. In fact, as noted above, Cu was the only metal which appeared, in flooded soils, to be consistently more available to rice during its early than its later growth.

The release of soil Cu into solution appears to be dominated by its associations with soil organic constituents (Hodgson *et al.* 1965; Beckwith 1973). Hence, as well as the level of Cu in soil, conditions affecting distribution of organic moieties between soil surfaces and solution, such as temperature, drying and rewetting of soils, presence of competing inorganic anions (Beckwith 1973) and factors causing modification to sorbing surfaces in soils, are all able to affect availability of soil Cu to plants. Until more is known about the molecular size and chemical nature of organic ligands under various soil conditions, and about the ease with which roots obtain metals from these different complexes, our understanding of availability of Cu seems likely to remain equally incomplete for flooded and drained soils.

The low yield under upland conditions and exceptional pattern of Zn uptake by rice on the Yanco cereal soil (Fig. 2, top right) cannot be certainly explained. Low Cu could have limited the yield from flooded pots but this soil, in fact, showed the largest stimulus of yield from flooding. Another nutrient(s), possibly Zn and/or Mn, appears to have been more severely limiting under upland conditions. If this were so, analogy with wheat and oats suggests that the Cu-deficient, flooded rice would translocate larger than normal proportions of Zn and Mn to its shoots (Beckwith 1973) and so cause contents of these metals to be large compared with plants not suffering Cu deficiency, i.e. upland rice in this soil. Analysis of roots might have confirmed this interpretation but, unfortunately, this soil was not included among those selected for root recovery.

There are several possible reasons why flooding did not generally increase rice yields in this experiment. The plants matured in Adelaide in July when light intensity, day length and temperature were possibly sub-optimal under glasshouse conditions. Also, any temperature differential between flooded and upland pots would have favoured the latter. And although we applied recommended field rates of fertilizers, these may have been sub-optimal for pot cultures, as indeed appears to be so for S. Under these circumstances, flooded rice may have been inhibited slightly more than the upland plants.

Despite any such limitations, the low levels of Cu, Zn and S in some of our plants suggest that these are nutrients which might limit rice yields in some parts of the MIA in future years. Alkaline soils, such as our sample from near Yanco, may be most liable to the metal deficiencies. It is possible that "Calrose" rice has an unusually small requirement for all these nutrients (IRRI 1972) but, even so, shortages may be imminent in some MIA soils unless fertilizer supplements are provided.

As noted previously in this paper, the concentrations of Fe and Zn in solutions from our flooded soils appear to be especially low compared with most Asian rice soils. The choice of urea as N fertilizer, although amply justified by its widespread use on rice crops, could have been at least a part cause of these low values (IRRI 1971). However, the possibility must be admitted that the unusual field treatment of MIA soils, in comparison with Asian paddy soils, may also have contributed to their differences in content of soluble metals; this question can only be resolved by further studies.

ACKNOWLEDGEMENTS

We thank Mr. J.T. Hutton for S analyses and Mr. P.M. Clayton and Miss L.R. Spouncer for assistance with other analytical work. We are indebted to Dr. D.J. McDonald, N.S.W. Department of Agriculture, for his help in locating and collecting soil samples and advice on culturing rice in pots, and to Mr. A.G. Constantine, CSIRO Division of Mathematical Statistics, for statistical analysis of data.

REFERENCES

BECKWITH, R.S. (1973) - Ph.D. Thesis, University of Western Australia

CANDIDO, P., RICARDO, P., ARRIAGA, J.M. & CUNHA, E. (1968) - *Agronomia lusit.* *29* : 57

CLARK, F.E. & RESNICKY, J.W. (1956) - *Proc. 6th Int. Congr. Soil Sci.* *C* : 545

HODGSON, J.F., GEERING, H.R. & NORVELL, W.A. (1965) - *Proc. Soil Sci. Soc. Am.* *29* : 665

HODGSON, J.F., LINDSAY, W.L. AND TRIERWEILER, J.F. (1966) - *Proc. Soil Sci. Soc. Am.* *30* : 723

IRRI (1964) - Annual Report Int. Rice Res. Inst., Los Banos, Philippines

IRRI (1970) - Annual Report Int. Rice Res. Inst., Los Banos, Philippines

IRRI (1971) - Annual Report Int. Rice Res. Inst., Los Banos, Philippines
IRRI (1972) - Annual Report Int. Rice Res. Inst., Los Banos, Philippines
JOHNSTON, E.J. (1953) - CSIRO (Aust.) Soil Publ. No. 1
JOSHI, N.V. & JOSHI, S.G. (1957) - *J. Indian Soc. Soil Sci.* *5* : 21
KARIM, A.Q.M.B. & VLAMIS, J. (1962) - *Pl. Soil 16* : 347
KUBOTA, J., LEMON, E.R. & ALLAWAY, W.H. (1963) - *Proc. Soil Sci. Soc. Am. 27* : 679
LINDSAY, W.L. & NORVELL, W.A. (1969) - *Agron. Abstr. 1969* : 84
NG, S.K. & BLOOMFIELD, C. (1962) - *Pl. Soil 16* : 108
OSTROVSKAYA, L.K. (1956) - *Fiziologiya Rast. 3* : 73
PONNAMPERUMA, F.N. (1972) - *Adv. Agron. 24* : 29
RAHAJA, P.C., YAWALKAR, K.S. & SINGH, R. (1959) - *Ind. J. Agron. 3* : 254
RAJAN, S.V.G. & RAO, G.H.G. (1964) - *J. Indian Soc. Soil Sci. 12* : 355
TANAKA, A. & YOSHIDA, S. (1970) - "Nutritional Disorders of the Rice Plant in Asia". IRRI Tech. Bull. No. 10
TILLER, K.G. (1958) - *J. Soil Sci. 9* : 225
TILLER, K.G., HONEYSETT, J.L. & DE VRIES, M.P.C. (1972) - *Aust. J. Soil Res. 10* : 165
VAN DIJK, D.C. (1961) - CSIRO (Aust.) Soils and Land Use Series No. 40
YAMASAKI, T. (1965) - *In* "The Mineral Nutrition of the Rice Plant". Published for IRRI, Johns Hopkins Press : Baltimore

MOBILITY AND PLANT UPTAKE OF MICRONUTRIENTS IN RELATION TO SOIL WATER CONTENT

E.K.S. Nambiar

Department of Agronomy
Waite Agricultural Research Institute
The University of Adelaide
Glen Osmond, South Australia 5064

INTRODUCTION

Most experiments dealing with the effects of soil water on micronutrient supply to plants compare well-drained (aerobic) soils with flooded (anaerobic) soils, the effects being measured with or without growing plants (e.g. Ng Siew Kee & Bloomfield 1962; Kubota *et al.* 1963; Giordano & Mortvedt 1972). Results from such experiments show gross effects but are often difficult to interpret, because, when soils are flooded, their physical, chemical and biological properties change greatly.

This paper will discuss the mobility and plant uptake of micronutrients in relation to soil water content within the range of 0.3 bar to 15 bar suction, in which soils are usually well drained and aerobic. Within this range, the ratio of maximum to minimum volumetric water content varies from 2 to 10, depending on the moisture characteristics, influencing directly and/or indirectly the mobile pool of micronutrients, diffusion and convection, processes at the root-soil interface, and extension and uptake ability of plant roots. Such effects will be briefly considered here with respect to micronutrient cations. Emphasis will be given to effects of topsoil drying on Mn and Zn uptake.

TABLE 1

Effect of soil water content and incubation time on ammonium acetate (0.4N, pH 4.8) extractable Mn and Fe (ppm air-dry soil) (Nambiar 1970)

Soil	pH (H_2O)	Organic carbon (%)	Element	Air-dry soil	Water content and time of incubation			
					Field capacity		50% Field capacity	
					7 days	28 days	7 days	28 days
Sandy	5.6	1.11	Mn	9.5	< 2.0	< 2.0	< 2.0	< 2.0
(Herantals)			Fe	30.0	17.5	12.5	15.5	15.5
Clay loam	5.8	0.64	Mn	36.5	2.0	2.0	18.5	7.0
(Glab-Zurbende)			Fe	10.0	5.5	< 5.0	< 5.0	< 5.0
Calcareous clay	7.4	0.74	Mn	25.0	3.0	< 2.0	3.5	< 2.0
(Lede)			Fe	7.5	5.0	< 5.0	5.0	< 5.0

MOBILITY OF MICRONUTRIENTS

Effect of Soil Water Content on Microbial Activity and on Available Manganese and Iron

When dry soils are wetted, extractable Mn, Fe and Co often decrease, whereas Zn and Cu remain unaffected (Nambiar 1970). Table 1 shows results from an experiment in which three soils of contrasting properties were incubated at 50 and 100% field capacity and at laboratory temperature. Despite appreciable differences in pH, organic matter status and the amounts extracted initially, extractable amounts in all soils converged to 2 ppm or less for Mn and 5 ppm or less for Fe. This was entirely due to biological re-oxidation of Mn^{2+} and Fe^{2+}, as shown by other experiments (Nambiar 1970) (Fig. 1). When the clay loam soil was incubated at field capacity, after suppressing biological oxidation by cell poisoning, chemical reduction of Mn occurred at nearly the same rate as in water-saturated soils.

In order to test the effects of biological oxidation on Mn supply to plants, a clay loam (Glab-Zurbende), which released

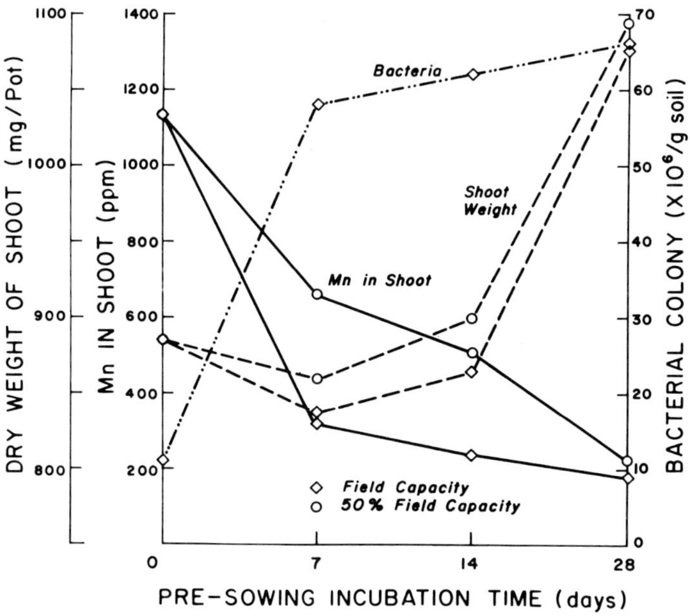

Fig. 1. Effect of soil water content on the microbial oxidation of Mn (Nambiar 1970; Nambiar & Cottenie 1971).

toxic amounts of Mn to plants when allowed to dry, was pre-incubated at different water content. The amount of Mn available to barley seedlings was then assessed using the Neubauer technique (Fig. 1). During the uptake period of 15 days all treatments were maintained at field capacity. Details of this experiment have been reported elsewhere (Nambiar & Cottenie 1971). Barley seedlings grown in soils which were not pre-incubated showed yellowing, followed by gradual and progressive necrosis of the upper one-third of the leaves. Plant growth was severely retarded after 12 days and the shoot had accumulated 1130 ppm Mn at harvest. In contrast, pre-incubation of the soil greatly decreased the Mn content of the shoot at harvest and increased the yield significantly (Fig. 1). The number of colonies of mould in air-dry soil was 0.2×10^4/g soil, which on incubation increased to 10×10^4, 10×10^4 and 40×10^4/g soil after 7, 14 and 28 days. At the peak of biological activity, soluble Mn was at its lowest (Fig. 1) (Nambiar & Cottenie 1971). Similarly, Grasmanis & Leeper (1966) reported that Mn toxicity in fruit trees grown in a near-neutral soil was aggravated by drying of the topsoil.

If the Mn^{2+} in solution were in equilibrium with the Mn^{2+}-MnO_2 system in the soil, the activity of Mn^{2+} could be calculated from the equation:

$$\log (Mn^{2+}) = \frac{E^o}{0.0295} - 33.9\ Eh - 4\ pH \quad .$$

However, Geering *et al.* (1969) found that predicted Mn^{2+} values were considerably higher than measured values. This discrepancy was attributed to the formation of MnO_2 through microbial oxidation. Soil water regulates the kinetics of precipitation and dissolution of Mn and Fe in systems such as Mn^{2+}-MnO_2 and Fe^{2+}-$Fe(OH)_3$ by controlling the chemical as well as microbiological redox processes.

Effect of Wetting and Drying Cycles

In nature, soils generally are subjected to wetting and drying (W-D) cycles of varying intensity and duration and do not maintain a constant water content for any appreciable length of time. This is particularly so in the topsoil, where most of the micronutrients are present.

In general, cycles of wetting and drying tend to decrease the amount of micronutrients in solution and in weakly adsorbed forms. This may be due to adsorption and chelation with organic matter as shown for Zn (Estepp & Keefer 1969) and Cu (Ng Siew Kee & Bloomfield 1962), entrapment in the clay lattice of the expanding clay minerals as shown for Zn (Reddy & Perkins 1974), or chemical oxidation and precipitation as shown for Fe (Table 2) (Nambiar 1970).

TABLE 2

Extractable Mn and Fe in a clay loam (Glab-Zurbende) after re-drying the incubated soil (ppm air-dry soil) (Nambiar 1970)

	Water content during incubation		
	Field capacity (0.3 bar)	Saturation (0.01 bar)	
	Mn	Mn	Fe
Without decomposed peat:			
Before drying	7.0	189.0	95.0
After drying	15.0	191.5	10.0
With 2% decomposed peat:			
Before drying	18.5	140.0	155.0
After drying	25.0	140.0	7.5

Effect of wetting and drying on Mn mobility will be influenced by the state of oxidation/reduction to which the soil has been brought during the wet phase, since this will influence not only the concentration of Mn in solution at the commencement of drying but also the ensuing microbial activity. For example, the effect of re-drying (re-oxidation) on the acid ammonium acetate-extractable Mn in the clay loam was considerably influenced by the wetness of the soil prior to drying (Table 2). Soil was incubated for 28 days at laboratory temperature and then dried at 35°C for 24 hr. After a further period of 48 hr at laboratory temperature, subsamples were extracted with ammonium acetate (0.4 N, pH 4.8). The drying caused a rapid re-oxidation of Fe but not Mn. This indicates that the re-oxidation of Mn^{2+} was mainly biological, while that of Fe^{2+} was mainly chemical (Table 2).

The fate of Mn in soils which are marginally Mn deficient, and in which the Mn is bound predominantly in organic forms, contrasts with that discussed so far. One such soil was studied from the "Ninety Mile Desert" of South Australia. The surface soil used was acidic (pH 6.3), sandy (95% > 0.02 mm) and contained moderate amounts of organic matter (0.74% organic carbon). The topsoil was underlain by more than 100 cm of pale yellow sand. The soil was incubated for 45 days at 20 ± 2°C. Samples were either maintained wet (10% w/w) or subjected to three wetting and drying (10% to 1% w/w) cycles.

Sub-samples were extracted with 1 M NH_4OAc (pH 7.0) to obtain exchangeable Mn. Residual soil was further extracted with 1 M NH_4OAc + 0.2% hydroquinone ("active" Mn), or 1 M NH_4OAc + 0.01 M EDTA (pH 7.0).

Figure 2 shows that fluctuations in soil water content had negligible effect on the various Mn forms measured over a wide range of Mn concentration. The extent of reversion of the exchangeable Mn, regardless of the water regime, was surprisingly high for an acid sandy soil. Other experiments (Table 3) revealed that this soil showed Mn deficiency when the topsoil was allowed to dry. This contrasts strongly with the behaviour of Mn in the soils described in Table 1 and Figure 1.

The effects of wetting and drying on exchangeable Zn and Cu in the sandy soil from the "Ninety Mile Desert" differed from those reported by other investigators and noted earlier in this section. W-D cycles did not affect the exchangeable Zn, but tended to increase exchangeable Cu. The transforma-

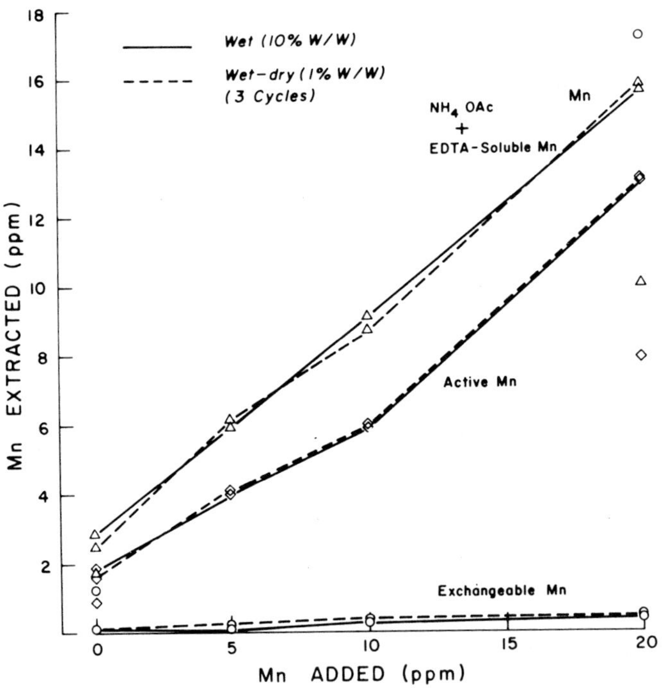

Fig. 2. The fractionation of added Mn in an acid sandy soil. Unconnected points show the amounts extracted at '0' time.

tion of Zn and Cu from solution or weakly adsorbed forms to chelated or specifically adsorbed forms does not necessarily impair availability to plants, as such forms remain isotopically exchangeable (Tiller *et al.* 1972; McLaren & Crawford 1974).

Micronutrients in Soil Solution

In contrast to non-reactive ions like Cl^- and NO_3^-, for which the concentration in the soil solution varies inversely with water content, the concentration of Fe, Zn and Cu may not be expected to change with moderate and transient changes in soil water content. For example, the concentration of Zn in displaced solutions does not change in soils equilibrated within the suction range 0.3-10 bar (Wilkinson *et al.* 1968; Warncke & Barber 1972).

Further arguments can also be advanced in support of the view that soil water content as such has little effect on the solution concentration of micronutrients other than Mn: (i) most of the labile ions (Zn, Co, Cu) are specifically adsorbed on the lattice sites of the silicate minerals, organic matter and oxides of Fe and Mn (Tiller *et al.* 1972; McKenzie 1972; McLaren & Crawford 1974); (ii) the presence of relatively high amounts of isotopically exchangeable micronutrients in solid phase compared to solution phase tends to buffer changes in solution concentration. For Mn, under well-drained conditions, the effect of water content on Mn in solution is essentially an indirect effect of soil water on microbial activity (Fig. 1).

However, if soils are inundated for sufficiently long periods so that Eh and pH are affected, the concentrations of Fe, Mn, Cu, Zn, Co, Mo and B in the equilibrium extract generally increase (Ng Siew Kee & Bloomfield 1962; Kubota *et al.* 1963).

EFFECTS OF TOPSOIL DRYING

When effects of soil drying on ion uptake are evaluated, physiological effects of water stress on the plant must be closely considered. However, in many soils and climates, drying of the topsoil, where most of the micronutrients and P are located, proceeds while the subsoil generally retains available water. In these circumstances there may be little if any water stress in the plant.

Experiments were conducted to study the effect of topsoil drying on root growth and micronutrient uptake, using the sandy topsoil from the "Ninety Mile Desert". In a glasshouse experiment, a 10 cm deep layer of topsoil and a subjacent layer of acid washed sand (15 cm) containing no detectable amounts of Mn, Zn and Cu were separated with a wax membrane. Ryegrass (*Lolium multiflorum* Lam) was planted in the initially wet topsoil and roots were allowed to grow through the membrane and into both layers. Nutrients other than those being studied and water were supplied plentifully in the subjacent layer. The topsoil was allowed to dry from the 30th day after planting (Table 3). The ability of the ryegrass to regrow after defoliation enabled harvests to be taken after the topsoil had become dry. Plants were defoliated after 43, 62 and 80 days. Growth rate of shoots decreased soon after the initiation of topsoil drying due to Mn deficiency. Table 3 shows that drying of the topsoil reduced the Mn uptake severely, but mulching the soil and so conserving a very small amount of water (15 ml/pot) increased the Mn uptake appreciably. This effect of mulching cannot be attributed either to an increase in available Mn, since incubation enhanced Mn reversion in this soil regardless of water content (Fig. 2), or to an increase in root growth (Table 3).

The experiment also showed that topsoil drying had little effect on Zn and Cu uptake, and that plant roots adsorbed significant amounts of Zn and Cu from nearly air-dry topsoil when the roots had access to subsoil water.

In a separate short-term experiment, the effect of water content of a sandy mallisol on Zn uptake by oats was examined.

TABLE 3

Cumulative uptake of Mn from an acid sandy topsoil by the shoot of ryegrass, when defoliated successively at 43, 62 and 80 days. The subsoil was plentifully supplied with water at all times.

Topsoil water supply	Topsoil water at day 43 (% w/w)	Mn uptake (μg/pot) Days 43	62	80	Rooting density (L_V) in topsoil at day 80 (cm^{-2})
Mulch removed- No water added after 30 days	0.8	27.2	36.1	39.1	10.7
Mulch retained- No water added after 30 days	1.1	31.1	50.2	59.2	13.8
Mulch retained- Watered daily	10.0	38.6	71.4	103.1	36.2

Figure 3 shows the effect of initial soil water content of the topsoil on root growth and ^{65}Zn uptake over 15 days. The uptake layer consisted of a 1.5 cm thick layer of light sandy loam previously equilibrated with carrier-free ^{65}Zn. This layer was sandwiched between two wax membranes to prevent movement of water from a top layer of non-radioactive wet soil (1.5 cm) and a Zn-free wet sand layer (13 cm) below.

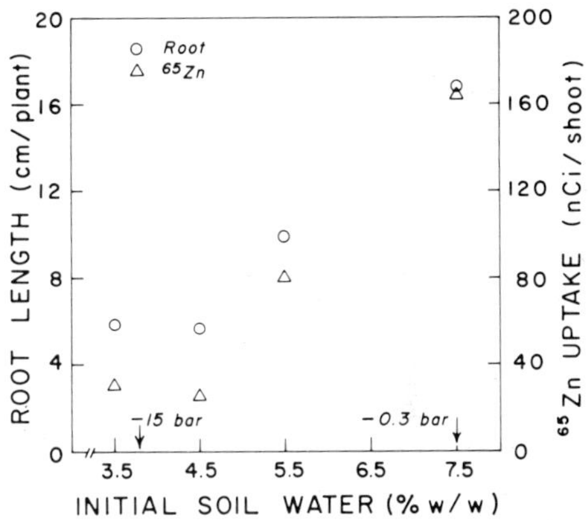

Fig. 3. Root growth and ^{65}Zn uptake by oats from topsoil initially wetted to different water content and allowed to dry during the uptake period.

Plants were grown in a growth cabinet maintained at 20 ± 0.5°C, 50% R.H., and with a 12 hr day/night cycle. During the uptake period no net outflow of water occurred from the root into the relatively dry uptake layer. Figure 3 shows the similar effects of soil water content on root growth and ^{65}Zn uptake. Root growth and ^{65}Zn uptake were significantly correlated (r = 0.89). Even when the soil was initially drier than wilting point, roots grew through the layer and absorbed appreciable amounts of ^{65}Zn (Fig. 3).

In a further experiment with the mallisol, ^{65}Zn absorption by unbranched seminal roots of oats was measured over 4 days. Seminal roots were allowed to grow through the uptake layer at volumetric water contents of 0.04 (> 15 bar), 0.06, 0.07 and 0.1 (0.3 bar). Mean rates of ^{65}Zn absorption were 1.37, 1.68, 2.12 and 3.13 nCi per cm of root per day. The rate of Zn absorption by oat seedlings from the driest soil, containing 0.1 ppm exchangeable Zn, ranged from 0.04 to 0.06 ng per cm of root per day. These rates are comparable with those reported by Carroll & Loneragan (1969) from solutions containing moderate to optimum concentrations of Zn. Examination of the roots at harvest showed that where the roots had grown in dry soil an annulus of soil remained adhering to the roots after washing, whereas the roots grown in wet soil could readily be washed clean.

MICRONUTRIENT TRANSFER TO ROOTS

In many cultivated soils the convective transfer of Fe, Mn, Zn and Cu to the roots is likely to be small (Oliver & Barber 1966; Wilkinson *et al.* 1968). Measurements of Zn diffusion coefficients in soil show that bulk diffusion coefficients may range from 10^{-11} to 10^{-8} cm^{-2} sec^{-1} at 13% water content to 10^{-9} to 10^{-8} cm^{-2} sec^{-1} at near saturation (45% v/v), depending on the soil properties (Warncke & Barber 1972).

The transfer of micronutrient cations to the roots is predominantly diffusive and is likely to occur from a zone within 1-2 mm from the root. This zone is influenced by the presence of the root through shifts in pH and Eh, changes in soil water potential and bulk density, exudates and presence of mycorrhizas. Such changes have special significance in relation to micronutrient uptake (Barley 1970; Wilkinson 1972). Data on Zn uptake reported by Wilkinson *et al.* (1968) for wheat in loamy sand (pH 5.0) and Tiller *et al.* (1972) for subterranean clover in alkaline soils (pH > 7.0) showed that release of specifically bound Zn resulting from root activity may account for 70% or more of the Zn absorbed by plants. In certain instances, ion transfer may be confined to an extremely narrow annulus of soil around the root. This has been shown for Fe (Glauser & Jenny 1960) and for Mn (Passioura & Leeper 1963).

The pathway for micronutrient cation transfer to the root proposed by Jenny (1966), namely surface diffusion of cations from clays through root mucilage, appears to be of added importance when plants absorb Mn and Zn from relatively dry soils (Fig. 3, Tables 3 and 4).

When the soil is dry, liquid continuity between the water in the root cell wall and soil matrix is likely to depend on the infiltration of mucilage, because, even at 1 bar suction, soil water menisci will retreat to voids of radius < 1.5 μ, which are too narrow for the entry of root hairs (radius, 5 μ) (Barley 1970). There are reports of increased exudation of mucilage and other rhizosphere products when plants experience water stress (Hale *et al.* 1971). The author recently observed increased exudation of mucilage in response to localized soil drying, though the plant as a whole had a plentiful supply of water.

Table 4 shows the width of the root-soil cylinder remaining after washing, when roots of various plant species had grown through a layer of dry soil (4% v/v) into wet sand, as described in the previous section. The width of the shell of soil retained around the root after washing varied 5-fold between species. However, no direct relationship appears to

TABLE 4

Width of the root and of the root-soil cylinder and uptake of ^{65}Zn from a dry mallisol by different genotypes

Genotype	Root	Diameter (mm)		^{65}Zn uptake ($nCi\ cm^{-1}$)
		Root	Root-soil cylinder	
Lucerne	Axis	0.66	0.79	2.74
Clover	Axis	0.88	0.95	2.50
Oats	Seminal	0.87	2.08	3.89
Wheat	Seminal	1.10	3.50	2.71

exist between the width of the cylinder and ^{65}Zn absorption per unit root length (Table 4). Qualitative differences in the exudates of plant species may also need to be considered, particularly their chelating properties.

Following Jenny (1966), Barley (1971) described the microscopic paths to ion transfer. Considering Barley's description further, the following path lengths may be recognized:

Path	*Path length (mμ)* Wet soil	Dry soil
Mucilage	1×10^3-1×10^4	1×10^4-1×10^6
Cell wall	5×10^2	4×10^2

In the dry soil, the radial extent of diffusion path for immobile micronutrients is unlikely to exceed 10^4 mμ and the thickness of the solution film linking the clay matrix and mucilage is likely to be 10 mμ or less. It is clear that in dry soil the mucilage path is very important. For micronutrients, the bulk properties of the soil, including such parameters as bulk diffusion coefficient, are likely to be less influential than the nature and lengths of the microscopic paths in the immediate vicinity of the root.

SUMMARY AND CONCLUSIONS

Soil water content, within the range permitting aerobic processes, may influence directly or indirectly the forms and size of labile pools of micronutrients, diffusion and convection of ions, extension of the root, processes at the root-soil interface, uptake ability of roots, and plant demand.

Effects of soil water content on microbial activity need to be considered closely when studying the changes in Mn supply. Depending on the nature of the soil, drying may induce either a net release of Mn in significant and sometimes toxic amounts or reversion of Mn sometimes leading to deficiency.

Concentration of Zn or Cu in solution is not likely to be significantly altered by moderate and transient changes in soil water content.

Evidence obtained recently shows that roots having access to subsoil water are able to absorb Mn, Zn and Cu from relatively dry topsoils, with negligible concomitant uptake of water. For Zn, absorption from dry topsoil has been shown to occur at rates which are significant in plant nutrition. The process that enables this surprisingly efficient transfer and absorption needs to be explored by enquiring into the nature and properties of the microscopic pathways in the immediate vicinity of the root.

ACKNOWLEDGEMENTS

The author wishes to thank Mr. K.P. Barley for helpful comments, and the Australian Research Grants Committee for financial support.

REFERENCES

BARLEY, K.P. (1970) - *Adv. Agron.* *22* : 159

BARLEY, K.P. (1971) - *In* "Towards a Model of Plant Nutrition", Modelling Workshop Proceedings, CSIRO Division of Land Research, Canberra

CARROLL, M.D. & LONERAGAN, J.F. (1969) - *Aust. J. agric. Res.* *20* : 457

ESTEPP, R. & KEEFER, R.F. (1969) - *Proc. W. Va Acad. Sci.* *41* : 85

GEERING, H.R., HODGSON, J.F. & SADANO, C. (1969) - *Proc. Soil Sci. Soc. Am.* *33* : 81

GIORDANO, P.M. & MORTVEDT, J.J. (1972) - *Agron. J.* *64* : 521

GLAUSER, R. & JENNY, H. (1960) - *Agrochimica* *4* : 263

GRASMANIS, V.O. & LEEPER, G.W. (1966) - *Pl. Soil* *25* : 41

HALE, M.G., FOY, C.L. & SHAY, F.G. (1971) - *Adv. Agron.* *23* : 89

JENNY, H. (1966) - *Pl. Soil* *25* : 265

KUBOTA, J., LEMON, E.R. & ALLAWAY, W.H. (1963) - *Proc. Soil Sci. Soc. Am.* *17* : 679

McKENZIE, R.M. (1972) - *Z. Pfl-Ernähr. Bodenk.* *131* : 221

McLAREN, R.G. & CRAWFORD, D.V. (1974) - *J. Soil Sci.* *25* : 111

NAMBIAR, E.K.S. (1970) - Ph.D. Thesis, State University of Ghent, Belgium

NAMBIAR, E.K.S. & COTTENIE, A. (1971) - *Proc. Int. Symp. Soil Fert. Evaln, New Delhi,* p. 157

NG SIEW KEE & BLOOMFIELD, C. (1962) - *Pl. Soil 16* : 108

OLIVER, S. & BARBER, S.A. (1966) - *Proc. Soil Sci. Soc. Am. 30* : 468

PASSIOURA, J.B. & LEEPER, G.W. (1963) - *Nature, Lond. 200* : 29

REDDY, M.R. & PERKINS, H.F. (1974) - *Proc. Soil Sci. Soc. Am. 38* : 229

TILLER, K.G., HONEYSETT, J.L. & DE VRIES, M.P.C. (1972) - *Aust. J. Soil Res. 10* : 151

WARNCKE, D.D. & BARBER, S.A. (1972) - *Proc. Soil Sci. Soc. Am. 36* : 39

WILKINSON, H.F. (1972) - *In* "Micronutrients in Agriculture" (J.J. Mortvedt *et al.*, eds.). Soil Sci. Soc. Am. Inc. : Madison, Wisc.

WILKINSON, H.F., LONERAGAN, J.F. & QUIRK, J.P. (1968) - *Proc. Soil Sci. Soc. Am. 32* : 831

SELENIUM IN SOILS AND PLANTS: CONTRASTS IN CONDITIONS PROVIDING SAFE BUT ADEQUATE AMOUNTS OF SELENIUM IN THE FOOD CHAIN

C.M. Johnson

Department of Soils and Plant Nutrition
University of California
Berkeley, California, U.S.A. 94720

INTRODUCTION

Among the elements of consequence in soil-plant-animal relationships, few have had a more varied and interesting evolution than Se. Direct implication of Se in alkali disease and blind staggers of livestock was reported only 40 years ago by scientists of the U.S. Department of Agriculture and the South Dakota and Wyoming State Agricultural Experiment Stations. Interest in the role of Se in animal health continued to be centred around the toxic effects of the element for grazing animals until 1957 when several workers demonstrated beneficial effects of trace amounts of selenite on rats and chicks. Very quickly reports were published on the use of Se in treatment of the so-called Se-responsive disorders of sheep and poultry.

Comprehensive descriptions of the manifestations of Se excess and inadequacy in animals have been presented elsewhere (Muth 1967; Rosenfeld & Beath 1964; Underwood 1971). The role and metabolism of Se in the animal and the diagnosis and treatment of Se inadequacies in animals will be discussed by others in this Symposium.

This review will not be exhaustive but rather will be selective in order to illustrate general principles of behaviour in soil-plant relationships. Large economic losses of animals from both Se excesses and inadequacies have been reported. Therefore, attention will be paid to origin and availability of the element in diverse areas of toxicity and inadequacy in the soil-plant-animal complex.

SELENIUM IN ROCKS AND SOILS

Selenium is one of the dispersed elements, occurring in minute amounts in nearly all materials of the earth's crust but rarely concentrated in any material above 100 ppm (Lakin 1961; Lakin & Davidson 1967). Fleischer (1972) notes that averages are not very useful when considering relationships of trace elements to disease. Rather it is of more interest to know about rocks in which these elements are concentrated or depleted. As geochemical knowledge increases, it may become possible to generalize about potential areas of Se excess and inadequacy.

Selenium, a member of Group VI in the periodic table, exhibits some of the chemical and physical behaviour of S, with some significant points of departure. As with S, Se is not an essential constituent of the rock-forming silicate minerals. Selenium tends to separate from cooling igneous rocks in association with sulphides of Fe, Co and Ni. Volatile Se escapes along with volatile S from cooling rocks and volcanoes. Thus igneous rocks and soils derived therefrom tend to be low in Se as well as S. Volatilization of Se may provide a means of dispersal of the element so that large areas of soils may be provided with small concentrations of the element when the volatile selenides come to earth in rain and snowfall.

The disintegration of rocks by the actions of wind, water, carbon dioxide, plants and animals results in soils of diverse characteristics. Variations in pH, Eh and temperature may also modify soil properties.

Of the rocks exposed to weathering and soil development, only about 20% are of igneous origin. Soils of such origin are generally regarded as uniformly low in Se. When these rocks decompose to form soils, the plants grown thereon are apt to be low in Se. The more humid the area and the more acid the soil, the greater the possibility of inadequacy of plant Se for animals.

Sedimentary rocks cover more than three-fourths of the land surface of the earth and the majority of agricultural soils are formed from such rocks. Shales generally have the highest concentrations of Se and are the principal sources of the Se-toxic soils of the Great Plains and Rocky Mountain foothills of the United States (Lakin 1961; Lakin & Davidson 1967), Ireland (Fleming 1962), Australia (McCray & Harwood 1963) and probably other countries of the world (Rosenfeld & Beath 1964).

Selenium contents of sandstones are quite variable. The element is often concentrated in organic debris in the sandstones.

Limestones are generally low in Se, but here again there is substantial variability.

Selenium concentrations in phosphate rocks are variable. Robbins & Carter (1970) analysed samples of phosphate rock from Florida and from western North America and found that generally western phosphate contained higher concentrations of Se. They suggested that normal and concentrated superphosphates made from phosphate rock containing 100 ppm Se can be expected to contain 60 and 40 ppm respectively. Selenium contents of phosphate fertilizers of course depend on the particular processing of the phosphate rock. Much of the Se may be removed from the sulfuric acid and the rock by volatilization during high-temperature processing according to Rader & Hill (1935). Gissel-Nielsen (1971*a*) analysed components of phosphate fertilizer through every step of production and found an overall 44% loss of Se. Losses of Se from the pyrites used in sulfuric acid manufacture accounted for nearly 50% of the total loss of Se.

Fossil fuels frequently contain Se. Spoil areas from coal mining may be a source of Se in drainage waters. Selenium volatilized in coal- and oil-burning areas can be a source of the element from particulate soot and dissolved Se compounds in rain and snow. Dust from air-conditioner filters has been found to contain from 0.05 to 10 ppm Se. Selenium from these sources is probably in the form of insoluble oxides or in the elemental form and may not be of immediate value to plants and animals.

Not all soils formed on seleniferous parent materials produce toxic vegetation. Total Se in a soil, as with many other soil-borne elements, is not a useful index of plant-available Se.

Pyritiferous shales, containing from 6 to 10 ppm Se, weather in high rainfall climates, as for example in Puerto Rico and Hawaii, to soils of high total Se content of low plant availability. Vegetation on these soils normally contains less than 1 ppm of the element. Lateritic soils of some parts of the United States contain 0.5-2.4 ppm Se but do not produce toxic vegetation. Lakin & Davidson (1967) state that these soils contain basic ferric selenite or selenite ions strongly adsorbed on ferric oxides. Geering *et al.* (1968), in a study of solubility and redox criteria for the possible forms of Se in soils, present evidence that the insolubility of selenite in soils containing reactive Fe oxides may be adequately explained by the formation of ferric selenite-ferric hydroxide complexes. Oxidation of selenite to the soluble selenate in alkaline soils is favoured by a decrease in the stability of the ferric hydroxide-selenite complex. Thus under alkaline, well-aerated conditions the Se is present as selenate, which forms neither highly insoluble salts nor stable adsorption complexes with the sesquioxides. For a given total Se content under acid and reducing conditions, as in humid areas, the selenite-ferric oxide complexes provide Se to protect livestock from Se inadequacy but not enough of the element to be toxic. Under well-drained conditions, in sub-humid areas, the Se will be principally in the selenate form and vegetation may be toxic to animals.

If Se from soils on upper levels enters the drainage channels and is deposited in lower and poorly drained areas, toxic plants are likely to be produced, especially on alkaline soils. Lakin & Davidson (1967) suggest that this is the reason for the occurrence of the very seleniferous soils of old lake bottoms of Limerick, Tipperary and Mean counties of Ireland, as described by Fleming (1962).

McCray & Harwood (1963) described a highly selenized "poison strip" located on an outcrop of Cretaceous marine clay limestone shales in Queensland.

Olson (1967) has prepared a diagram illustrating the Se cycle especially for cases of excessive Se (Fig. 1). With

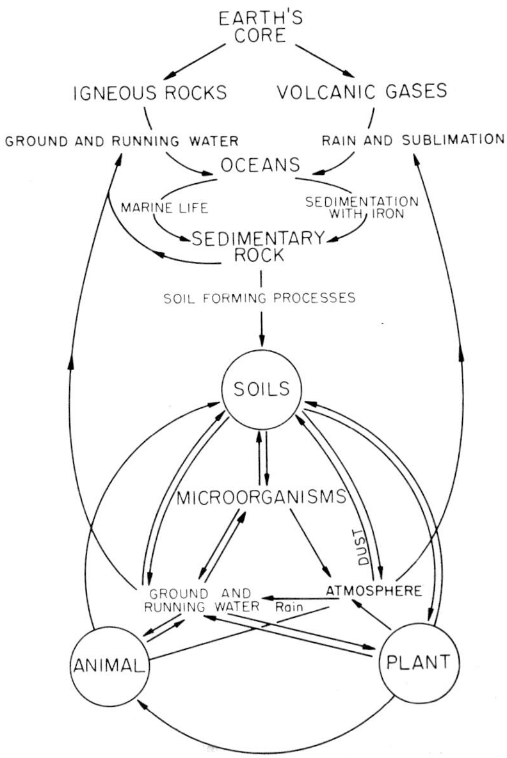

Fig. 1. Cycling of Se in nature. [From Olson, O.E., "Selenium in Biomedicine", Chapter 18, Avi Publishing Co. Inc. (1967)]

small modifications this cycle is appropriate not only for toxic amounts of the element but also for adequate and inadequate situations.

Many agriculturally important areas of the world produce crop plants and forages which are very low in Se. Here Se-responsive diseases, such as white muscle disease of sheep and cattle, are prevalent. Kubota *et al.* (1967) have prepared a map showing the regional distribution of Se concentrations in crops of the United States (Fig. 2). In addition to the Great Plains and Rocky Mountain foothill areas which

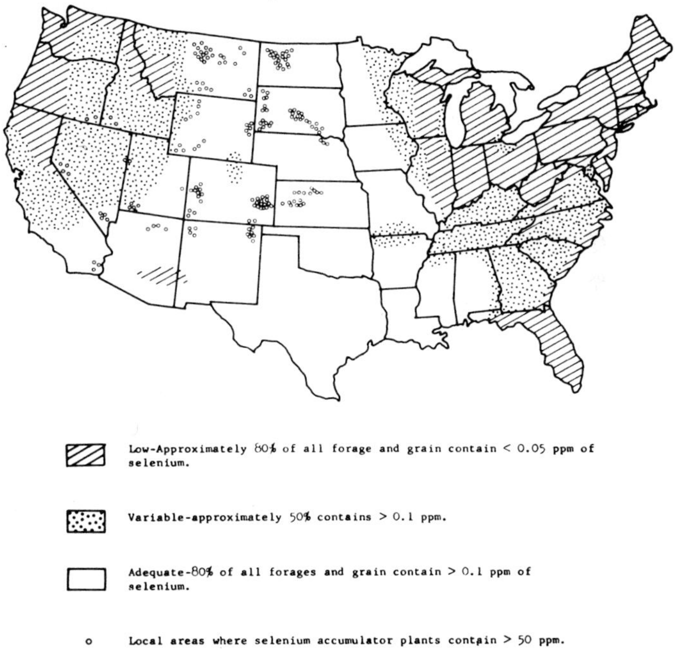

Fig. 2. Geographic distribution of low, variable and adequate Se areas in the U.S.A. [From Kubota, J. & Allaway, W.H., "Micronutrients in Agriculture", Chapter 21; reprinted with permission of the Soil Sci. Soc. of America Inc. (1972)]

produce high to acutely toxic forages, they were able to delineate large areas of apparent adequacy and where Se toxicity is only occasionally and locally a problem. Three major areas of the country are of interest because of the incidence of Se-responsive diseases. Parts of the Pacific Northwest, including portions of Washington, Oregon and northern California, produce forages which are very low in Se (less than 0.05 ppm) and Se-responsive diseases of animals are often observed. This is an area of recent volcanic deposits and the soils formed on them or transported from them are inherently very low in Se. In addition, the low native Se is plant-unavailable because of low pH and high sesquioxides, as discussed earlier.

In the Northeast of the United States, acid soils derived from very old sedimentary rocks that pre-date the major Cretaceous period of selenization of North America produce crops which are only slightly higher than those of the Pacific Northwest. Selenium-responsive diseases are reported in this area but the incidence is low because of intermingling of feeds from adequate areas.

A third low Se area is located along the South Atlantic seaboard, where a combination of soil parent material of low total Se and low availability to plants appears to account for the low Se crops.

In Denmark several soils of low native Se have been reported by Bisbjerg & Gissel-Nielsen (1969).

Davies & Watkinson (1966*a*) described a soil in New Zealand where lambs responded to oral supply of Se. This soil was formed on a rhyolitic pumice alluvium. Here low total Se and low pH (5.6) resulted in low plant availability.

In Western Australia, white muscle disease has been observed in the Great Southern and South Coastal districts. These highly leached soils were developed on granites and sedimentary rocks (Gardiner 1962).

Allaway (1971) has succinctly summarized current knowledge of relationships among Se in rocks, soils and plants as follows:

1. Where rocks with a high Se content decompose to form well-drained soils in sub-humid areas (less than 20 inches annual rainfall), selenides and other insoluble forms of Se will be converted to selenates and organic Se compounds. Vegetation produced on such soils will contain potentially toxic concentrations of Se.
2. Where rocks with a high content of Se weather to form soils in humid areas, slightly soluble complexes of ferric oxide or hydroxide and selenite ions will be formed. These soils will also be slightly to strongly acid, and the plants produced on them will not contain toxic concentrations of Se, but they may contain sufficient Se to protect livestock consuming them from Se deficiency.
3. Where rocks with a high content of Se weather to form poorly drained soils, or where Se from higher lying areas is deposited in poorly drained areas by alluvial action, and the soils are alkaline, plants containing toxic levels of Se are likely to be produced. This will be especially probable if the aeration of these soils is improved by

artificial drainage. The more acid the soils in an area, the less the likelihood that vegetation will contain toxic levels of Se.

4. Where rocks with a low content of Se decompose to form soils under either humid or dry conditions, the plants produced are likely to contain insufficient Se to protect animals from Se deficiency. The more humid the area and the more acid the soil, the greater the likelihood of extremely low Se concentrations in the plants.

MANAGEMENT PRACTICES

Management practices of value in control of Se toxicity in the Great Plains area of the United States have been the subject of investigations which are discussed by Rosenfeld & Beath (1964) and by Anderson (1961). These writers recommend mapping of seleniferous areas based on geological surveys and on occurrences of plant species known to be Se indicators and accumulators. Plant species differ markedly in their ability to accumulate Se in both high and low Se areas. These differences will be discussed later.

As the result of soil and plant surveys and observation of animal toxicities on some lands, the United States Government has withdrawn large areas from wheat production with conversion of the areas to controlled grazing (Anderson 1961).

The use of various soil amendments, especially gypsum, has been ineffective in reducing Se absorption by plants. The soils were probably already moderately high in sulphates and other salts in this sub-humid area. Further, many of the native plants are deep-rooted and sulphate probably does not penetrate deeply in the soils.

The effects of phosphate and sulphur fertilizers have been studied. Davies and Watkinson (1966*b*) found that mixed forages from a low Se soil frequently contained less Se from superphosphate-fertilized areas than when monocalcium phosphate was used. There may be two explanations: (1) sulphur additions stimulated clover growth, and clover in mixed pastures contains less Se than the grasses; (2) where yield was markedly increased in the superphosphate plants, there would be the further effect of dilution of Se taken up by the roots. Carter *et al.* (1972) considered other explanations for the cause of phosphate fertilization increasing Se concentrations in alfalfa. Possibly phosphate may compete with Se for adsorption sites in the soil. However, they could not detect differences in Se extractable by 0.01 N $CaCl_2$. They considered the effect of phosphate application on root proliferation to be the principal reason for increase in plant Se because phosphate applications to these low Se phosphate-responsive soils caused an increase in Se in the plant from both native and applied Se.

Allaway (1970) concludes that S fertilizers are not effective in preventing Se toxicity of plants in Se excess areas. High levels of available S may lead to decreased Se concentrations in plants growing on low Se soils, but Se deficiencies may occur in livestock in some places without regard to S status. If the S concentration in the animal diet is constant, the Se contained high S plants will be of nearly equal value to animals as the Se in low S plants. He also states that good practice requires satisfying S and Se requirements of both plants and animals.

Irrigation removes large quantities of Se, as noted in drainage water analyses (Lakin & Davidson 1967). However, most of the Se is removed from surface layers of the soil and not from lower profiles where deep-rooted plants still can accumulate toxic amounts of Se.

PLANT SPECIES DIFFERENCES

In the high Se areas of the United States the most seleniferous plants include species of *Astragalus*, *Machaeranthera*, *Haplopappus* and *Stanleya* (Rosenfeld & Beath 1964). Some of these plants are less palatable to grazing animals than are range grasses and other plants which are lower in Se. Animals eat the former only under conditions of over-grazing. However, some more palatable native plants such as *Aster* spp. and saltbush (*Atriplex* spp.) may still accumulate enough Se to present a hazard in open-range grazing.

In the three seleniferous areas of Ireland (Fleming 1962), alkaline highly organic lacustrine soils produce toxic vegetation in all species analysed. In contrast to the Great Plains situation, where native grasses rarely accumulate large excesses of Se, grasses in the Irish areas contained nearly as much Se as other species. Cruciferae had the highest Se content, followed by legumes and cereals.

Hamilton & Beath (1964) also found that cruciferae contained the greatest amount of Se when grown on a soil containing variable levels and forms of Se. Of 18 vegetables in this experiment, all had the highest Se concentrations of the element in the inedible parts, stems, pods, peels, etc. In another study, Hamilton & Beath (1963*a*) compared a number of crop plants for their ability to accumulate and store Se. Of these, sunflower possessed the highest Se-absorbing ability, with flaxseed, safflower seed and the root of sugar beets absorbing and storing the least. Hamilton & Beath (1963*b*) tabulated the analytical values obtained for 20 species of range plants, permitting comparisons of their Se-accumulating capacity when grown on the same soil in the greenhouse.

Peterson & Butler (1962) found that seeds of *Neptunia amplexicaulis*, an endemic Australian legume confined to seleniferous soils in central Queensland, contained 123 μg Se/seed when grown on the Australian soil. However, seeds of plants grown from the Australian seed on Manawatu soil in New Zealand contained only 0.45 μg Se/seed, all of which could have been derived from the Australian seed.

Management of Se-inadequate areas seems more feasible than in the case of the Se-toxic areas. Although the former are larger in extent, they are usually in areas of intensive agriculture and amenable to established agricultural practices such as plant species selection, fertilization and irrigation. Moreover, in some areas, supplemental feeding of grains and forage from adequate or even toxic areas can serve as a temporary but not very economic stopgap. Dosing of individual animals in areas of severe Se inadequacy has been practised, as will be discussed by others in this Symposium.

Factors affecting the concentration of Se in plants have been reviewed by Johnson *et al.* (1967). Widespread instances of conditions in animals which respond to Se, e.g. "ill-thrift" and white muscle disease of lambs, prompted studies of Se content of pasture plants by Davies & Watkinson (1966*a*, *b*). They found lower values for Se in clover and ryegrass than in other species of the mixed pasture. This may be significant in the light of observed animal responses to Se being usually on improved pastures containing clover and ryegrass rather than on unimproved pastures dominated by browntop. Browntop, ryegrass and cocksfoot all accumulated more Se from 4 and 16 ounce Se per acre applications than did clover, and the differences persisted for at least 342 days after application.

Bisbjerg & Gissel-Nielsen (1969), in pot trials using six low Se Danish soils, found a decreasing Se content as follows: crucifers, ryegrass, legumes and cereals. The order was unchanged by variations in concentration and oxidation state of applied Se.

Ehlig *et al.* (1968) consider the differences among plant species in Se accumulation from soils low in available Se to be relatively small. This is in sharp contrast to the findings of Hamilton & Beath (1963*a*,*b*, 1964), where large species differences were shown on soils of high available Se. Ehlig *et al.* (1968) suggest that species differences are of minor importance in the incidence of Se disease syndromes except in marginal cases where a shift from alfalfa or trefoil to red clover or certain grasses might cause a higher incidence of Se-responsive diseases. They conclude that addition of 0.1-0.3 ppm Se to soils is feasible for production of alfalfa, clovers and grasses without causing toxic concentrations in other species, including weeds, present in the hay or pastures.

Allaway (1971) states that, in the United States, alfalfa accumulates more Se than red clover, timothy or brome grass from soils having moderately low levels of Se. Differences among species have not been consistent on soils having very low levels of available Se.

FORMS OF SELENIUM APPLIED TO SOILS

The forms of Se applied to soils have been studied by Butler & Peterson (1961), Peterson & Butler (1962), Carter *et al.* (1969) and Gissel-Nielsen & Bisbjerg (1970). Butler & Peterson found that, when ^{75}Se-selenious acid was introduced into the rumen of sheep, 51% of the ^{75}Se was recovered in the dung, mostly in insoluble form. The labelled Se of the dung was only slightly available, less than 0.10% to perennial ryegrass in pot trials. Peterson & Spedding (1963), upon feeding red clover grown on nutrient solutions containing ^{75}Se to sheep, found also that ^{75}Se from the dung was only slightly (less than 0.3%) available to pasture species. In agreement with the findings of Davies & Watkinson (1966*a*), uptake of Se by browntop was greater than that by ryegrass and red clover.

Carter *et al.* (1969) studied the effect of laboratory-prepared Se salts on the concentration of Se in alfalfa grown on a low Se silt loam of pH 7.8. Barium selenate, even when mixed with barium sulphate, provided toxic plant levels of Se. Cupric selenite provided toxic concentrations in the first alfalfa cuttings. Ferric-hydroxy bi-selenites provided safe and adequate Se levels in the alfalfa. Elemental Se provided adequate and safe levels of Se for a year but then availability decreased. Cost of elemental Se (at 2-4 kg/hectare) may be excessive for use in grazing areas.

Gissel-Nielsen & Bisbjerg (1970) applied elemental Se and a series of selenites and selenates to soils in field and pot experiments. During a two-year experiment with mustard, the total uptake as a percentage of the added Se was 0.01% of Se^{O}, 4% of K_2SeO_3 and 30% of K_2SeO_4 and $BaSeO_4$. With lucerne, barley and sugar beet the uptake was one-third of this or less. They conclude that elemental Se has such a small effect that it may be ruled out immediately as a Se source.

The use of soluble selenates has the advantage that a substantial part is taken up by plants, but has the disadvantage of resulting in toxic crops if dosage is not controlled. Selenite has a more protracted action. However, Se accumulation in the soil may present a problem with continued use. Cary & Allaway (1969) caution that additions of any form of Se to the soil to protect animals from Se deficiency will require monitoring of Se concentrations in each of the crops produced for the first 2 or 3 years following treatment. They also note that, in soils of acid and neutral reaction, the concentration of Se soluble in dilute $CaCl_2$ or K_2SO_4 solutions may be used as an index of concentrations of Se to be expected in alfalfa growing on such soils.

When elemental Se is added to soils, some is quite rapidly oxidized to selenite, which then reacts to form an adsorption complex with Fe of the soil. The availability of Se from such addition-complexes resembles that from synthetic $Fe(OH)_3$-$HSeO_3$ complexes prepared in the laboratory by these workers. Selenates do not appear to be present in appreciable quantities in acid and neutral soils except possibly in the first few weeks after addition of selenites or selenates to soil. Equilibrium values for the different forms of the element and the kinetics involved need further study.

Gissel-Nielsen (1971*b*) studied the relationship between pH, clay content and time on uptake of Se by ryegrass. Increasing clay contents of soils decreased uptake of Se. Uptake of Se increased with increasing pH. Both these effects became less obvious with time. He concluded that lesser amounts of Se should be used on sandy soils than on loams. He also agreed with the warning of Allaway *et al.* (1967) that addition of Se may be initially much more hazardous on alkaline soils than on acid soils.

VOLATILE SELENIUM

Volatile Se compounds have a characteristic and offensive odor. This odor has been stated to be a factor in the rejection of Se accumulator plants by grazing animals in favour of less seleniferous plants such as grasses (Rosenfeld & Beath 1964).

Moxon & Rhian (1938) reported Se losses of 4 to 73% when grains of barley, maize and wheat were stored from 3 to 5 years. Beath *et al.* (1935) found that analytical values for Se decreased during storage at 20°C. The element could be trapped by passing the air from over the drying plant material through concentrated sulfuric acid.

Soils also liberate volatile Se (Abu-Erreish *et al.* 1968). Evolution of volatile Se was almost completely inhibited by autoclaving the soil or by removal of oxygen by sweeping with N. On this evidence, these workers suggested that mould growth was responsible for loss of Se from these soils. These findings confirmed the earlier report of Ganje & Whitehead (1958) of losses of ^{75}Se (as either selenite or selenate) previously added to a shale-loam-vermiculite mixture.

Lewis *et al.* (1966) measured losses of ^{75}Se-labelled compounds from intact alfalfa in greenhouse experiments. Evolution of volatile Se by these plants was a function of age and Se content of the plants, of temperature, and of light intensity. A biological process was suggested. Not surprisingly, the accumulator species, *Astragalus*, also produced volatile Se.

Ehlig *et al.* (1968) dried samples of a number of agronomic species containing ^{75}Se for from 2 to 30 hr at 70°C with losses only within the experimental error of ± 5% observed. Asher *et al.* (1967) and Evans *et al.* (1968) collected volatile Se compounds from intact plants of *Medicago sativum* and *Astragalus racemosus* as well as from tops and roots of these plants on drying at 70°C for 48 hr. The plants were supplied with ^{75}Se to facilitate analytical procedures. The compounds were characterized by solubility studies and by gas chromatographic patterns. Four Se compounds were collected from the volatiles of *Astragalus racemosus*. One of these was dimethyl diselenide. The remaining three compounds were not identified. Alfalfa produced four volatile Se compounds, none of which was identified. Neither dimethyl diselenide nor dimethyl selenide was produced by alfalfa. The amount of volatile Se in each case was small, representing 0.3-7.0% of the total Se of the plants.

Lewis *et al.* (1971,1974) have identified a volatile compound from cabbage as dimethyl selenide. They also reported isolation of an enzyme which catalyzes the cleavage of dimethyl selenide from Se-methylselenomethionine selenonium salt. Thus cabbage produces a volatile Se compound different from those produced by either *Astragalus* or alfalfa.

Gissel-Nielsen (1970) reported losses of up to 20% of Se from ^{75}Se-labelled barley, ryegrass, lucerne, mustard, radish, onion and sugar beet upon drying at 60°C. He also reported smaller losses of the element on storage of hay species.

Except in certain extreme cases, it does not seem that the small reported losses of Se compounds from plants are significant either in diminishing the incidence of toxicity for animals or in increasing occurrences of Se-responsive diseases.

Most ordinary cooking techniques probably do not result in major losses of Se from foods. Higgs *et al.* (1972) found little or no loss of the element as the result of broiling meats or seafoods, frying eggs or boiling cereals. Dry heating of cereals, however, led to 7-23% losses. Boiling two vegetables, asparagus and mushrooms, resulted in 29 and 44% losses respectively.

SELENIUM AS A PLANT NUTRIENT

Evidence for a Se requirement by accumulator species, *Astragalus racemosus* and *A. beathii*, has been presented by Trelease and Trelease (1938,1939). Shrift (1969,1973) has provisionally supported this concept but stated that the evidence may not be unequivocal. Broyer *et al.* (1972*a*,*b*) concluded that the medium used in the Trelease studies contained high levels of phosphate, at least for these leguminous species. Photos published in the Trelease works suggested phosphate toxicity symptoms in the plants. Broyer's experiments seem to verify this idea. At low maintained phosphate supply the plants showed no symptoms, and growth responses were not obtained with increased Se supply. They observed a complex interaction between selenite, phosphate and micronutrient elements. Selenite depressed sorption of phosphate and the micronutrient metals; phosphate tended to restrict selenite sorption. Broyer *et al.* concluded that, if Se is required for these *Astragalus* species, the critical concentration of the element in the dry whole plant tops must be less than 0.08 ppm.

In earlier work with the agronomic species, alfalfa (*Medicago sativa* L.) and subterranean clover (*Trifolium subterraneum* L. var. Mt. Barker), Broyer *et al.* (1966) had found no beneficial effect of additions of selenite to highly purified cultures on which these plants were grown. The yields of neither alfalfa or subterranean clover were significantly affected by additions of selenite from 0.02 to 0.2 ppm Se but a depressing effect was observed at the 2 ppm Se rate. They concluded that, if Se is essential, the critical level in tops must be below 0.08 ppm.

It should be noted that analytical values for tops of many grazing species on low selenium soils where Se-responsive diseases are observed are often as low as 0.02 ppm. Yield responses to top-dressing of selenite have not been reported in these instances. Bisbjerg & Gissel-Nielsen (1969) noted a yield depression in some plants in soil experiments when between 0.5 and 2.5 ppm selenate Se was added.

SUMMARY

Soils derive their Se from the parent material rocks. Igneous rocks weather to soils which are likely to be low in Se. Soils formed by weathering of sedimentary rocks are generally variable in Se content but higher than those on igneous rocks. Shales are invariably higher in Se than other rocks and, in addition, make up a larger proportion of rocks exposed to weathering processes.

The availability of soil Se to plants depends on rainfall, drainage, aeration, pH and sesquioxide content of the soil.

The Se status of vegetation may be classified as (a) excessive or toxic to animals, (b) adequate for animal health and (c) low or inadequate for animal health. Selenium content of plants depends on plant species, plant available Se, fertilizer and cultural practices.

No plant requirement for Se has been established.

REFERENCES

ABU-ERREISH, G.M., WHITEHEAD, E.I. & OLSON, O.E. (1968) - *Soil Sci. 106* : 415

ALLAWAY, W.H. (1970) - *J. Sulphur Inst. 6* : 3

ALLAWAY, W.H. (1971) - *In* "Selenium in Nutrition". Report of the Sub-committee on Selenium, Committee on Animal Nutrition, Agricultural Board, National Research Council. National Academy of Sciences

ALLAWAY, W.H., CARY, E.E. & EHLIG, C.F. (1967) - *In* "Symposium: Selenium in Biomedicine" (O.H. Muth, ed.), p. 273. Avi Publishing Co. Inc. : Westport, Conn.

ANDERSON, M.S. (1961) - *In* "Selenium in Agriculture", Agriculture Handbook No. 200, Agricultural Research Service, U.S. Department of Agriculture in cooperation with the Geological Survey, U.S. Department of Interior : Washington, D.C.

ASHER, C.J., EVANS, C.S. & JOHNSON, C.M. (1967) - *Aust. J. biol. Sci. 20* : 737

BEATH, O.A., EPPSON, H.F. & GILBERT, C.S. (1935) - *Wyoming Agr. Expt. Sta. Bull. 206* : 1

BISBJERG, B. & GISSEL-NIELSEN, G. (1969) - *Pl. Soil 31* : 287

BROYER, T.C., JOHNSON, C.M. & HUSTON, R.P. (1972*a*) - *Pl. Soil 36* : 635

BROYER, T.C., JOHNSON, C.M. & HUSTON, R.P. (1972*b*) - *Pl. Soil 36* : 651

BROYER, T.C., LEE, D.C. & ASHER, C.J. (1966) - *Pl. Physiol. 41* : 1425

BUTLER, G.W. & PETERSON, P.J. (1961) - *N.Z. Jl agric. Res. 4* : 484

CARTER, D.L., BROWN, M.J. & ROBBINS, C.W. (1969) - *Proc. Soil Sci. Soc. Am. 33* : 715

CARTER, D.L., ROBBINS, C.W. & BROWN, M.J. (1972) - *Proc. Soil Sci. Soc. Am. 36* : 624

CARY, E.E. & ALLAWAY, W.H. (1969) - *Proc. Soil Sci. Soc. Am. 33* : 571

DAVIES, E.B. & WATKINSON, J.H. (1966*a*) - *N.Z. Jl agric. Res. 9* : 317

DAVIES, E.B. & WATKINSON, J.H. (1966*b*) - *N.Z. Jl agric. Res. 9* : 641

EHLIG, C.F., ALLAWAY, W.H., CARY, E.E. & KUBOTA, J. (1968) - *Agron. J. 60* : 43

EVANS, C.S., ASHER, C.J. & JOHNSON, C.M. (1968) - *Aust. J. biol. Sci. 21* : 13

FLEISCHER, M. (1972) - *Ann. N.Y. Acad. Sci. 199* : 3

FLEMING, G.A. (1962) - *Soil Sci. 94* : 28

GANJE, T.J. & WHITEHEAD, E.I. (1958) - *Proc. S. Dak. Acad. Sci. 37* : 81

GARDINER, M.R. (1962) - *Aust. vet. J. 38* : 387

GEERING, H.R., CARY, E.E., JONES, L.H.P. & ALLAWAY, W.H. (1968) - *Proc. Soil Sci. Soc. Am. 32* : 35

GISSEL-NIELSEN, G. (1970) - *Pl. Soil 32* : 242

GISSEL-NIELSEN, G. (1971*a*) - *J. agric. Fd Chem. 19* : 564

GISSEL-NIELSEN, G. (1971*b*) - *J. agric. Fd Chem. 19* : 1165

GISSEL-NIELSEN, G. & BISBJERG, B. (1970) - *Pl. Soil 32* : 382

HAMILTON, J.W. & BEATH, O.A. (1963) - *Agron. J. 55* : 528

HAMILTON, J.W. & BEATH, O.A. (1963*b*) - *J. Range Mgmt 16* : 261

HAMILTON, J.W. & BEATH, O.A. (1964) - *J. agric. Fd Chem. 12* : 371

HIGGS, D.J., MORRIS, V.C. & LEVANDER, O.A. (1972) - *J. agric. Fd Chem. 20* : 678

JOHNSON, C.M., ASHER, C.J. & BROYER, T.C. (1967) - *In* "Symposium: Selenium in Biomedicine" (O.H. Muth, ed.), p. 57. Avi Publishing Co. Inc. : Westport, Conn.

KUBOTA, J., ALLAWAY, W.H., CARTER, D.L., CARY, E.E. & LAZAR, V.A. (1967) - *J. agric. Fd Chem. 15* : 448

LAKIN, H.W. (1961) - *In* "Selenium in Agriculture", Agriculture Handbook No. 200, p. 3. Agricultural Research Service, U.S. Department of Agriculture in cooperation with the Geological Survey, U.S. Department of Interior : Washington, D.C.

LAKIN, H.W. & DAVIDSON, D.F. (1967) - *In* "Symposium: Selenium in Biomedicine" (O.H. Muth, ed.), p. 27. Avi Publishing Co. Inc. : Westport, Conn.

LEWIS, B.G., JOHNSON, C.M. & BROYER, T.C. (1971) - *Biochim. biophys. Acta 237* : 603

LEWIS, B.G., JOHNSON, C.M. & BROYER, T.C. (1974) - *Pl. Soil* *40* : 107

LEWIS, B.G., JOHNSON, C.M. & DELWICHE, C.C. (1966) - *J. agric. Fd Chem.* *14* : 638

McCRAY, C.W.R. & HARWOOD, I.S. (1963) - *Qd J. agric. Sci.* *20* : 475

MOXON, A.L. & RHIAN, M. (1938) - *Proc. S. Dak. Acad. Sci.* *18* : 20

MUTH, O.H. (ed.) (1967) - "Symposium: Selenium in Biomedicine". Avi Publishing Co. Inc. : Westport, Conn.

OLSON, O.E. (1967) - *In* "Symposium: Selenium in Biomedicine" (O.H. Muth, ed.), p.297. Avi Publishing Co. Inc. : Westport, Conn.

PETERSON, P.J. & BUTLER, G.W. (1962) - *Aust. J. biol Sci.* *15* : 126

PETERSON, P.J. & SPEDDING, D.J. (1963) - *N.Z. Jl agric. Res.* *6* : 13

RADER, L.F. & HILL, W.L. (1935) - *J. agric. Res.* *51* : 1071

ROBBINS, C.W. & CARTER, D.L. (1970) - *Proc. Soil Sci. Soc. Am.* *34* : 506

ROSENFELD, I. & BEATH, O.A. (1964) - "Selenium - Geobotany, Biochemistry, Toxicity, and Nutrition". Academic Press : New York.

SHRIFT, A. (1969) - *A. Rev. Pl. Physiol.* *20* : 475

SHRIFT, A. (1973) - *In* "Organic Selenium Compounds: Their Chemistry and Biology" (D.L. Klayman and W.H.E. Gunther, eds.). John Wiley & Sons : New York

TRELEASE, S.F. & TRELEASE, H.M. (1938) - *Am. J. Bot.* *25* : 372

TRELEASE, S.F. & TRELEASE, H.M. (1939) - *Am. J. Bot.* *26* : 530

UNDERWOOD, E.J. (1971) - "Trace Elements in Human and Animal Nutrition", 3rd Edn. Academic Press : New York and London

THE FUNCTIONS OF TRACE ELEMENTS IN PLANTS

D.J.D. Nicholas

Department of Agricultural Biochemistry & Soil Science
Waite Agricultural Research Institute
The University of Adelaide
Glen Osmond, South Australia 5064

INTRODUCTION

At the present time the following trace elements are known to be essential for the growth of plants: Fe, Cu, Zn, Mn, Mo, Co, B, Cl, Na and possibly V. Not all of these elements are universally required by plants but all of them have been shown to be required by some plant. In this paper I will not only consider higher plants, but will also draw freely on examples from lower forms, including algae and fungi as well as bacteria. The classification of essential mineral elements into (a) major or macro-element and (b) minor element, trace element or oligo-element reflects a large or a small requirement of these by plants. Although this may be a convenient subdivision, it is arbitrary and has obvious limitations. Thus, for example, Ca and S are essential in large amounts for higher plants, whereas for microorganisms they are required at much smaller levels and could therefore be regarded as trace elements.

Two main approaches have been used to establish a trace element requirement for microorganisms and plants: (a) nutrition-type experiments in sand or solution cultures, often confirmed under field conditions, whereby a trace element is shown to be specifically required for growth (Wallace 1961; Hewitt 1966), and (b) biochemical studies to determine the functions of the elements in metabolism (Nicholas 1961).

This paper will be primarily concerned with examples of the latter approach, although the nutritional work has often preceded and guided metabolic studies. A good example of this was the observation that Mo deficiency is more severe when fungi or higher plants are grown with nitrate rather than ammonia as a sole source of N (Nicholas *et al.* 1954; Nicholas 1959,1963; Hewitt 1959). This led to the discovery that Mo is indeed a component of nitrate reductase enzymes in fungi and green plants.

CHEMISTRY OF TRANSITION METALS

A number of the essential trace elements for the growth of microorganisms and plants are transition metals, viz. Fe, Cu, Zn, Mn, Mo, Co and V, which have electronic structures with distinct physical and chemical properties. These include variable oxidation states, complex formation, metallic nature and catalytic activity. While it is not suggested that the individual properties are necessarily unique to transition metals, collectively they add up to behaviour quite distinct from that shown by any other type of element (Earnshaw & Harrington 1973). Transition metals are involved in electron transfer processes as follows:

$$A^+ H^- \longrightarrow A^+ + H^+ + 2e$$
$$2e + B^+ \longrightarrow B^-$$
$$B^- + H^+ \longrightarrow B^+ H^-$$

These are usually single one-electron reactions and it is likely that even when a two-electron transfer occurs, it does so in two consecutive one-electron reactions.

Two one-electron transfers in biological systems are usually fast and here transition metals are quite important. It is well established that enzymes or proteins which contain metals often assume the reactive characteristics of the metals themselves. Some transition metals are not restricted to single electron changes but can react by multi one-electron steps, e.g. Mn^{II} to Mn^{IV}, Mo^{III} to Mo^{VI}. Metals are involved in electron transfer and O atom transfer but not usually in H^- atom transfer. Low oxidation states of metals, however, are not normally used by biological systems since they are unstable and chemically very reactive. Metals often bridge the gap between redox couples of non-metal compounds which have a relatively small span of potentials, e.g. those of N and S, whereas transition metals cover a much wider range of potentials. Thus for Fe alone the range is from -0.3 to +1.0 volts at pH 7.0 and this can be extended further as a result of complex formation with ions and ligands of proteins. Electron transfer reactions of transition metals can often be made with a low overall energy change.

I will now consider briefly some of the main functions of

the trace elements in microorganisms and plants. Since Dr. Boardman will be describing the role of trace elements in photosynthesis and Drs. Jackson & Chapman of this department will be discussing the possible role of B in RNA metabolism of plants, I will, in this brief review, concentrate on some other aspects of trace element function.

IRON

Non-haem Iron Proteins

Proteins that possess Fe-S clusters as active sites have been isolated from a wide variety of bacteria, plant and animal sources and are usually known as ferredoxins. These non-haem Fe-S proteins, which have small molecular weights, are involved in key metabolic events such as N_2 fixation, photosynthesis, electron transfer enzymes and hormone biosynthesis (Buchanan & Arnon 1970). The main feature of these non-haem Fe-S proteins is their very negative potential often at or near the H electrode (E_o^1 at pH 7.0, 0.4 volts). Ferredoxin was first isolated from *Clostridium pasteurianum* and is also involved in electron transfer in chloroplasts of green plants, to be discussed by Dr. Boardman. The amino acid sequences of the various ferredoxins have been determined and detailed spectroscopic work with electron spin resonance, Mössbauer and nuclear magnetic resonance have elucidated their mechanism of action. Cysteine residues usually bind the Fe-S clusters to the polypeptide chain. Ferredoxin from *Clostridium* has 8 Fe atoms associated with 8 cysteine residues and 8 sulphides, with a molecular weight of 12,000 daltons for the dimer. A variant known as rubredoxin prepared from the same bacterium contains 1 Fe atom/4 cysteine residues with a molecular weight of 6100 daltons. Recent X-ray work at 3 Å resolution has resolved the crystal structure of these proteins (Jensen *et al.* 1973).

In ferredoxin from bacteria the close association of tyrosine with each of the Fe complexes suggests that the two aromatic groups are involved in the mechanism by which these molecules transfer electrons. Ferredoxins may be likened to capacitator proteins in that they store reducing equivalents at the various Fe-S centres (Williams 1973).

Haem-iron Compounds

In green plants, components of the respiratory chain located in the mitochondria are similar to those found in the animal system. Various haem-iron compounds, the cytochromes, are important respiratory electron carriers but these differ in detail from those in animal mitochondria. The reduced absorption bands of the cytochromes are sharpened considerably at low temperature (77°K). Using this technique, three *b*-type cytochromes compared with two in the animal mitochondria and two *c*-type cytochromes that are different from cytochromes *c* and c_1 in the animal system have been identified. A possible general scheme of electron transfer in plant mitochondria is as follows:

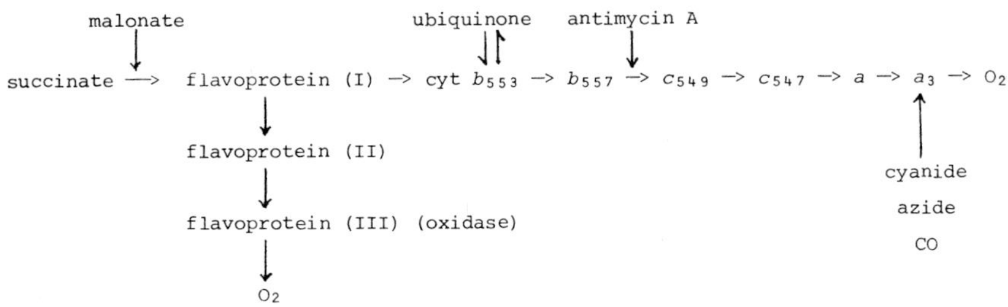

(subscripts are the reduced α-bands of the cytochromes)

It should be noted that there is a cytochrome *b*, *c*, *a*, a_3 pathway from substrate succinate to O_2 which is inhibited by antimycin A between b_{557} and c_{549}, whereas cytochrome a_3 is inhibited by cyanide, azide and carbon monoxide (Storey 1970; Ikuma 1972).

Special cytochromes are present in chloroplasts. These include cytochromes b_{563}, b_{559} and f_{554}, as well as non-haem Fe ferredoxin. The light-induced electron transfer involving these components will be discussed later By Dr. Boardman and reference should also be made to a review by Bendall & Hill (1968).

Numerous groups have identified a ferredoxin-linked nitrite reductase in higher plants (Hewitt 1959; Paneque *et al.* 1964; Hucklesby & Hewitt 1966; Joy & Hageman 1966; Ramirez *et al.* 1966) which reduces nitrite to ammonia. Recently a ferredoxin-linked nitrite reductase of spinach, an enzyme that catalyses the six-electron reduction of nitrite to ammonia, was shown to contain sirohaem, a new type of haem prosthetic group (Murphy *et al.* 1974). The absorption bands of this haem compound are as follows: Soret 380-340 nm; α-band 589. Sirohaem is also a component of sulphite reductase, both assimilatory and dissimilatory types, which also mediate the six-electron reduction of sulphite to sulphide. The prosthetic group of the sulphite and nitrite reductases is an iron tetrahydroporphyrin of the isobacteriochlorin type (adjacent pyrrole rings reduced) with eight carboxylate side-chains (Fig. 1).

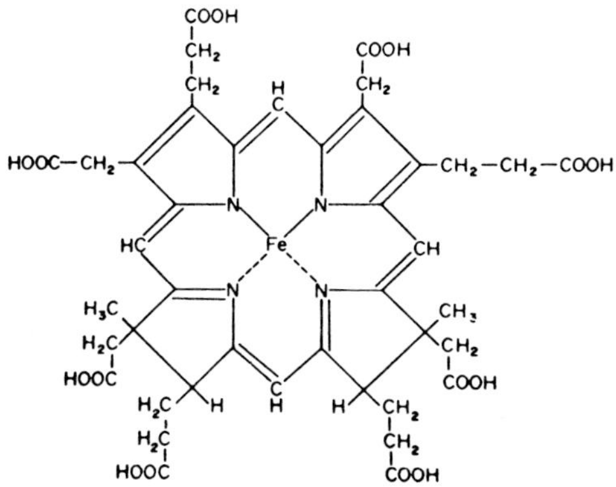

Fig. 1. Postulated structural formula for the sirohaem prosthetic group.

The main reactions for nitrite and sulphite reductases may be summarized as follows:

$$NO_2^- + [6\ H\bullet] \longrightarrow NH_4^+ + 2OH^-$$

$$SO_3^{2-} + [6\ H\bullet] \longrightarrow S^{2-} + 3H_2O$$

The enzymic reductions of nitrite and sulphite occur in two different physiological pathways. Thus, in green plants, fungi and many bacteria the reduction of sulphite to sulphide and nitrite to ammonia are steps in the assimilation of sulphate and nitrate respectively for the synthesis of S- and N-containing cellular constituents. In some dissimilatory bacteria, sulphite and nitrite are associated with anaerobic respiration, utilizing sulphate and nitrate as terminal acceptors instead of O_2 and producing H_2S from sulphite and the oxides of N and N_2 gas from nitrite.

The association of sirohaem with reductive processes involved in the metabolism of two major elements of the biosphere, N and S, suggests a key role for this novel haem compound in the evolution of redox metabolism (Murphy *et al.* 1974).

MOLYBDENUM

Molybdenum is a component of at least five distinct enzymes that catalyse diverse and unrelated reactions, namely nitrogenase, nitrate reductase, xanthine oxidase, aldehyde oxidase and sulphite oxidase. These enzymes are complex and each has additional non-protein groups besides Mo that are also redox carriers. Thus the nitrogenase enzyme contains Fe-S proteins in addition to a Mo-Fe-S protein, nitrate reductase is a molybdoflavoprotein, xanthine and aldehyde oxidases contain flavin and Fe-S protein, and sulphite oxidase has a haem component in addition to Mo. Each enzyme, which has two substrates, the oxidant and the reductant, facilitates a low energy pathway for electron flow from the reducing equivalents to the oxidants. In each case the Mo interacts directly with the oxidant.

Nitrogenase

This is an enzyme complex composed of two distinct components that combine to reduce N_2 gas to ammonia or acetylene to ethylene:

$$N_2 + 6H^+ + 6e^- \longrightarrow 2NH_3$$
$$C_2H_2 + 2H^+ + 2e^- \longrightarrow C_2H_4$$

The two enzyme components prepared from various N_2-fixing organisms have the compositions described in Table 1. Thus component I contains Fe, Mo and $S^=$ and is referred to as molybdoferredoxin (MoFd), and the smaller protein II, or azoferredoxin (AzoFd), contains Fe and sulphide. The AzoFd, which contains two identical subunits, is very O_2 sensitive and is cold labile and salt sensitive, whereas the MoFd contains four subunits, two of which are different (Eady & Postgate 1974). The tetramer, which has two atoms of Mo, 20 Fe atoms and 20 acid-labile sulphide groups, is less affected by O_2 than AzoFd. The Mg-ATP complex binds to the AzoFd component. By using $^{13}N_2$-labelled gas produced in a linear accelerator, it was shown that the MoFd protein chemisorbs the N_2 gas, as shown in Table 2.

A possible mechanism of action for nitrogenase, based on the biochemical work, has been presented by Leigh (1971). He proposes that both Mo and Fe are involved in activating N_2 gas. The change from the coordinated dinitrogen molecule to a coordinated nitridoanion requires an oxidation-state change of three units and Mo is more likely to be involved here since Fe has not a wide range of oxidation states. Reduction involves

TABLE 1

Composition of purified enzymes (nitrogenase) for N_2 fixation

Bacterium	Enzyme	Molecular weight	Mo : Fe : Labile $S^=$
Azotobacter vinelandii	I	270,000	2 : 32 : 25
	II	39,000	0 : 2 : 2
Clostridium pasteurianum	I	200,000	2 : 15 : 15
	II	40,000	0 : 2 : 2
Rhizobium	I	180,000	1 : 9 : -
	II	50,000	0 : 1 : -

a nitridation of Mo and a protonation of the N atom attached to the Fe. Ammonia is not a good ligand for Fe and subsequent hydrolysis would remove it as well as the nitride ligand:

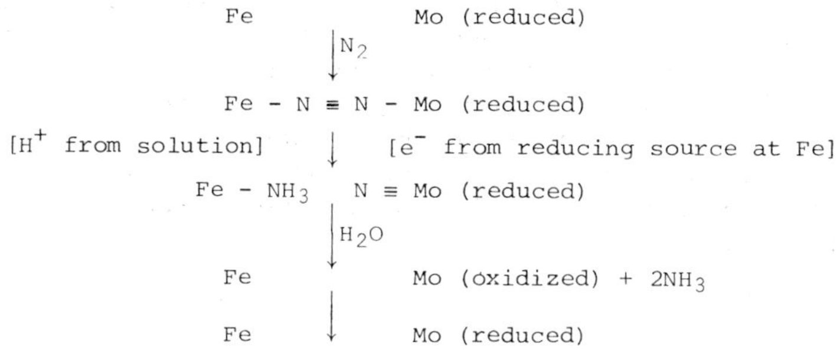

TABLE 2

Incorporation of $^{13}N_2$ *by nitrogenase from* Azotobacter vinelandii *(purified enzymes I and II of the nitrogenase complex)*

Reaction mixtures contained in 10 ml Warburg flasks in a final volume of 0.6 ml: 12.5 μmoles Tris-HCl buffer (pH 7.5); ATP generating system, 15 μmoles creatine phosphate, 2 μmoles ATP, 4 μmoles $MgCl_2$ *and 0.125 mg creatine phosphokinase; 2 mg* $Na_2S_2O_4$ *and 0.05 ml Enzyme I and/or 0.05 ml Enzyme II (Deering & Nicholas 1974). The* $^{13}N_2$ *gas was generated in a 3 MeV van der Graaf Linear Accelerator and flushed from the target area in He gas (Nicholas* et al. *1961; Ritchie 1968; Stanley & Nicholas 1969). The gas was bled off into a series of 10 ml Erlenmeyer flasks fitted with Quickfit* (B_{10}) *necks through which entry and exit tubes were fitted as sealed units. The inlet tube was drawn into a fine jet which was just clear of the reaction mixture (usually not more than 1 ml) in each flask. A sideport on each flask fitted with a Suba-seal was used to inject the enzyme and a solution of* $Na_2S_2O_4$ *(sparged with He) into the reaction mixture under strictly anaerobic conditions. The reaction mixtures as described were dispensed under anaerobic conditions and exposed to* $^{13}N_2$ *gas in He for 10 min and then He gas was used to flush through the reaction mixtures to remove excess labelled gas. The samples were then radioassayed in a well-type scintillation counter.*

Enzyme	Components of reaction mixture	^{13}N counts/100 sec
I (Mo-Fe protein)	ATP + $Na_2S_2O_4$	23,260
II (Fe protein)	ATP + $Na_2S_2O_4$	2,380
I + II	ATP + $Na_2S_2O_4$	67,730
I + II	*omit* ATP + $Na_2S_2O_4$	2,380
I + II	ATP *omit* $Na_2S_2O_4$	900

The scheme provides a specific function for both Mo and Fe, and it suggests an asymmetric reduction of dinitrogen which avoids intermediates such as diimide and hydrazine.

Stiefel (1973), on the other hand, considers that after the substrate, either acetylene or dinitrogen, is bound to enzyme I, then enzyme II (Fe•S protein) donates two electrons to the $Mo^{(VI)}$ in enzyme I, reducing it to $Mo^{(IV)}$, which in turn transfers two electrons to the substrate. In addition there is a proton transfer from the coordinated nitrogens on enzyme I to the substrate. Thus acetylene is reduced to ethylene by this scheme:

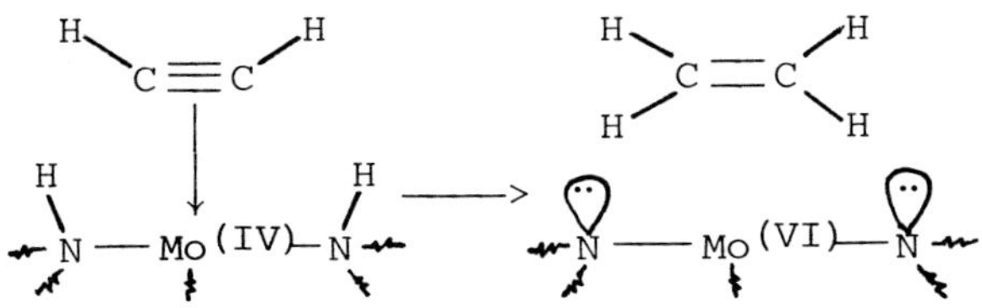

Hardy *et al.* (1971) have proposed a two-site hypothesis for the reduction of dinitrogen whereby the gas is coordinated end-on with Fe as well as side-on with Mo on enzyme I. The dinitrogen is then reduced by a coupled two-electron reduction [$Mo^{(IV)}$ to $Mo^{(VI)}$] and proton transfer from the coordinated nitrogens to form diimide, which then reduced to ammonia:

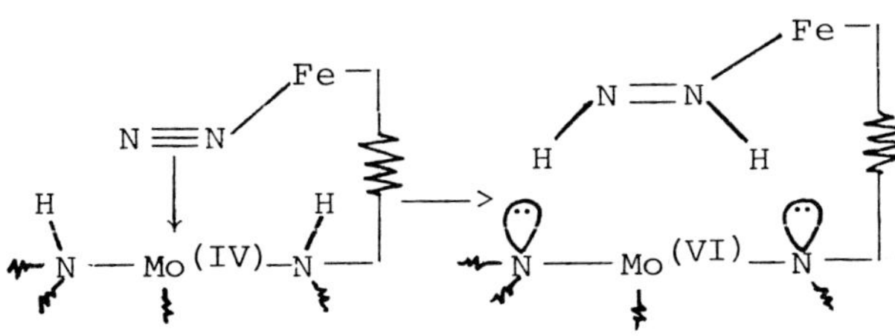

In the legume nodules it is of interest that haemoglobin functions as an oxygen-regulating system controlling the amount of O_2 available to the bacteroids. Excess O_2 would of course compete effectively with N_2 for reducing equivalents and thus inhibit N_2 fixation.

Nitrate Reductase

This enzyme is known to require Mo and flavin for its activity and for the reduction of nitrate to nitrite.

$$NO_3^- + 2H^+ + 2e^- \longrightarrow NO_2^- + H_2O \quad ..... \text{(i)}$$

$$NO_3^- + H^+ + 2e^- \longrightarrow NO_2^- + OH^- \quad ..... \text{(ii)}$$

Equation (i) is based on the acidic half reaction, whereas (ii) allows for OH^- participation at physiological pH.

The mechanism of electron flow proposed by Nicholas & Nason (1954*a,b*), based on their work with the fungus *Neurospora crassa* and higher plants, is as follows:

$$NAD(P)H \longrightarrow FAD \longrightarrow Mo \longrightarrow NO_3^- \longrightarrow NO_2^-$$

Evidence from chromatographic separation of the valency states of Mo indicates that during enzyme action the Mo undergoes a valency change from $Mo^{(V)}$ to $Mo^{(VI)}$ (Nicholas & Stevens 1955), and recently the possibility that even an $Mo^{(IV)}$ to $Mo^{(VI)}$ change might occur has also been considered (Stiefel 1973). It is of interest that $Mo^{(III)}$ reduces nitrate to nitrite non-enzymically (Nicholas & Stevens 1956). In some cases, non-haem Fe or cytochromes are present as additional carriers in the electron transfer sequence to nitrate (Nason 1963; Garrett & Nason 1969). When W is substituted for Mo in the nitrate reductase enzyme, it then has no catalytic activity (Notton & Hewitt 1971), presumably because $W^{(VI)}$ is too stable to undergo a valency change. It is of interest that Re approaches the catalytic activity of Mo but its abundance in nature is extremely low and so it is not a likely candidate for an enzyme system as common as nitrate reduction or N_2 fixation.

Sulphite Oxidase

This enzyme, which contains Mo, is found in microorganisms (Peck 1961; Hempfling *et al.* 1967), plants (Tager & Rautanen 1955; Arrigoni 1959; Fromageot *et al.* 1960) and animals (Cohen & Fridovich 1971*a,b*; Cohen *et al.* 1971). It catalyses the following reaction:

$$SO_3^{2-} + H_2O \longrightarrow SO_4^{2-} + 2H^+ + 2e^-$$

The enzyme contains Mo and haem Fe in the ratio of 1:1 and, in addition, haem-iron is also implicated in electron transfer from the reduced enzyme to oxygen as follows:

$$\begin{array}{l} \text{sulphite} \longrightarrow \underset{\downarrow}{\text{Mo}} \longrightarrow \text{haem} \longrightarrow O_2 \\ \qquad\qquad\quad \text{ferricyanide} \\ \qquad\qquad\quad \textit{or}\ \text{cytochrome } c \end{array}$$

Evidence from electron-spin resonance suggests that the Mo is involved in a one-electron change from valence states V to VI.

COBALT

Cobalt is essential for N_2 fixation by free-living bacteria, blue-green algae and symbiotic systems (Evans *et al.* 1965; Nicholas *et al.* 1962). In root nodules the metal requirement is usually associated with the production of cobamide compounds by the rhizobia. Cobamide coenzymes are required for metabolic processes of the bacteroids, especially for the enzyme methyl malonyl CoA isomerase (Evans & Kliewer 1964; Evans *et al.* 1965). It is now known that cobamide compounds are involved in several enzymic reactions involving group transfer. Some examples of these are:-

Methylmalonyl-CoA isomerase

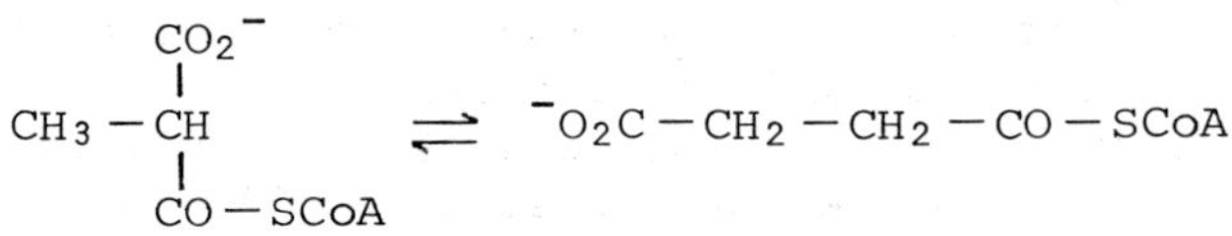

Glutamate mutase

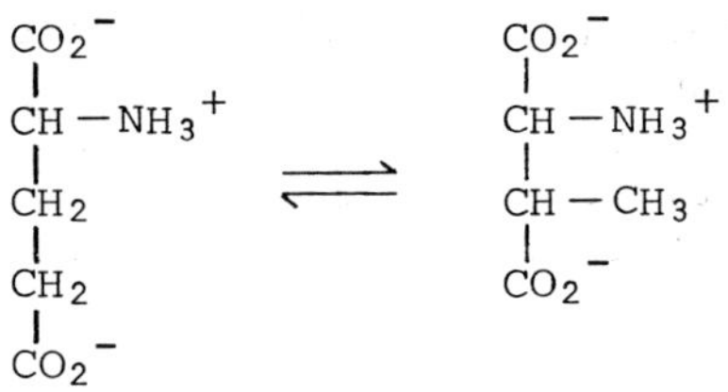

Glycerol dehydratase

$HO-CH_2-CH(OH)-CH_2-OH \longrightarrow HO-CH_2-CH_2-CHO$

Diol dehydratase

$CH_3-CH(OH)-CH_2-OH \longrightarrow CH_3-CH_2-CHO$

$HO-CH_2-CH_2-OH \longrightarrow CH_3-CHO$

Ethanolamine deaminase

$H_3N^+-CH_2-CH_2-OH \longrightarrow CH_3-CHO + NH_4^+$

β-*Lysine mutase*

$H_3N^+-CH_2-[CH_2]_3-CH(NH_3^+)-CO_2^-$

$\longrightarrow CH_3-CH(NH_3^+)-[CH_2]_2-CH(NH_3^+)-CO_2^-$

The current status of the metabolism of cobamide coenzymes has been discussed recently by Abeles (1973). Although the enzyme reactions above are diverse, they all involve H transfer concomitant with group migration, which may be illustrated as follows:

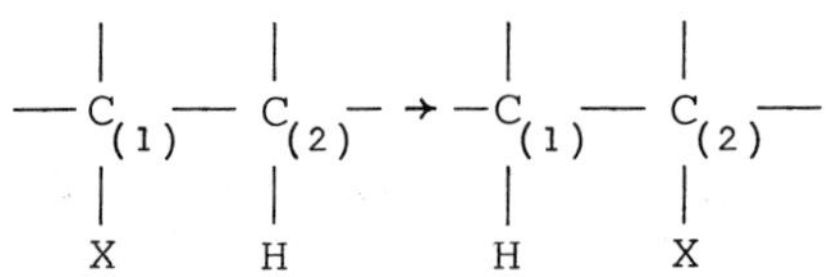

The migration of the H atom can be readily followed by isotopic labelling. Thus, when the vitamin B_{12} coenzyme containing labelled ^{3}H in the C-5' position is added to the apoenzyme, the label is transferred to the reaction products. The partial structure of vitamin B_{12} coenzyme is as follows:

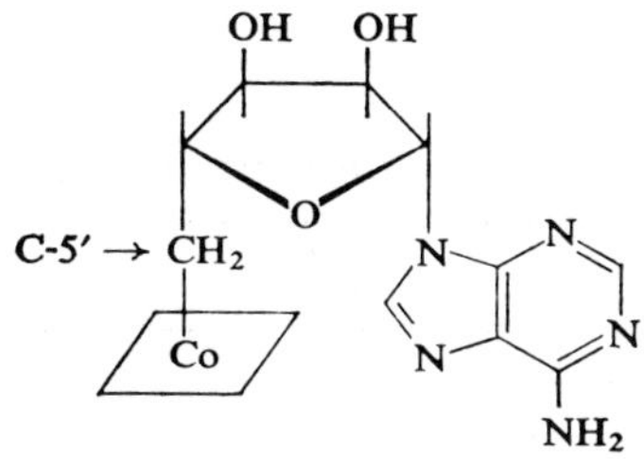

It is now clear that both C-5' hydrogen atoms, although chemically non-equivalent, participate in the reaction. Essenberg *et al.* (1971) consider that vitamin B_{12} coenzyme functions as an intermediate H carrier in all these reactions.

Relatively little is known about the effect of the micronutrient on the non-nodulated plants. A Co deficiency based on growth measurements has been reported in rubber and tomatoes (Bolle-Jones & Mallikarjuneswara 1957) and in non-nodulating subterranean clover (Hallsworth *et al.* 1965; Wilson & Hallsworth 1965), but symptoms were not recorded. Traces of cobamide coenzymes have been detected in plants, but it is not known whether these compounds might have originated from microorganisms associated with the plants (Fries 1962).

Cobalt deficiency has been produced in non-nodulated subterranean clover, *Trifolium subterraneum*, grown with nitrate or urea nitrogen, and in wheat, *Triticum durum*, supplied with nitrate (Wilson & Nicholas 1967). These plants were grown under aseptic conditions in highly purified culture solutions. Cobamide coenzymes were not detected in these plants. Indeed infiltration of ^{60}Co-labelled cyanocobalamin was degraded very readily by the plants. Over 90% of a ^{60}Co-labelled salt supplied to the plants was associated with small molecular weight complexes that were separated by gel filtration techniques (Wilson & Nicholas 1967).

ZINC

Knowledge of the functional role of Zn as a component of an enzyme is incomplete. It differs in many respects from Fe. Thus Zn enzymes are colourless, and the atom has no unpaired electrons at ambient temperature so that it has one stable state in solution. The metal is usually tightly bound to the apoenzyme and can be removed only with severe chemical treatment. It forms strong complexes with radicals of polar groups containing oxygen, nitrogen and sulphur. The main Zn-containing enzymes are listed in Table 3.

TABLE 3

Some enzymes containing or requiring Zn^{2+}, Cu^{2+} and Mn^{2+}

Zn^{2+}	Cu^{2+}	Mn^{2+}
Glutamic dehydrogenase	Tyrosinase	Arginase
Alcohol dehydrogenase	Laccase	Phosphotransferase
Lactic dehydrogenase	Cytochrome oxidase	
Carbonic anhydrolase	Plastocyanin	
Carboxypeptidase		
DNA polymerase I		

Carbonic anhydrase from bovine erythrocytes was first shown to contain Zn by Keilin & Mann (1940); this enzyme, which equilibrates CO_2 with H_2CO_3, is also present in plants, although its function in chloroplasts is not clear. The enzyme, which has a molecular weight of about 30,000, contains 0.2-0.3% Zn.

Numerous dehydrogenases contain Zn (Vallee 1955,1960). Thus alcohol dehydrogenase (ADH) from yeast, with a molecular weight of 150,000, contains 0.18% Zn. It appears that four atoms of Zn bind nicotinamide adenine dinucleotide (NAD) to the enzyme. Zinc is firmly bound to the ADH protein and NAD is reversibly bound to Zn:

$$[(ADH)Zn_4] + 4(NAD) \longrightarrow [(ADH)Zn_4]\ (NAD)_4$$

Other dehydrogenases, including the NAD-linked glutamic dehydrogenase and lactic dehydrogenase, are Zn metallo-enzymes.

Deoxyribonucleic acid polymerase I from *E. coli* contains 0.15 g atom of Zn per mole of enzyme. Enzyme-bound Zn^{2+} probably functions by coordinating the 3'-OH group of the terminal deoxyribose of the growing strand of DNA; Zn^{2+} would facilitate the deprotonation of this group, an essential step in the nucleotidyl transfer reaction which elongates the primer (Springgate *et al.* 1973). The reverse transcriptase has also been found to contain Zn^{2+} (Stengle & Baldeschwieler 1966). These findings are relevant to nutritional-type experiments where a deficiency of Zn^{2+} interferes with DNA replication *in vivo.*

COPPER

Some of the enzymes containing Cu are listed in Tables 3 & 4. The oxidases catalysing reactions that reduce both atoms of molecular oxygen to water, namely cytochrome *c* oxidase, ascorbate oxidase and laccase, all contain Cu (Beinert & Palmer 1965). Plastocyanin (E_0' +0.30 volts), a blue non-autoxidizable Cu-containing protein present to the extent of 1 molecule/400 chlorophyll molecules, is a component of the photosynthetic electron transfer chain; this will be considered later by Dr. Boardman. A summary of some of the main characteristics of these Cu-containing proteins are given in Table 4.

The catalytic unit of cytochrome *c* oxidase is thought to contain two distinct forms of Cu in addition to cytochrome *a*. Evidence from electron spin resonance measurements suggests that Cu undergoes a valence change during electron transfer to O_2. Malkin & Malmström (1970) suggest that there are various forms of Cu present in laccase, i.e. two Cu^{2+} types as well as a diamagnetic form consisting of a strongly coupled Cu^{2+}-Cu^{2+} pair. They propose that reducing equivalents enter the enzyme molecule through Cu^{2+} and thence to the two-electron acceptor Cu^{2+}-Cu^{2+} pair to O_2.

TABLE 4

Properties of some Cu-containing proteins

Proteins	Source	Molecular weight	% Cu in protein	Atoms metal/MW
Ascorbic acid oxidase	*Cucumis sativus* (squash)	146,000	0.26	6
Phenol oxidase	*Psalliota campestris* (mushroom)	100,000	0.25	4
Laccase	*Rhus vernicifera*	60,000	0.25	4
Ceruloplasmin	Mammalian blood	160,000	0.34	8

MANGANESE

Brief mention only will be made of the role of Mn in plant metabolism since this will be discussed more fully by Dr.

Boardman. In Table 3 are listed two Mn-dependent enzymes, arginase and phosphotransferase, and it is also known that this element substitutes for Mg in many of the ATP-dependent enzymes of glycolysis. In addition, Mn appears to participate in the O_2-evolving system of photosynthesis and there are reports that the element is required for glycollate formation in green algae (Tanner *et al.* 1960). The electron spin resonance (ESR) spectrum of Mn^{2+} is very characteristic and is readily recognized in biological materials by its six hyperfine lines centred around $g = 2$ (Beinert & Palmer 1965). ESR measurements have been used to determine the binding constants of Mn^{2+} to enzymes such as creatine and pyruvic kinase, enolase and 3-phosphoglycerate kinase.

A very thorough study has been made by Cohn (1963) of kinase enzyme systems using Mn^{2+} as a probe to study the enzyme-metal-substrate interactions of these systems. The most interesting results were obtained by studying the influence of free and bound Mn on the proton relaxation rate of the nuclear resonance of water.

VANADIUM

Bortels (1933) proposed that V can substitute for Mo in the fixation of dinitrogen by *Azotobacter* species. Arnon & Wessel (1953) claimed that V was essential for the growth of the alga *Scenedesmus obliquus*. Subsequently Allen (1956) and Holm-Hansen (1968) were unable to establish a V requirement for nitrogen fixation in blue-green algae. It is likely that the early effects for V reported by Bortels (1933) might have been due to a contamination of Mo in the relatively impure V salts used at that time. Recently it has been shown that V can replace Mo in the nitrogenase enzyme, but under these conditions the fixation of dinitrogen was only about 10% of the Mo-containing enzyme (McKenna *et al.* 1970; Burns *et al.* 1971; Benemann *et al.* 1972). Vanadium has been found to exert an inhibitory effect on the fixation of dinitrogen by the blue-green alga *Anabaena cylindrica* and its presence in Mo-deficient cultures enhanced Mo deficiency (Fay & de Vasconcelos 1974).

SODIUM

As early as 1860, before agricultural chemistry was a defined science, many believed Na to be an essential element for plant growth. In subsequent years, however, Na was regarded as an element that could partially substitute for K in many plants, e.g. sugarbeet. In 1955, Allen & Arnon showed that Na was essential for the blue-green alga, *Anabaena cylindrica*, and Brownell & Wood (1957), here in Adelaide, demonstrated its essentiality for a higher plant, *Atriplex vesicaria* of the family *Chenopodiaceae*.

The functions of Na in plant metabolism are not known with any certainty. Experiments with radioactive ^{24}Na indicate that the element is present in various combined forms in plants; one form is readily exchangeable with all cations, whereas the other type exchanges with Na but only with difficulty with Ca, K and Mg. Brownell & Nicholas (1967) found that a Na deficiency in *Anabaena cylindrica* resulted in a higher nitrate reductase activity and that nitrite accumulation in the culture solution resulted in chlorosis. By contrast, N_2 fixation was reduced in Na-deficient cells.

Brownell & Crossland (1972) found that plants utilizing the C_4 dicarboxylic acid pathway of CO_2-fixation respond to Na. This pathway is restricted to a few diverse angiosperm families and involves a variant of the Calvin cycle, whereby $^{14}CO_2$ is initially incorporated in the carbon-4 of the C_4 acids, oxalacetate, malate and aspartate (Hatch & Slack 1970; Hatch 1974). This fixed carbon is then released as CO_2 and refixed, in accordance with the operation of the Calvin cycle. The precise role of Na in this pathway remains to be elucidated.

CONCLUSIONS

An understanding of the functions of trace elements in metabolism, whether associated with microorganisms, plants or animals, has advanced considerably during the last ten years. It is clear that trace elements function in key metabolic events such as respiration, photosynthesis, N_2 fixation and the assimilation of nitrates and sulphates. The main effort, of course, has concentrated on the transition metals which are involved in oxidation-reduction reactions in living systems. The main reason for this has been a renaissance in a basic understanding of the chemical structures and mechanism of action of metals and their coordination compounds. These new concepts and techniques have flowed on to biological systems. In particular, new spectroscopic methods, including electron spin resonance, Mössbauer and nuclear magnetic resonance, have been effectively employed to study the mechanism of electron transfer in metallo-enzyme systems.

Despite these advances there are many challenging gaps in

our knowledge of trace element function. Notably these include the non-transition metals, Na, B and Cl, since as yet there is no definitive information about their functions in plant metabolism.

REFERENCES

ABELES, R.H. (1973) - *Biochem. Soc. Trans. 1* : 39

ALLEN, M.B. (1956) - *Sci. Monthly 83* : 100

ALLEN, M.B. & ARNON, D.I. (1955) - *Physiologia Pl. 8* : 653

ARNON, D.I. & WESSEL, G. (1953) - *Nature, Lond. 172* : 1039

ARRIGONI, O. (1959) - *Ital. J. Biochem. 7* : 181

BEINERT, H. & PALMER, G. (1965) - *Adv. Enzymol. 27* : 105

BENDALL, D.S. & HILL, R. (1968) - *A. Rev. Pl. Physiol. 19* : 167

BENEMANN, J.R., McKENNA, C.E., LIE, R.F., TAYLOR, T.G. & KAMEN, M.D. (1972) - *Biochim. biophys. Acta 264* : 25

BOLLE-JONES, E.W. & MALLIKARJUNESWARA, V.R. (1957) - *J. Rubb. Res. Inst. Malaya 15* : 128

BORTELS, H. (1933) - *Zbl. Bakt., II. Abt. 87* : 476

BROWNELL, P.F. & CROSSLAND, C.J. (1972) - *Pl. Physiol. 49* : 794

BROWNELL, P.F. & NICHOLAS, D.J.D. (1967) - *Pl. Physiol. 42* : 915

BROWNELL, P.F. & WOOD, J.G. (1957) - *Nature, Lond. 179* : 635

BUCHANAN, B.B. & ARNON, D.I. (1970) - *Adv. Enzymol. Relat. Areas Mol. Biol. 33* : 119

BURNS, R.C., FUCHSMAN, W.H. & HARDY, R.W.F. (1971) - *Biochem. biophys. Res. Commun. 42* : 353

COHEN, H.J. & FRIDOVICH, I. (1971*a*) - *J. biol. Chem. 246* : 359

COHEN, H.J. & FRIDOVICH, I. (1971*b*) - *J. biol. Chem. 246* : 367

COHEN, H.J., FRIDOVICH, I. & RAJAGOPALAN, K.V. (1971) - *J. biol. Chem. 246* : 374

COHN, M. (1963) - *Biochemistry 2* : 623

DEERING, J.V. & NICHOLAS, D.J.D. (1974) - *Proc. Aust. Biochem. Soc. 7* : 52

EADY, R.R. & POSTGATE, J.R. (1974) - *Nature, Lond. 249* : 805

EARNSHAW, A. & HARRINGTON, T.J. (1973) - "The Chemistry of the Transition Elements". Clarendon Press : Oxford

ESSENBERG, M.K., FREY, P.A. & ABELES, R.H. (1971) - *J. Am. chem. Soc. 93* : 1242

EVANS, H.J. & KLIEWER, M. (1964) - *Ann. N.Y. Acad. Sci. 112* : 735

EVANS, H.J., RUSSELL, S.A. & JOHNSON, G.V. (1965) - *In* "Non-haem Iron Proteins: Role in Energy Metabolism" (A.S. Pietro, ed.), p. 303. Antioch Press : Ohio

FAY, P. & DE VASCONCELOS, L. (1974) - *Arch. Mikrobiol. 99* : 221

FRIES, L. (1962) - *Physiologia Pl. 15* : 566

FROMAGEOT, P., VAILLANT, R. & PEREZ-MILAN, H. (1960) - *Biochim. biophys. Acta 44* : 77
GARRETT, R.H. & NASON, A. (1969) - *J. biol. Chem. 244* : 2870
HALLSWORTH, E.G., WILSON, S.B. & ADAMS, W.A. (1965) - *Nature, Lond. 205* : 307
HARDY, R.W.F., BURNS, R.C. & PARSHALL, G.W. (1971) - *Adv. Chem. Ser. 100* : 219
HATCH, M.D. (1974) - *Proc. Aust. Biochem. Soc. 7* : 1
HATCH, M.D. & SLACK, C.R. (1970) - *A. Rev. Pl. Physiol. 21* : 141
HEMPFLING, W.P., TRUDINGER, P.A. & VISHNIAC, W. (1967) - *Arch. Mikrobiol. 59* : 149
HEWITT, E.J. (1959) - *Biol. Rev. 34* : 333
HEWITT, E.J. (1966) - Commonwealth Bur. Hort. Tech. Comm. No. 22, East Malling, England, p. 156
HOLM-HANSEN, O. (1968) - *A. Rev. Microbiol. 22* : 47
HUCKLESBY, D.P. & HEWITT, E.J. (1966) - *Biochem. J. 119* : 615
IKUMA, H. (1972) - *A. Rev. Pl. Physiol. 23* : 419
JENSEN, L.H., SIEKER, L.C., WATENPAUGH, K.D., ADMAN, E.T. & HERRIOTT, J.R. (1973) - *Biochem. Soc. Trans. 1* : 27
JOY, K.W. & HAGEMAN, R.H. (1966) - *Biochem. J. 100* : 263
KEILIN, D. & MANN, T. (1940) - *Biochem. J. 34* : 1163
LEIGH, G.J. (1971) - *In* "Chemistry and Biochemistry of Nitrogen Fixation" (J.R. Postgate, ed.), p. 19. Academic Press : New York
MALKIN, R. & MALMSTRÖM, B.G. (1970) - *Adv. Enzymol. Relat. Areas Mol. Biol. 33* : 177
McKENNA, C.E., BENEMANN, J.R. & TAYLOR, T.G. (1970) - *Biochem. biophys. Res. Commun. 41* : 1501
MURPHY, M.J., SIEGEL, L.M., TOVE, S.R. & KAMIN, H. (1974) - *Proc. natn. Acad. Sci. U.S.A. 71* : 612
NASON, A. (1963) - *In* "The Enzymes" (P.D. Boyer, H. Lardy & K. Myrback, eds.), Vol. 7, p. 587. Academic Press : New York
NICHOLAS, D.J.D. (1959) - *In* "Utilization of Nitrogen and Its Compounds by Plants" (H.K. Porter, ed.), Vol. 8, p. 1. Cambridge University Press : Cambridge
NICHOLAS, D.J.D. (1961) - *A. Rev. Pl. Physiol. 12* : 63
NICHOLAS, D.J.D. (1963) - *Biol. Rev. 38* : 530
NICHOLAS, D.J.D., KOBAYASHI, M. & WILSON, P.W. (1962) - *Proc. natn. Acad. Sci. U.S.A. 48* : 1537
NICHOLAS, D.J.D. & NASON, A. (1954*a*) - *J. biol. Chem. 207* : 353
NICHOLAS, D.J.D. & NASON, A. (1954*b*) - *J. biol. Chem. 211* : 183
NICHOLAS, D.J.D., NASON, A. & McELROY, W.D. (1954) - *J. biol. Chem. 207* : 341
NICHOLAS, D.J.D., SILVESTER, D.J. & FOWLER, J.F. (1961) - *Nature, Lond. 189* : 634

NICHOLAS, D.J.D. & STEVENS, H.M. (1955) - *Nature, Lond.* *176* : 1066

NICHOLAS, D.J.D. & STEVENS, H.M. (1956) - *In* "Inorganic Nitrogen Metabolism" (W.B. McElroy & B. Glass, eds.), p. 178. Johns Hopkins University Press : Baltimore

NOTTON, B.A. & HEWITT, E.J. (1971) - *Biochem. biophys. Res. Commun.* *44* : 702

PANEQUE, A., RAMIREZ, J.M., DEL CAMPO, F.F. & LOSADA, M. (1964) - *J. biol. Chem.* *239* : 1737

PECK, H.D., Jr. (1961) - *Biochim. biophys. Acta* *49* : 621

RAMIREZ, J.M., DEL CAMPO, F.F., PANEQUE, A. & LOSADA, M. (1966) - *Biochim. biophys. Acta* *118* : 58

RITCHIE, A.I.M. (1968) - *Nucl. Instr. Meth.* *64* : 181

SPRINGGATE, C.F., MILDVAN, A.S., ABRAMSON, R., ENGLE, J.L. & LOEB, L.A. (1973) - *J. biol. Chem.* *248* : 5987

STANLEY, P.E. & NICHOLAS, D.J.D. (1969) - *Atom. Energy in Australia* *12* : 11

STENGLE, T.R. & BALDESCHWIELER, J.D. (1966) - *Proc. natn. Acad. Sci. U.S.A.* *55* : 1020

STIEFEL, E.J. (1973) - *Proc. natn. Acad. Sci. U.S.A.* *70* : 988

STOREY, B.T. (1970) - *J. Theor. Biol.* *28* : 233

TAGER, J.M. & RAUTANEN, N. (1955) - *Biochim. biophys. Acta* *18* : 111

TANNER, H.A., BROWN, T.E., EYSTER, C. & TREHAME, R.W. (1960) - *Biochem. biophys. Res. Commun.* *3* : 205

VALLEE, B.L. (1955) - *Adv. Protein Chem.* *10* : 318

VALLEE, B.L. (1960) - *In* "The Enzymes" (P.D. Boyer, H. Lardy & K. Myrback, eds.), Vol. 3, p. 497. Academic Press : New York

WALLACE, T. (1961) - "The Diagnosis of Mineral Deficiencies in Plants by Visual Symptoms". H.M. Stationery Office : London

WILLIAMS, R.J.P. (1973) - *Biochem. Soc. Trans.* *1* : 1

WILSON, S.B. & HALLSWORTH, E.G. (1965) - *Pl. Soil* *22* : 260

WILSON, S.B. & NICHOLAS, D.J.D. (1967) - *Phytochemistry* *6* : 1057

TRACE ELEMENTS IN PHOTOSYNTHESIS

N.K. Boardman

Division of Plant Industry, CSIRO
Canberra, A.C.T. 2601, Australia

INTRODUCTION

Photosynthesis is divisible into two main phases : a light phase in which quanta of radiation are absorbed by chlorophyll and other photosynthetic pigments and converted into chemical energy in the form of adenosine triphosphate (ATP) and the reduced form of nicotinamide adenine dinucleotide phosphate (NADPH); and a dark phase in which ATP and NADPH are used for the conversion of CO_2 to carbohydrate. The chlorophylls and carotenoids are localized in the internal membranes (termed thylakoids) of the chloroplast, and the light reactions of photosynthesis are associated with these membranes. Electrons are driven from water to $NADP^+$ *via* an electron transport chain and oxygen is evolved, according to equation 1. ATP formation accompanies the electron flow (equation 2).

$$2NADP^+ + 4H_2O \rightarrow 2NADPH + 2H_2O + O_2 + 2H^+ \quad \ldots (1)$$

$$ADP + P_i \rightarrow ATP \quad \ldots (2)$$

The trace elements Fe, Cu, Mn and Cl are essential components of the electron transport chain. Iron is complexed to porphyrin and protein in the cytochromes, and to sulphur and protein in the ferredoxins. Copper is present in the chain as a protein complex. Little is known about the nature of the manganese complex, although present evidence suggests that a protein is involved. Chlorine is essential for the water-splitting sequence of reactions, but its specific role has not been elucidated.

TRACE ELEMENT COMPLEXES

Cytochromes

The early work of Hill and co-workers (cf. Bendall *et al.* 1971) showed that chloroplasts contain a *c*-type cytochrome (cytochrome *f*) with an α-band at 554 nm and a midpoint potential of +0.36 volts at pH 7, and an autoxidizable *b*-type cytochrome (cytochrome b_6) with an α-band at 563 nm and a potential below zero. Subsequent studies (Lundegardh 1962; Boardman & Anderson 1967) established the presence in chloroplasts of a non-autoxidizable *b*-type cytochrome with an α-band at 559 nm and a potential close to that of cytochrome *f*. Recently, it has been shown that chloroplasts also contain an autoxidizable *b*-type cytochrome with an α-band at 559 nm (Bendall 1968). Its redox potential is somewhat more positive than that of cytochrome b_6, but considerably more negative than the non-autoxidizable cytochrome (Table 1). Fractiona-

TABLE 1

Properties of chloroplast cytochromes

	α-peak (nm)	E_0' (Volt)
Cytochrome *f* (*c*-type)	554	+0.36
Cytochrome b_6 (*b*-563)	563	-0.18
Cytochrome *b*-559, high potential form	559	+0.37
Cytochrome *b*-559, low potential form	559	+0.065

tion of chloroplasts with detergents has shown that cytochromes *f*, b_6 and *b*-559 (low potential) are associated with photosystem 1, whereas cytochrome *b*-559 (high potential) is localized in photosystem 2 (Boardman & Anderson 1967; Anderson & Boardman 1973). None of the chloroplast cytochromes is extracted by aqueous solvents. Cytochrome *f* is extracted into ammoniacal ethanol (Bendall *et al.* 1971), but the *b*-cytochromes are tightly associated with the thylakoid membranes and they are only released on disruption of the membranes with detergents. After purification, cytochrome *f* has a molecular weight of 250,000 and contains 4 haem groups per protein molecule (Forti *et al.* 1965). Cytochrome *f*, extracted by a detergent method, showed a molecular weight of 31,000 or 62,000, with an average of 1 haem per 62,000-molecule (Singh & Wasserman 1971), which suggests that it readily aggregates in aqueous media.

Recently, Wasserman and co-workers have also reported on

the extraction and purification of cytochrome *b*-559_{HP} and cytochrome b_6 in the presence of a Triton X-100-urea mixture (Stuart & Wasserman 1973; Garewal & Wasserman 1974). Electrophoretically homogeneous preparations of cytochrome *b*-559 contained 56% lipid, including chlorophyll *a* and β-carotene, and 46,000 g protein/mole of haem. Each molecule appears to contain eight polypeptide chains of molecular weight 5600 ± 1000. Cytochrome b_6 migrated as a single band on disc gel electrophoresis, but its spectral properties at liquid nitrogen temperature suggest that it may have contained some low potential cytochrome *b*-559 (Anderson & Boardman 1973). Cytochromes *b*-559 and b_6 are typical of membrane proteins and they readily aggregate on removal of the detergent, Triton X-100.

Ferredoxins

Ferredoxins are stable Fe-S proteins, but unlike the cytochromes they do not contain haem. The Fe is coordinated to the S atoms of cysteine and also to inorganic S (Fig. 1).

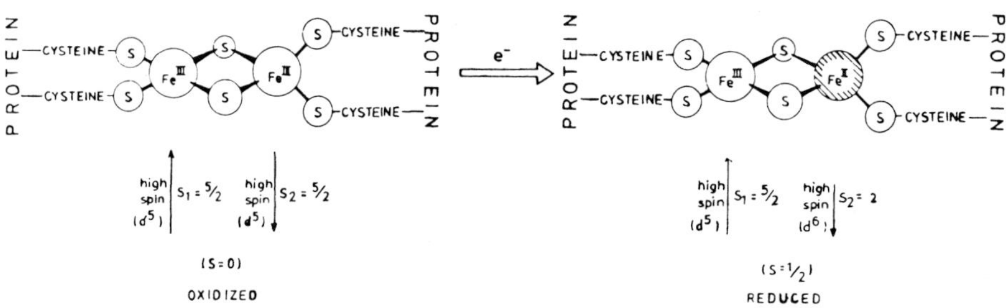

Fig. 1. Proposed model of the Fe-S group in soluble ferredoxin. [Reproduced by kind permission from Hall et al. *1972*]

Chloroplasts contain a soluble ferredoxin which is readily released from the chloroplasts in aqueous media, and one or more membrane-bound ferredoxins. Soluble ferredoxin has a molecular weight of 12,000 and contains 2 atoms of Fe and 2 atoms of inorganic sulphide per molecule (Arnon 1967). Soluble ferredoxin acts as a one-electron carrier, and only one of the two Fe^{3+} ions is reduced by illuminated chloroplasts. Soluble ferredoxin has a redox potential of -0.43 volts at pH 7.5, about 0.1 volts more negative than the pyridine nucleotides.

The presence in chloroplasts of one or more bound ferredoxins was established by the use of electron paramagnetic resonance (EPR) at very low temperatures (Malkin & Bearden 1971; Evans *et al.* 1972; Ke *et al.* 1973). Malkin *et al.*

(1974) extracted and purified bound ferredoxin, and showed that it had a molecular weight of 8000 with 4 atoms of Fe and 4 atoms of inorganic sulphide per molecule. It differs from soluble ferredoxin in its absorption spectrum and its redox potential is considerably more negative (Table 2).

TABLE 2

Properties of chloroplast ferredoxins

	Soluble ferredoxin	Bound ferredoxin
Molecular weight	12,000	8,000
Fe (per mole)	2	4
Major spectral bands (nm)	280, 330, 420, 463	280, 400
Redox potential (volt)	-0.43	-0.53

Plastocyanin

Plastocyanin, the Cu-containing protein of chloroplasts, is readily extracted on sonication of chloroplasts or by detergents. It has been purified from several higher plants and shown in most instances to have a molecular weight of about 10,500 and to contain 1 atom of Cu per molecule (Ramshaw *et al.* 1973). Plastocyanin from spinach has a molecular weight of 21,000 with 2 Cu atoms per molecule (Katoh *et al.* 1962). Each Cu atom appears to be bound to a sulphydryl group of cysteine and possibly to another amino acid. The oxidized form of plastocyanin is intensely blue with a main absorption band at 597 nm, and it shows a characteristic EPR signal at low temperatures (Blumberg & Peisach 1966). Its redox potential (+0.37 volts at pH 7) is close to those of cytochromes *f* and b-559_{HP}.

Manganese Complex

The Mn which functions in electron transport is not released by repeated washing of chloroplasts (Possingham & Spencer 1962), by treatment with a variety of chelating agents, or by extraction with lipid solvents (Cheniae & Martin 1966). It does not exchange with Mn^{2+} or MnO_4^-, and it lacks the characteristic EPR signal of Mn^{2+}. It is released from chloroplasts by digestion with the proteolytic enzyme, pronase, by chaotropic agents, low concentrations of hydroxylamine, 0.8 M Tris buffer pH 8.0, and temperature treatment at 50°C (Homann 1967; Cheniae 1970). These characteristics suggest that Mn is coordinated to a protein in the chloroplast thylakoid membrane, but attempts to isolate a Mn-protein complex have not met with any success.

ROLE OF TRACE ELEMENTS IN ELECTRON TRANSPORT

Chloroplasts contain two types of photosynthetic units [photosystem 1 (PS-1) and photosystem 2 (PS-2)] which cooperate in a sequential manner to transfer electrons from water to $NADP^+$ (Fig. 2). PS-2 generates a strong oxidant Y^+, and a weak reductant Q^-. The strong oxidant is estimated to have a redox potential of approximately +1.0 volt, which is more oxidizing than the potential of the H_2O/O_2 couple. The weak reductant with a potential of about zero volts is incapable of reducing $NADP^+$, without an input of light energy into PS-1. Quanta absorbed by PS-1 are transferred to P-700, a special complex of chlorophyll absorbing at 700 nm. On excitation, P-700 donates an electron to an acceptor Z, resulting in the formation of P-700^+ (a weak oxidant with a redox potential of +0.43 volt) and Z^- (a strong reductant with a potential in the vicinity of -0.6 volt). P-700^+ interacts with the weak reductant Q^-, generated by PS-2, *via* an electron transport chain, and the reducing power of Z^- is used to reduce $NADP^+$. Electron flow from Q^- to P-700^+ is accompanied by the formation of ATP from adenosine diphosphate and inorganic phosphate. Work in the last few years provides evidence for two sites of ATP formation on the electron transport pathway between PS-2 and PS-1 (only one site is shown in Fig. 2), in addition to a site of ATP formation on a cyclic pathway around PS-1.

Cytochrome f *and Plastocyanin*

Cytochrome *f* and plastocyanin are electron carriers on the pathway from Q^- to P-700^+. Duysens *et al.* (1961) first reported the photooxidation of cytochrome *f* in red algae by 680 nm light and photoreduction by green light absorbed by PS-2. The photooxidation of cytochrome *f* by far-red light absorbed by PS-1 was subsequently observed in a range of photosynthetic organisms, including higher plants and isolated chloroplasts. Photooxidation is also observed in red light if electron flow from PS-1 is inhibited by addition of the herbicide 3-(3,4-dichlorophenyl)-1,1-dimethylurea (DCMU). It seems likely that cytochrome *f* also functions on a cyclic pathway around PS-1 (Larkum & Bonner 1972; Biggins 1973) and that cytochrome *f* is common to both the non-cyclic and cyclic pathways.

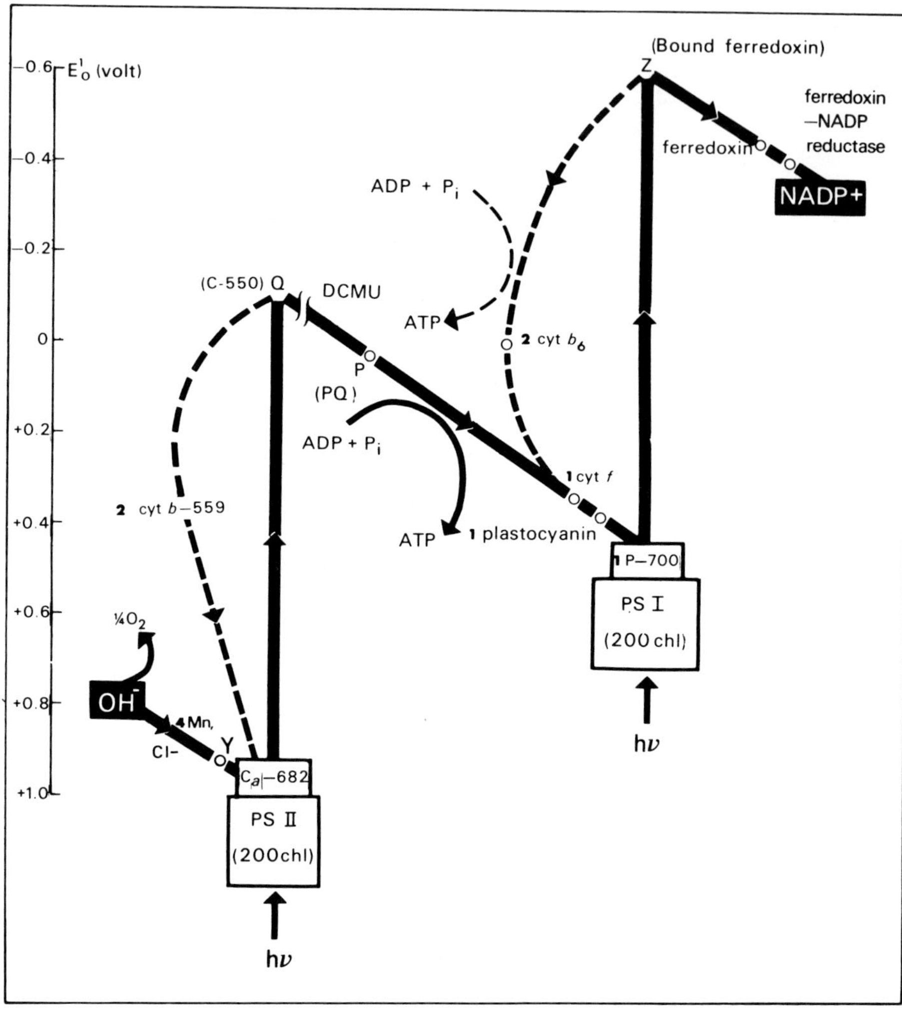

Fig. 2. Scheme for photosynthetic electron transport and photophosphorylation in chloroplasts (see text for explanation). The number beside a component indicates the number of molecules of that component per photosynthetic unit of 400 chlorophyll molecules. P appears to be identical to plastoquinone A (PQ), and Z may be identical to bound ferredoxin. Two sites of ATP formation are located on the pathway between PS-2 and PS-1, but only one is shown in the figure. A scale of redox potentials is shown on the left.

Plastocyanin, which has a redox potential *in situ* slightly more positive than cytochrome *f* (Malkin *et al.* 1973), acts at a site close to that of cytochrome *f*. Light-induced absorbance measurements at 597 nm have not provided clear evidence for the photooxidation of plastocyanin, but recent studies, using low temperature (25°K) EPR measurements to identify oxidized plastocyanin, have shown that plastocyanin is photooxidized by PS-1 and photoreduced by PS-2 (Malkin & Bearden 1973; Visser *et al.* 1974). Like cytochrome *f*, plastocyanin also appears to function in a cyclic pathway around PS-1 (Visser *et al.* 1974). The photoreduction of plastocyanin by PS-2 is inhibited by DCMU. Treatments which inactivate plastocyanin by complexing with the Cu (KCN or $HgCl_2$), or remove plastocyanin from the membrane (sonication, detergents or polyene antibiotics) or complex with it (polylysine, histone), generally inhibit the photooxidation of cytochrome *f*, as well as electron flow to $NADP^+$ from water and from artificial electron donors which feed into PS-1. A mutant of *Chlamydomonas* which lacked plastocyanin behaved in a similar way to the treated chloroplasts (Gorman & Levine 1966). In contrast, the photoreduction of cytochrome *f* is enhanced by the inactivation of plastocyanin with $HgCl_2$ (Kimimura & Katoh 1972).

Cytochromes-b

Cytochrome b_6 appears to function on a cyclic electron transport pathway around PS-1, and it may play a role in cyclic phosphorylation *in vivo* (Arnon *et al.* 1967; Böhme & Cramer 1972; Amesz *et al.* 1972; Knaff 1972). Our recent work with intact spinach chloroplasts, capable of high rates of CO_2 fixation, is consistent with the view that cytochrome b_6 functions on a cyclic pathway around PS-1, but we have observed an antagonistic effect of red *vs* far-red light on the reduction and oxidation of cytochrome b_6 in chloroplasts at 0°C under anaerobic conditions (U. Heber, N.K. Boardman & J.M. Anderson, unpublished). We conclude from this result that cyclic flow is sensitive to the relative poising of the redox levels of cytochrome b_6 and components of the non-cyclic pathway.

There are conflicting views concerning the role of cytochrome b-559_{HP}. This cytochrome is photooxidized by PS-2 at liquid nitrogen temperature in leaves and chloroplasts, and at room temperature if chloroplasts are treated with carbonyl cyanide m-chlorophenylhydrazone (CCCP), an uncoupler of phosphorylation, or by treatments which inhibit electron flow between water and PS-2 (Knaff & Arnon 1969; Boardman *et al.* 1971; Hiller *et al.* 1971). These results suggest that cytochrome b-559_{HP} is either on a side pathway from PS-2 or on a cycle around PS-2 (Fig. 2). On the other hand, Cramer *et al.* (1971) concluded that CCCP converts cytochrome b-559_{HP} to a form of lower potential which then functions as an electron carrier between the photosystems, being oxidized by PS-1 and reduced by PS-2. In our recent work with intact chloroplasts, we observed an efficient photooxidation of cytochrome b-559_{HP} by PS-2 at room temperature in the presence of carbonyl cyanide-p-trifluoromethoxyphenyl hydrazone (FCCP). Light absorbed by PS-1 was much less efficient. The quantum requirement in red light (655 nm) was 5.7 per molecule of cytochrome b-559_{HP} oxidized, compared with 142 in far-red light (714 nm). We have also studied the synthesis of cytochrome b-559_{HP} during greening of dark-grown barley seedlings. Cytochrome b-559_{HP} is absent from the plastids of etiolated tissue, and its synthesis during greening does not correlate with the onset of oxygen evolution, which precedes the synthesis of the cytochrome (Henningsen & Boardman 1973). Cytochrome b-559_{HP} does not appear to be essential for electron flow from water to $NADP^+$. Its function in mature chloroplasts may be concerned with the removal of excess oxidizing power from PS-2 under conditions of high light intensity.

The function of cytochrome b-559_{LP} in photosynthesis is unknown at present.

Ferredoxins

Soluble ferredoxin mediates the transfer of electrons from PS-1 to $NADP^+$ (Arnon 1967). If ferredoxin is added to chloroplasts in the absence of $NADP^+$, a photoreduction of ferredoxin is observed. Addition of $NADP^+$ results in a rapid reoxidation of ferredoxin, which is inhibited by the antibody to the flavoprotein, ferredoxin-$NADP^+$ reductase (Avron & Chance 1966).

EPR spectroscopy indicated that bound ferredoxin is photoreduced at very low temperatures (25°K) both in red and far-red light. PS-1 subchloroplast fragments, prepared either by the digitonin procedure or by passage of chloroplasts through the French Press, were found to be enriched in bound ferredoxin, and to show an enhanced EPR signal on illumination with far-red light (Bearden & Malkin 1972; Evans *et al.* 1972). Because of the temperature insensitivity of the photoreduction of bound ferredoxin, Bearden & Malkin (1972) proposed that bound ferredoxin is the primary electron acceptor of PS-1. However, the situation is complicated by the finding that PS-1 subchloroplast particles contain two or three bound ferredoxins with slightly different redox potentials (Evans *et al.* 1974; Ke 1974).

Manganese

Pirson (1937) first showed in *Ankistrodesmus* that Mn deficiency decreased photosynthesis considerably without affecting to a significant extent chlorophyll content or respiration. Kessler (1955) found that Mn deficiency had no effect on the photoreduction of hydrogen gas by *Ankistrodesmus* and concluded that Mn is required for O_2 evolution. Spencer & Possingham (1960,1961) isolated chloroplasts from Mn-deficient higher plants. Electron transport from water was inhibited by more than 50%, but cyclic phosphorylation was not affected. Possingham & Spencer (1962), from a study of the light saturation characteristics of Mn-deficient chloroplasts, concluded that Mn deficiency causes an impairment in the water-splitting sequence of reactions.

Cheniae & Martin (1968) measured the yield of O_2 per flash of light in Mn-deficient *Scenedesmus* and found that the O_2 yield was related linearly over a 20-fold range with the amount of bound Mn. From their measurements, they calculated that there are 3-4 functional Mn atoms per PS-2 photosynthetic unit, which is in agreement with the analyses of Anderson *et al.* (1964) for chloroplasts and PS-2 subchloroplast fragments of spinach.

Good evidence for locating the site of Mn on the electron pathway from water to PS-2 has come from the use of artificial donors which feed electrons to PS-2. Yamashita & Butler (1968) showed that treatment of chloroplasts with 0.8 M Tris buffer (pH 8.0) inhibits O_2 evolution, but electron flow to $NADP^+$ could be restored by addition of electron donors, including ascorbate, p-phenylenediamine and hydroxylamine. Tris treatment is effective in removing at least two-thirds of the bound Mn (Homann 1968; Cheniae 1970). Heath & Hind (1969) observed that hydroxylamine restored electron flow in Mn-deficient *Euglena* chloroplasts to the control level provided the chloroplasts were not too deficient.

Manganese deficiency has been shown to have an effect on chloroplast structure, resulting in fewer grana and disorganization of intergrana lamellae (Possingham *et al.* 1964), but it seems unlikely that the loss of O_2 evolution capacity in Mn-deficient plants can be correlated with structural changes in the chloroplast.

Restoration of O_2 evolution in Mn-deficient algae requires light (Cheniae & Martin 1966,1969; Homann 1967). An action spectrum for photorestoration, as well as the action of inhibitors such as DCMU, indicated that restoration is driven by PS-2 (Gerhardt & Wiessner 1967).

Chloride

Warburg & Lüttgens (1946) first showed that photosynthetic oxygen evolution by chloroplast fragments is dependent on chloride ions, and they concluded that Cl^- was essential for photosynthetic electron flow. Subsequent work (Gorham & Clendenning 1952; Bové *et al.* 1963; Hind *et al.* 1969) cast some doubt on this conclusion because a consistent effect of Cl^- on photosynthetic electron flow was observed only at alkaline pH. Izawa *et al.* (1969) reported a reproducible effect of Cl^- depletion on electron transport over a range of pH (5.7-8.3) if the chloroplasts were first washed with EDTA, a treatment which results in a loss of the capacity of the chloroplasts to make ATP.

The studies of Izawa *et al.* (1969) localized the site of action of Cl^- in the water-splitting sequence of reactions close to the functional site of Mn. Electron flow from donors to PS-2, such as ascorbate and hydroxylamine, was not affected by Cl^- depletion of chloroplasts, in contrast to electron flow from water, which was inhibited 4-10 fold. Fluorescence studies on isolated chloroplasts confirmed that Cl^- acts on the oxidizing side of PS-2 (Izawa *et al.* 1969; Cheniae 1970).

POSSIBLE ROLE OF TRACE ELEMENTS IN DARK PHASE OF PHOTOSYNTHESIS

Zinc is an essential element of carbonic anhydrase, the enzyme which catalyses the reversible reaction of CO_2 with water.

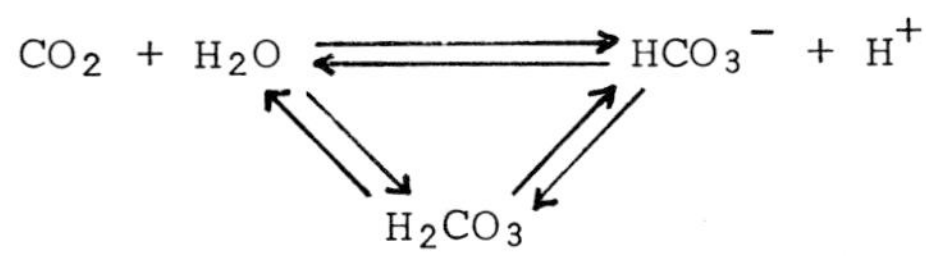

Enns (1967) and Suchdeo & Schultz (1974) showed that carbonic anhydrase facilitates the diffusion of CO_2 across liquid artificial membranes. Carbonic anhydrase is present in plants, and it has been suggested that it is involved in photosynthesis by facilitating the diffusion of CO_2 through the liquid phase of the cell to the chloroplast (Hatch & Slack 1970). Since CO_2 is the substrate for the carboxylation enzyme (ribulose diphosphate carboxylase), carbonic anhydrase could also act in photosynthesis by catalysing the dehydration of carbonic acid at the site of carboxylation.

When *Chlorella* or *Chlamydomonas* is grown in high CO_2 concentrations (5% in air), the cells lack carbonic anhydrase (Graham *et al.* 1971). On transfer to low CO_2 (0.03% in air), the cells exhibit very poor rates of photosynthesis. An induction period of up to 2 hr is needed to establish normal photosynthetic rates and during this time there is a rapid increase in the level of carbonic anhydrase. Everson (1970) found that CO_2 fixation in isolated spinach chloroplasts was partially inhibited by Diamox (5-acetamido-1,3,4-thiadiazole-2-sulphonamide), a specific inhibitor of carbonic anhydrase from animal sources.

Recently, Randall & Bouma (1973) measured photosynthetic rates and carbonic anhydrase levels in spinach plants grown under conditions of Zn deficiency. The deficient plants contained about one-half of the total Zn per unit area of leaf of normal plants, and carbonic anhydrase activity was drastically reduced to 1-13% of that of normal plants. However, photosynthesis rates of the deficient plants were usually only slightly affected, and it was only in leaves with very low levels of carbonic anhydrase that photosynthesis was reduced appreciably (to 85% of control leaves). When photosynthesis was measured over a range of CO_2 levels, Zn-deficient leaves were the same as controls at lower CO_2 levels (< 0.02% in air), but they exhibited lower photosynthetic rates at high levels of CO_2 (0.03-0.05%). These results indicate no close relationship between photosynthesis and carbonic anhydrase. They argue against the hypothesis that carbonic anhydrase facilitates the supply of CO_2 to the chloroplast, although the possibility that small amounts of carbonic anhydrase may be sufficient cannot be excluded.

Sodium was shown to be essential for the growth of the blue-green alga *Anabaena cylindrica* (Allen & Arnon 1955) and the higher plant *Atriplex vesicaria* (Bladder saltbush) (Brownell 1965). More recently, Brownell & Crossland (1972) have examined the Na requirement of over 30 species of higher plants. Only certain species were shown to respond to Na and these were characterized by having the C_4-dicarboxylic pathway of photosynthesis. C_4-species, grown in the absence of Na, exhibited chlorosis and necrosis, whereas C_3-species, which possess only the Calvin cycle, were free of lesions. It would appear that species with the C_4 pathway of photosynthesis have a requirement for Na, but the role of Na is unknown.

SUMMARY

Several trace elements are required for photosynthesis, a process essential not only to plants but indirectly to all living organisms. The role of Fe and Cu as electron carriers in photosynthetic electron transport is reasonably well understood. Fe is an essential component of the cytochromes and the ferredoxins and Cu is complexed to protein in plastocyanin. The roles of Mn and Cl are less well understood, although both elements are essential for the oxidation of water to oxygen. Mn appears to be complexed to protein and requires light for activation, but nothing is known about the nature of the chloride complex. Zinc is a component for carbonic anhydrase, but there are doubts as to whether carbonic anhydrase is essential for photosynthesis. Sodium appears to be required for the C_4 pathway of photosynthesis.

REFERENCES

ALLEN, M.B. & ARNON, D.I. (1955) - *Physiologia Pl. 8* : 653

AMESZ, J., PULLES, M.P.J., VISSER, J.W.V. & SIBBING, F.A. (1972) - *Biochim. biophys. Acta 275* : 442

ANDERSON, J.M. & BOARDMAN, N.K. (1973) - *FEBS Lett. 32* : 157

ANDERSON, J.M., BOARDMAN, N.K. & DAVID, D.J. (1964) - *Biochem. biophys. Res. Commun. 17* : 685

ARNON, D.I. (1967) - *Physiol. Rev. 47* : 317

ARNON, D.I. TSUJIMOTO, H.Y. & McSWAIN, B.D. (1967) - *Nature, Lond. 214* : 562

AVRON, M. & CHANCE, B. (1966) - *Brookhaven Symp. Biol. 19* : 149

BEARDEN, A.J. & MALKIN, R. (1972) - *Biochem. biophys. Res. Commun. 46* : 1299

BENDALL, D.S. (1968) - *Biochem. J. 109* : 46P

BENDALL, D.S., DAVENPORT, H.S. & HILL, R. (1971) - *In* "Methods in Enzymology" (A. San Pietro, ed.), Vol. 23, Part A, p. 327. Academic Press : New York

BIGGINS, J. (1973) - *Biochemistry 12* : 1165

BLUMBERG, W.E. & PEISACH, J. (1966) - *Biochim. biophys. Acta 126* : 269

BOARDMAN, N.K. & ANDERSON, J.M. (1967) - *Biochim. biophys. Acta 143* : 187

BOARDMAN, N.K., ANDERSON, J.M. & HILLER, R.G. (1971) - *Biochim. biophys. Acta 234* : 126

BÖHME, H. & CRAMER, W.A. (1972) - *Biochim. biophys. Acta 283* : 302

BOVÉ, J.M., BOVÉ, C., WHATLEY, F.R. & ARNON, D.I. (1963) - *Z. Naturf. 186* : 683

BROWNELL, P.F. (1965) - *Pl. Physiol. 40* : 460

BROWNELL, P.F. & CROSSLAND, C.J. (1972) - *Pl. Physiol. 49* : 794

CHENIAE, G.M. (1970) - *A. Rev. Pl. Physiol. 21* : 467

CHENIAE, G.M. & MARTIN, I.F. (1966) - *Brookhaven Symp. Biol. 19* : 406

CHENIAE, G.M. & MARTIN, I.F. (1968) - *Biochim. biophys. Acta 153* : 819

CHENIAE, G.M. & MARTIN, I.F. (1969) - *Pl. Physiol. 44* : 351

CRAMER, W.A., FAN, H.N. & BÖHME, H. (1971) - *J. Bioenerg. 2* : 289

DUYSENS, L.N.M., AMESZ, J. & KAMP, B.M. (1961) - *Nature, Lond. 190* : 510

ENNS, T. (1967) - *Science 155* : 44

EVANS, M.C.W., REEVES, S.G. & CAMMACK, R. (1974) - *Proc. 3rd Int. Congr. Photosynthesis* (in press)

EVANS, M.C.W., TELFER, A. & LORD, A.V. (1972) - *Biochim. biophys. Acta 267* : 530

EVERSON, R.G. (1970) - *Phytochemistry 9* : 25

FORTI, G., BERTOLE, M.L. & ZANETTI, G. (1965) - *Biochim. biophys. Acta 109* : 33

GAREWAL, H.S. & WASSERMAN, A.R. (1974) - *Biochemistry 13* : 4063,4072

GERHARDT, B. & WIESSNER, W. (1967) - *Biochem. biophys. Res. Commun. 28* : 958

GORHAM, P.R. & CLENDENNING, K.A. (1952) - *Archs Biochem. Biophys. 37* : 199

GORMAN, D.S. & LEVINE, R.P. (1966) - *Pl. Physiol. 41* : 1648

GRAHAM, D., ATKINS, C.A., REED, M.L., PATTERSON, B.D. & SMILLIE, R.M. (1971) - *In* "Photosynthesis and Photorespiration" (M.D. Hatch, C.B. Osmond & R.O. Slatyer, eds.), p. 267. Wiley-Interscience : New York

HALL, D.O., CAMMACK, R. & RAO, K.K. (1972) - *In* "Proceedings of the Second International Congress on Photosynthesis Research" (G. Forti, M. Avron & A. Melandri, eds.), Vol. 3, p. 1707. Dr. W. Junk : The Hague

HATCH, M.D. & SLACK, C.R. (1970) - *A. Rev. Pl. Physiol. 21* : 141

HEATH, R.L. & HIND, G. (1969) - *Biochim. biophys. Acta 189* : 222

HENNINGSEN, K.W. & BOARDMAN, N.K. (1973) - *Pl. Physiol. 51* : 1117

HILLER, R.G., ANDERSON, J.M. & BOARDMAN, N.K. (1971) - *Biochim. biophys. Acta 245* : 439

HIND, G., NAKATANI, H.Y. & IZAWA, S. (1969) - *Biochim. biophys. Acta 172* : 277

HOMANN, P.H. (1967) - *Pl. Physiol. 42* : 997

HOMANN, P.H. (1968) - *Biochem. biophys. Res. Commun. 33* : 229

IZAWA, S., HEATH, R.L. & HIND, G. (1969) - *Biochim. biophys. Acta 180* : 388

KATOH, S., SHIRATORI, I. & TAKAMIYA, A. (1962) - *J. Biochem.* *51* : 32

KE, B. (1974) - *Proc. 3rd Int. Congr. Photosynthesis* (in press)

KE, B., HANSEN, R.E. & BEINERT, H. (1973) - *Proc. natn. Acad. Sci. U.S.A.* *70* : 2941

KESSLER, E. (1955) - *Archs Biochem. Biophys.* *59* : 527

KIMIMURA, M. & KATOH, S. (1972) - *Biochim. biophys. Acta* *283* : 279

KNAFF, D.B. (1972) - *FEBS Lett.* *23* : 92

KNAFF, D.B. & ARNON, D.I. (1969) - *Proc. natn. Acad. Sci. U.S.A.* *63* : 956

LARKUM, A.W.D. & BONNER, W.D. (1972) - *Biochim. biophys. Acta* *256* : 385

LUNDEGARDH, H. (1962) - *Physiologia Pl.* *15* : 390

MALKIN, R., APARICIO, P.J. & ARNON, D.I. (1974) - *Proc. natn. Acad. Sci. U.S.A.* *71* : 2362

MALKIN, R. & BEARDEN, A.J. (1971) - *Proc. natn. Acad. Sci. U.S.A.* *68* : 16

MALKIN, R. & BEARDEN, A.J. (1973) - *Biochim. biophys. Acta* *292* : 169

MALKIN, R., KNAFF, D.B. & BEARDEN, A.J. (1973) - *Biochim. biophys. Acta* *305* : 675

PIRSON, A. (1937) - *Z. Bot.* *31* : 193

POSSINGHAM, J.V. & SPENCER, D. (1962) - *Aust. J. biol. Sci.* *15* : 58

POSSINGHAM, J.V., VESK, M. & MERCER, F.V. (1964) - *J. Ultrastruct. Res.* *11* : 68

RAMSHAW, J.A.M., BROWN, R.H., SCAWEN, M.D. & BOULTER, D. (1973) - *Biochim. biophys. Acta* *303* : 269

RANDALL, P.J. & BOUMA, D. (1973) - *Pl. Physiol.* *52* : 229

SINGH, J. & WASSERMAN, A.R. (1971) - *J. biol. Chem.* *246* : 3532

SPENCER, D. & POSSINGHAM, J.V. (1960) - *Aust. J. biol. Sci.* *13* : 441

SPENCER, D. & POSSINGHAM, J.V. (1961) - *Biochim. biophys. Acta* *52* : 379

STUART, A.L. & WASSERMAN, A.R. (1973) - *Biochim. biophys. Acta* *314* : 284

SUCHDEO, S.R. & SCHULTZ, J.S. (1974) - *Biochim. biophys. Acta* *352* : 412

VISSER, J.W.M., AMESZ, J. & GELDER, B.F. (1974) - *Biochim. biophys. Acta* *333* : 279

WARBURG, O. AND LÜTTGENS, W. (1946) - *Biokhimiya* *11* : 303

YAMASHITA, T. & BUTLER, W.L. (1968) - *Pl. Physiol.* *43* : 1978

THE ROLE OF BORON IN PLANTS

J.F. Jackson and K.S.R. Chapman

Department of Agricultural Biochemistry & Soil Science
Waite Agricultural Research Institute
The University of Adelaide
Glen Osmond, South Australia 5000

INTRODUCTION

Boron deficiency in commercial crops does occur in Australia, being confined mainly to the slopes of the Great Dividing Range from northern Queensland to Tasmania, where it is quite widespread (Fig. 1). The deficiency affects a variety of crops, including lucerne, beetroot, celery, swedes, turnips, tobacco, apples and pears, in widely scattered areas of the Range, and generally can be corrected by application of borax or boric acid.

The biochemical role of B is as yet not well understood, and, unlike other generally recognized trace elements, it has not been shown to form part of an enzyme system. Speculations as to its role, then, can be deduced only indirectly from physiological experiments, usually with B-deficient plants. The earliest morphological symptom of B deficiency in, for example, *Phaseolus aureus* (mung bean) appears to be a slow-down in root extension, followed by a degeneration of meristematic tissue, possibly due to a repressive effect of B deficiency on cell division. A positive effect on cell expansion and many other morphological properties follow subsequently, depending on the plant concerned (Stiles 1961).

Physiological changes due to a B deficit may be more useful to us than morphological observations in that the underlying causes can be more easily related to biochemical events. Thus, it seems that the "browning" of tissue is a late effect of B deficiency, caused by the accumulation of polyphenolic compounds, and it has been postulated that B is involved in the synthesis of cell-wall components such as these polyphenolic

Fig. 1. Boron deficiency in commercial crops has been noted in the following areas of Australia: 1. Tobacco at Mareeba, Qld. 2. Beans in the Burdekin area, Qld. 3. Tomatoes at Bowen, Qld. 4. Beetroot along the Lockyer River valley, Qld. 5. Apples at Stanthorpe, Qld. 6. Lucerne at Tenterfield, Inverell, Tamworth; cauliflowers at Glen Innes; N.S.W. 7. Apples, pears, lucerne at Orange, Bathurst; cauliflowers at Bathurst; N.S.W. 8. Lucerne at Goulburn, Queenbeyan; N.S.W. 9. Lucerne at Tumut; apples, pears at Batlow; N.S.W. 10. Cauliflowers around Sydney, N.S.W. 11. Apples at Kiewa, Vic. 12. Clovers at Tolmie; lucerne at St. Arnaud; Vic. 13. Celery, redbeet, parsnips in areas south-east of Melbourne, Vic. 14. Swedes, turnips at Smithton, Tas. 15. Swedes, turnips at Strahan, Tas. 16. Swedes, turnips at Launceston, St. Marys; Tas. 17. Apples in Huon and Channel districts; swedes, turnips, celery, cauliflower at Hobart, New Norfolk, Huon, Ouse; Tas. 18. Poppies for alkaloid production in northern Tas. 19. Apples in the Tasman Peninsula area. 20. Boron deficiency has been reported at McLaren Vale, S.A. [see A. Walkley & H.H. Kemp (1939) - J. Aust. Inst. Agric. (continued on next page)

compounds (Slack & Whittington 1964). Similarly, B has been implicated in carbohydrate metabolism. In this case, however, changes are observed earlier than the usual morphological symptoms, although still several days after deficiency began. Thus, B-deficient tomato plants were found to translocate more sugar when 50 ppm B were added with sucrose through a cut petiole, compared with sucrose supplied alone (Sisler *et. al.* 1956). The ability of borate to complex with *cis*-hydroxyl groups was somehow thought to be involved in these effects (Dugger *et al.* 1957).

A physiological symptom to show up even earlier (within a few hours of transfer to a medium lacking B) is increased uptake of certain labelled precursors of RNA into root tips (Cory *et al.* 1966; Cory & Finch 1967; Chapman & Jackson 1974). On the assumption that this earliest known effect of B deficiency is closest to the fundamental biochemical site of action of B, a closer study of this aspect of RNA metabolism was undertaken.

NUCLEIC ACID PRECURSOR STUDIES

Radioactivity from ^{14}C-uridine was found to be incorporated into the RNA of 3 mm long root-tip segments from *Phaseolus aureus* at a linear rate over a 4 hr incubation period. When 5-day-old *P. aureus* seedlings were transferred to a B-deficient medium and root-tip segments incubated as above with the precursor at various times after the transfer, it was found that the amount of label incorporated into RNA increased in the tips from B-deficient seedlings over and above that for control plants. The specific activity of label in the RNA also increased in B-deficient root tips and, like total incorporation of label into RNA, continued to increase with time of culture in the B-deficient medium (Fig. 2). The increased specific activity of label in RNA was apparent within several hours (6-10 hr) of transfer to the deficient medium, and so

---

Sci. 5 : 229]; no other case has been reported in S.A. 21. An isolated case of deficiency in celery has been reported in the Perth area, W.A.

The following are acknowledged for contributing towards the above data: R.C. Bruce, Queensland Department of Primary Industries; A.N. Old & R.G. Weir, Department of Agriculture, New South Wales; Dr. R. Rowe, Department of Agriculture, Victoria; Prof. G. Wade, University of Tasmania; Prof. C.M. Donald, Waite Agricultural Research Institute, South Australia; L.T. Jones, Department of Agriculture, Western Australia.

represents a much earlier symptom of B deficiency than the slowdown in root elongation or swelling of the root tip. The latter symptoms are the earliest observable morphological effects of a B deficit, the inhibition of root elongation becoming noticeable within 24-40 hr after transfer (Fig. 3a) and a thickening of the root tips becoming measurable after 50-100 hr in the B-deficient medium (Fig. 3b). Cell counts

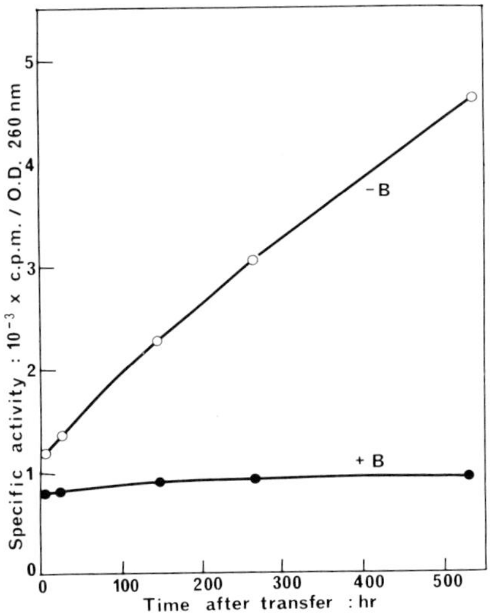

Fig. 2. The effect of B deficiency on the specific activity of labelled RNA was determined by incubating 400-600 root tips from 5-day-old P. aureus *seedlings with 2-^{14}C-uridine (2 mCi; 40 mCi/mmole) for 4 hr at 30°C in 2 ml of nutrient medium containing 2% (w/v) sucrose. RNA was extracted and its specific activity determined (Chapman & Jackson 1974).*

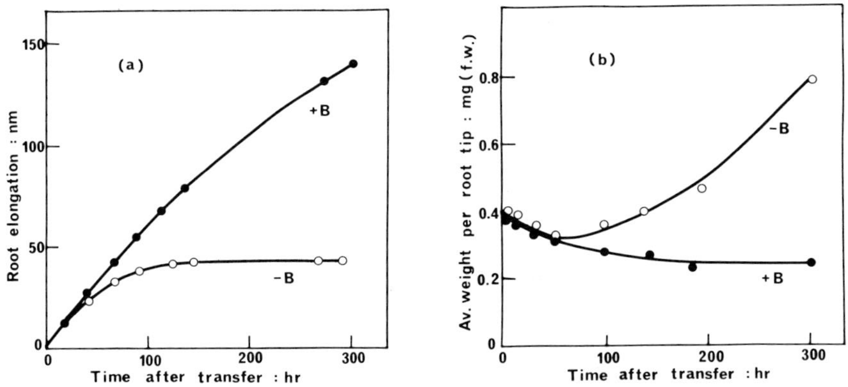

Fig. 3. Root elongation (a) was measured after marking the roots with a mixture of finely ground charcoal in lanolin at the time of transfer of the 5-day-old P. aureus *seedlings to another medium, either boron deficient (-B) or complete (+B) (Chapman & Jackson 1974). Thickening of the root tip (b) was estimated by weighing 200-300 of excised 3 mm root tips from* P. aureus *seedlings cultured for varying periods in boron-deficient (-B) or complete (+B) medium.*

carried out on the 3 mm long root tips suggested that the swelling of the tips was primarily due to cell enlargement rather than an increase in cell numbers.

Another precursor of RNA to show an increased incorporation as a result of B deficiency was ^{32}P-orthophosphate. Here again, the increase occurred before morphological changes were measurable. As with ^{14}C-uridine, the specific activity of label incorporated into the RNA of root-tip segments was increased by B deficiency and continued to increase with length of culture in the medium lacking B (Table 1). Cory *et al.* (1966) and Cory & Finch (1967) demonstrated that transfer of *Vicia faba* seedlings to a B-deficient medium resulted in an increase in specific activity of root-tip ^{32}P-RNA after incubation with ^{32}P-orthophosphate. The effect observed with *Vicia faba* was also apparent well before the more familiar morphological symptoms appeared.

TABLE 1

Effect of B deficiency on the specific activity of RNA following ^{32}P-orthophosphate incorporation

Root tips (400-800) were incubated in 2 ml of nutrient medium (diluted × 4) containing 2% (w/v) sucrose and 1 mCi ^{32}P-orthophosphate (3 mCi/μmole). Each sample of RNA extracted contained at least 10^5 cpm.

Time after transfer (hr)	Increase in specific activity (%)
24	53
48	69
144	100
312	202

The incorporation of some other nucleic acid precursors has been briefly investigated. Using root-tip segments from *P. aureus* seedlings grossly deficient in B (100 hr after transfer), it was shown that neither ^{14}C-orotic acid nor ^{14}C-thymidine was incorporated at a rate greatly in excess of control segments (Table 2). In the case of ^{14}C-thymidine, the result indicates that DNA synthesis is not increased as a result of B deficiency. This experiment is particularly relevant since both uridine and thymidine was thought to be incorporated into nucleic acids by "salvage" pathways. Although orotic acid, unlike thymidine, is a precursor of RNA, it was not incorporated at rates comparable with that for uridine in B-deficient segments. However, unlike uridine, orotic acid is thought to be incorporated into RNA *via* the

TABLE 2

Incorporation of precursors into grossly deficient root tips

The indicated numbers of 3 mm root tips from *P. aureus* seedlings grown for 100 hr in either a complete medium or a medium lacking B were incubated in 4 ml of medium containing 2% (w/v) sucrose and 4 μM ^{14}C-precursor (1.2 × 10^7 cpm/μmole) for 2 hr (Expt. 1) or 3 hr (Expt. 2). The tips were then washed, fixed, dried and counted.

Precursor	Experiment 1 (cp 10 min/20 tips)		Experiment 2 (cp 10 min/10 tips)	
	^{14}C-orotate	^{14}C-uridine	^{14}C-uridine	^{14}C-thymidine
+B	3,500	5,386	9,758	1,307
-B	3,974	9,635	14,863	1,055
% increase in -B	13	79	52	-29

"*de novo*" pathway, rather than the "salvage" pathway; this may explain the low rate of increase obtained with this precursor.

SPECIFICITY OF THE EFFECT

Many factors can affect RNA metabolism. For this reason, some other trace element deficiencies were investigated for an effect on the incorporation of precursors into RNA and compared with that obtained for B deficiency. Control plants were grown in the complete medium of Hoagland & Arnon (1950), while B deficiency was induced by transfer to this medium lacking the normally present boric acid; Fe-deficient medium lacked the Fe-EDTA complex of Jacobson (1951); Cu-deficient cultures lacked $CuCl_2$, and Mn deficiency was brought about by omitting $MnCl_2$ from the medium. All plant cultures were carried out in black plastic containers, in media prepared from double-distilled water, further distilled in a silica apparatus.

The results given in Table 3 show that, during the first 48 hr, Mn and Cu deficiencies do not result in any significant increase in the incorporation of ^{14}C-uridine into RNA, but do give an increase into the acid-soluble fraction. The latter represents acid-soluble nucleoside and nucleotide precursors of RNA synthesis. An increase into RNA is detectable at later stages of these deficiencies. Iron deficiency gives an

TABLE 3

Effect of trace element deficiencies on ^{14}C-uridine incorporation

Deficient and normal root tips (150 of each) were incubated with ^{14}C-uridine as described in Figure 1 and the RNA and acid-soluble fractions extracted, separated and counted. All samples incorporated at least 10^4 cpm into each fraction, and the results are expressed as the ratio:

$$\frac{\text{counts incorporated into 150 deficient root tips}}{\text{counts incorporated into 150 normal root tips}} .$$

(from Chapman & Jackson 1974)

Deficient element	Time after transfer to deficient medium (hr)				
	12	24	48	144	336
			RNA fraction		
Fe	0.75	0.55	0.60	0.62	1.06
Mn	-	0.98	0.92	1.30	1.56
Cu	-	0.88	1.10	1.32	1.08
B	1.07	1.24	1.50	1.52	3.23
			Acid-soluble fraction		
Fe	1.00	1.21	0.97	0.90	1.85
Mn	-	1.34	1.42	1.58	1.10
Cu	-	1.08	1.27	1.77	1.33
B	1.00	1.02	1.08	1.57	2.62

early inhibition of incorporation into RNA, without any apparent effect on the acid-soluble fraction. The outcome of deficiencies in these trace elements, then, was quite different to that for B and points to a relatively specific effect of B deficiency in giving an early increase in ^{14}C-uridine incorporation into RNA of *P. aureus* root-tip segments.

Although we have demonstrated a reproducible and apparently specific effect of the very early stages of B deficiency on ^{14}C-uridine incorporation into the RNA of root-tip segments of *P. aureus* seedlings, it is not clear how widespread this phenomenon is. Cory *et al.* (1966) and Cory & Finch (1967) have shown an increased incorporation of another precursor, ^{32}P-orthophosphate, to be an early symptom of B deficiency in *Vicia faba* radicles. In this case, intact plants were used, so the results cannot be strictly compared with the present studies. However, there seems little doubt that we are looking at the same phenomenon.

Shkol'nik & Kositsyn (1962), Sherstnev & Kurilenok (1962),

Timashov (1967) and Rapota (1970) grew pea and sunflower plants until they showed clear signs of B deficiency (at least 6 days), and then labelled the nucleic acids by culturing the plants a further 2-4 days in a culture medium (Knop's) containing either ^{14}C-adenine or ^{32}P-orthophosphate. In contrast to the work with *P. aureus* and *V. faba*, they report a decrease in the specific activity of labelled RNA from the roots and shoots (including the growing points), using a 40 min pulse of ^{32}P-orthophosphate. Sherstnev (1967) demonstrated the same effect on incorporation into the RNA of pea roots.

The lack of agreement between the result for *P. aureus* and *V. faba* on the one hand, and pea and sunflower on the other, could be due to the very different experimental conditions used. The latter work made use of grossly deficient plants, and incorporation into the total root and leaf systems was studied, while the former was carried out at the very earliest stages of B deficiency and on the root-tip region only. It could be that the extensive regions behind the root tip, where the bulk of the RNA is to be found, do in fact incorporate less precursor into RNA in grossly deficient plants, while the root-tip region, containing a small amount of RNA, could register an increased incorporation and yet not show up in the results for bulk RNA. It can be seen then that the two results are not necessarily in disagreement or mutually exclusive. We decided to study the root-tip region only, since this is where the first visible symptoms occur (inhibition of root elongation) and where the first physiological effects have been recorded (cell enlargement, and then disintegration).

NATURE OF RNA WITH INCREASED LABEL

To study further the increased incorporation of precursor into RNA during the early stages of B deficiency, the character of the labelled RNA in the *P. aureus* root tips was investigated using methylated albumin keiselguhr (MAK) chromatography, polyacrylamide gel electrophoresis and RNA-DNA hybridization. The first two techniques tell us something of the size of the RNA molecules, and the last, information on the degree of similarity between the repetitious base sequences of RNA labelled normally compared with that labelled under conditions of B deficiency.

The amount of label incorporated into all RNA fractions from the MAK column was increased during the early stages (15 hr) of B deficiency (Table 4). However, the increase was not

TABLE 4

Effect of B deficiency on the labelling of RNA fractions separated by MAK chromatography

Nucleic acid samples (0.1-0.5 mg) from ^{14}C-uridine-labelled *P. aureus* root-tip segments were applied to the column (0.65 × 10 cm) and the latter washed with 0.1 M NaCl-0.05 M phosphate, pH 6.7. RNA fractions were eluted with a linear gradient of NaCl (0.2-1.1 M) in the phosphate buffer, and the TB-RNA fraction collected with 0.5% (w/v) SDS.

Time after transfer to deficient medium (hr)	RNA fraction (label in fraction; % of control)		
	t-RNA	r-RNA	TB-RNA
15	125	140	210
30	145	170	200
100	285	315	275
170	335	290	265
260	395	260	330
515	470	110	250

uniform through all the fractions, but was greatest in the tenaciously bound (TB)-RNA fraction, where incorporation of ^{14}C-uridine was found to be more than double the control value. The specific activity of label in the three fractions showed a corresponding increase at this early stage of deficiency, and similar results were obtained for root tips harvested after 30 hr in deficient medium. Polyacrylamide gel electrophoresis of the RNA from root tips of *P. aureus* cultured for 24 hr in B-deficient medium verified that the amount of label (^{32}P from ^{32}P-orthophosphate) incorporated into the r-RNA fraction at this early stage was increased, with the 25 s and 18 s components showing an equal rise. The electrophoretic results also served to demonstrate that at no time was there appreciable radioactivity in the 16 s and 23 s regions, indicating that bacterial contamination was not contributing significantly to increased RNA labelling.

Grossly deficient *P. aureus* root tips (515 hr of B deficiency) showed a completely different trend. Although the label in all three major fractions obtained by MAK chromatography was increased by B deficiency, the t-RNA fraction now gave by far the largest increase (Table 4). By 515 hr, label in the r-RNA fraction was back to that shown by control plants, after having risen to approximately three-fold that of controls after 100 hr of B deficiency. The amount of label in the TB-RNA remained higher than controls at all times.

Hybridization of the labelled RNA (from normal or B-deficient root tips) with *P. aureus* DNA in the presence of varying amounts of competing unlabelled RNA (from normal or B-deficient root tips) gave the competition curves shown in Fig. 4 for RNA from plants subjected to 48 hr of B deficiency. The lack of divergence of the one curve from the other suggests

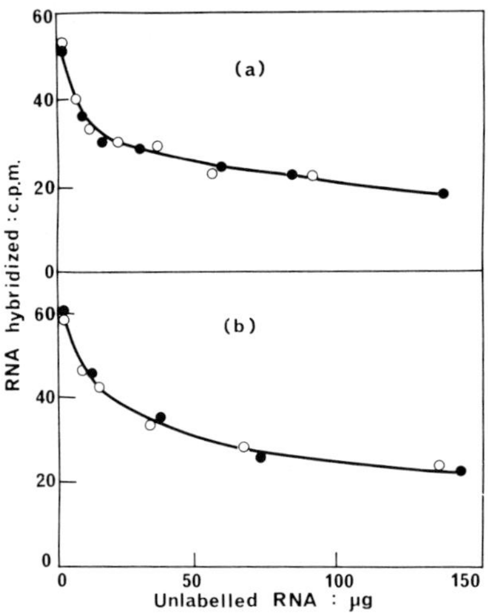

Fig. 4. Competitive DNA-RNA hybridization after 48 hr of B deficiency was carried out by annealing three Sartorius membrane discs, each containing 0.4 μg of DNA, with 11 μg of ^{32}P-RNA from (a) normal root-tip segments or (b) B-deficient segments. Hybrids were formed in the presence of various amounts of unlabelled competing RNA from normal (●) or B-deficient (○) root tips. The constant amount of unlabelled RNA used approached the previously determined saturation level.

that there was no significant difference between the repetitious sequences of labelled RNA from deficient or control root tips. In complex higher organisms such as *P. aureus*, it is only the repetitious sequences which will hybridize under the conditions of the experiment. Similar results were obtained for plants subjected to 24 hr of deficiency. Taken together with the results from MAK chromatography, this may suggest that increased labelling at the very earliest stages of B deficiency is not associated with any one sequence (new or old), but rather with sequences normally synthesized and found within the TB-RNA fraction.

In contrast to these early effects of a B deficit, competition curves for RNA from plants subjected to 144 hr of deficiency did show a degree of divergence and it was concluded that at these later stages of B deficiency there was a change in concentration of particular base sequences within the RNA population. This could be brought about by, for example, selective degradation of RNA. Extensive degradation of labelled RNA was suggested by the high concentration of small molecular weight RNA (t-RNA on MAK chromatography) found at these later stages of deficiency. An increased ribonuclease activity in root tips at the later stages of B deficiency reinforces this interpretation (Chapman & Jackson 1974).

CONCLUSIONS

Investigations with *P. aureus* and *V. faba* suggest that the earliest known response to the removal of B from a plant culture medium is an increased precursor incorporation into the RNA of the root-tip region. This response is greatest with ^{14}C-uridine and ^{32}P-orthophosphate, and it occurs without any apparent change in the total nucleotide precursor pool (Cory & Finch 1967; Chapman & Jackson 1974). The various RNA species were not equally affected; the fraction tenaciously bound to MAK (TB-RNA) showed the earliest substantial increase. The distribution of ^{32}P among the component nucleotides suggested a base composition for the increased TB-fraction higher in AMP content than r-RNA. This is consistent with the label being associated with a m-RNA type molecule, which is thought to make up a substantial amount of newly synthesized RNA in short-term precursor incorporation experiments. Hybridization studies suggested that there was no change in the nucleotide sequences of the RNA labelled as a result of B withdrawal.

These responses to B deficiency are markedly similar to some of the effects of the application to plant tissue of such hormones as auxin, gibberellic acid and cytokinin. Thus, Matthysse & Phillips (1969), working with tobacco tissue cultures, found that the incorporation of ^{32}P-orthophosphate into RNA was increased within as little as 15 min after auxin addition. An enhanced incorporation of precursor into RNA shortly after auxin application has also been shown for several other auxin-responsive tissues (Click & Hackett 1964; Masnda & Kamisaka 1969; Key 1969; Klambt 1974). As for B withdrawal, it is the tenaciously bound RNA fraction which is affected first (Tester & Dure 1967; O'Brien *et al.* 1968; Tanimoto & Masnda 1969). Similarly, treatment of plant tissue with gibberellic acid can give rise to increased precursor incorporation into RNA (Chandra & Varner 1965; Fletcher & Osborne 1966; Johri & Varner 1968; Broughton 1969; Poulson & Beevers 1970). In pea nuclei isolated from gibberellic acid-treated plants, the stimulation of RNA synthesis took place mainly in the TB-RNA fraction (Johri & Varner 1968). Cytokinin treatment can also result in enhanced incorporation of labelled precursor into RNA (Osborne 1962; Oota 1964; Carpenter & Cherry 1966; Burdett & Wareing 1968; Kulaeva *et al.* 1971). In this case, however, all RNA fractions from MAK chromatography were found to be equally affected.

In view of the general similarity of B deficiency symptoms (in so far as RNA metabolism is concerned) with the response of plant tissue to treatment by some hormones, it is tempting to interpret the metabolic action of B in terms of some common reaction(s) shared by these hormones. The concept of a second messenger, such as cyclic AMP, is now well established in animal tissue (Robison *et al.* 1971), and it is in this area that hormones may have a mechanism in common. In this system, the various hormones interact with a membrane-bound receptor in a target cell, causing the amplifier, adenylate kinase, to synthesize the second messenger, cAMP, within the cell, and the appropriate response is produced by cAMP. If, in plant cells, B were involved somewhere in this type of mechanism, in such a way that it held the production of the second messenger in check in the absence of hormone, then its removal could result in the overproduction of the second messenger, resulting in such responses as increased labelling of RNA, swelling of root tips, etc.

Cyclic AMP has been shown to mimic many of the effects of such plant hormones as gibberellic acid (Gilbert & Galsky 1972) and auxin (Hartung 1972), alluding to a second messenger role for this cyclic nucleotide. However, evidence has been presented which suggests that cAMP does not occur in higher plants (Amrhien 1974), despite the many claims purporting to show its presence in plant tissue (e.g. Kessler & Levinstein 1974). The existence of a second messenger in plants, then, has not been adequately proven, and until this is decided it is difficult to test this hypothesis, or to relate B to hormone action in other ways.

REFERENCES

AMRHEIN, N. (1974) - *Planta 118* : 241

BROUGHTON, W.J. (1969) - *Ann. Bot. 33* : 227

BURDETT, A.M. & WAREING, P.F. (1968) - *Planta 81* : 88

CARPENTER, W.J.G. & CHERRY, J.H. (1966) - *Biochim. biophys. Acta 114* : 640

CHANDRA, G.R. & VARNER, J.E. (1965) - *Biochim. biophys. Acta 108* : 583

CHAPMAN, K.S.R. & JACKSON, J.F. (1974) - *Phytochem. 13* : 1311

CLICK, R.E. & HACKETT, D.P. (1964) - *Fedn Proc. 23* : 525

CORY, S. & FINCH, L.R. (1967) - *Phytochem. 6* : 211

CORY, S., FINCH, L.R. & HINDE, R.W. (1966) - *Phytochem. 5*:625

DUGGER, W.M., HUMPHREYS, T.E. & CALHOUN, B. (1957) - *Plant Physiol. 32* : 364

FLETCHER, R.A. & OSBORNE, D.J. (1966) - *Can. J. Bot. 44* : 739
GILBERT, M.L. & GALSKY, A.G. (1972) - *Pl. Cell Physiol. 13* : 867
HARTUNG, W. (1972) - *Z. Pflanzenphysiol. 67* : 380
HOAGLAND, D.R. & ARNON, D.I. (1950) - *Calif. Agric. Expt Sta. Circ. no. 347*
JACOBSON, L. (1951) - *Pl. Physiol. 26* : 411
JOHRI, M.M. & VARNER, J.E. (1968) - *Proc. natn. Acad. Sci. U.S.A. 59* : 269
KESSLER, B. & LEVINSTEIN, R. (1974) - *Biochim. biophys. Acta 343* : 156
KEY, J.L. (1969) - *A. Rev. Pl. Physiol. 20* : 449
KLAMBT, D. (1974) - *Planta 118* : 7
KULAEVA, O.N., SELIVANKINA, S.Yu. & KUROEDOV, V.A. (1971) - *Soviet Pl. Physiol. 18* : 632
MASNDA, Y. & KAMISAKA, S. (1969) - *Pl. Cell Physiol. 10* : 79
MATTHYSSE, A.G. & PHILLIPS, C. (1969) - *Proc. natn. Acad. Sci. U.S.A. 63* : 897
O'BRIEN, T.J., JARVIS, B.C., CHERRY, J.H. & HANSON, J.B. (1968) - *Biochim. biophys. Acta 169* : 35
OOTA, Y. (1964) - *A. Rev. Pl. Physiol. 15* : 17
OSBORNE, D.J. (1962) - *Pl. Physiol. 37* : 595
POULSON, R. & BEEVERS, L. (1970) - *Pl. Physiol. 46* : 782
RAPOTA, V.V. (1970) - *Fiziol. Biokhim. Knl't. Rast 2* : 210
ROBISON, G.A., BUTCHER, R.W. & SUTHERLAND, E.W. (1971) - "Cyclic AMP". Academic Press : New York
SHERSTNEV, E.A. (1967) - *Dokl. Akad. Nauk SSSR 175* : 1190
SHERSTNEV, E.A. & KURILENOK, G.V. (1962) - *Dokl. Akad. Nauk SSSR 142* : 1201
SHKOL'NIK, M.Ya. & KOSITSYN, A.V. (1962) - *Dokl. Akad. Nauk SSSR 144* : 662
SISLER, E.C., DUGGER, W.M. & GAUCH, H.G. (1956) - *Pl. Physiol. 31* : 11
SLACK, C.R. & WHITTINGTON, W.J. (1964) - *J. exp. Bot. 15* : 495
STILES, W. (1961) - "Trace Elements in Plants". Cambridge University Press : Cambridge
TANIMOTO, E. & MASNDA, Y. (1969) - *Pl. Cell Physiol. 10* : 485
TESTER, C.F. & DURE, L.S. (1967) - *Biochemistry 6* : 2532
TIMASHOV, N.D. (1967) - *Dokl. Akad. Nauk SSSR 175* : 240

TRACE ELEMENTS AND THEIR PHYSIOLOGICAL ROLES IN THE ANIMAL

E.J. Underwood*

Institute of Agriculture
University of Western Australia
Nedlands, Western Australia 6009

THE ESSENTIAL TRACE ELEMENTS

At the present time 14 elements, commonly accepted as trace elements, are believed to be essential for animal life. In the general order of their discovery these are Fe, I, Cu, Mn, Zn, Co, Mo, Se, Cr, Sn, F, Si, Ni and V. Boron, one of the earliest trace elements shown to be required by the higher plants, has not yet been shown to be essential for animals.

An element is considered essential if its deficiency consistently results in impairment of function from optimal to sub-optimal. The function that is most obvious and most readily quantitated is somatic or body growth. Growth impairment in the young animal, responsive to dietary supplements of the element in question, is therefore the most satisfactory criterion of essentiality that we have. Unfortunately this tells us little of the element's physiological role in the tissues and still less of its metabolic function or functions at the cellular level. These have to come from more detailed studies of the concentration, distribution and chemical forms of the element in the cells and tissues and fluids of the animal body and of various structural and functional disorders which may precede, accompany or succeed the growth impairment. With several of the so-called 'newer' trace elements, which have only emerged as essential elements for rats and chicks in the last 3-4 years, no critical structural or functional disorders in the animal, other than growth impairment, have yet been observed. Their physiological roles thus remain obscure.

---

*Present address: CSIRO, Private Bag, P.O., Wembley, W.A. 6014

More than 30 years ago Green (1941) stated that "enzymic catalysis is the only rational explanation of how a trace of some substance can produce profound biological effects". Since that time the trace elements have been shown to participate in a wide range of enzymic processes involving many aspects of the intermediary metabolism and physiological functions of the organism. They act as catalysts or cofactors in enzyme systems with roles ranging from relatively weak, non-specific ion effects (metal-ion activated enzymes) to highly specific associations (metallo-enzymes) in which the metal is firmly attached to the protein in a fixed number of atoms per molecule. These atoms cannot be removed without loss of activity and in most cases cannot be replaced by any other metal. However, the specificity is not always complete. For instance, Vallee (1971) has shown that Co and Cd can be substituted, *in vitro*, for the native Zn atoms in several Zn metallo-enzymes while the enzyme remains active. Examples of metallo-enzymes are given in Table 1 to illustrate the diversity of functions in which trace elements are involved.

Some progress is being made in our knowledge of the three-dimensional structure and bonding of the metallo-enzymes, especially due to the work of Vallee (1971) in Boston. He has shown that in certain of these compounds some of the metal is critical to its catalytic step, while the rest seems to serve primarily in preserving the protein structure. Despite these encouraging advances, we are still largely ignorant of the precise mechanisms by which the trace elements exert their enzymatic roles and we can seldom express or explain their biological specificity in physico-chemical terms. An interesting attempt to do this for the five known Mo enzymes (nitrogenase, nitrate reductase, xanthine oxidase, aldehyde oxidase and sulphite oxidase), with their diverse functions, has recently been made by Stiefel (1973). The reactions catalysed by these enzymes each find the product differing from the substrate by two electrons and two protons, or some multiple thereof. A simple molecular mechanism embodying coupled electron-proton transfer to and from the substrate, compatible with the coordination chemistry of Mo, is presented and discussed for each of the Mo enzymes and reasons for the physiological use of Mo in these ways, as distinct from most other metals, are suggested. Fortunately our understanding of the nature of the enzymic and biochemical processes in which the trace elements participate and of the relation of defects in these processes to the functional and structural disorders that arise in the deficient animal is much more advanced and is rapidly unfolding. Some of these advances and some of the gaps in knowledge in this area will be discussed elsewhere (Smith & Gawthorne, this volume).

TABLE 1

Some metallo-enzymes and metallo-proteins, and their physiological functions

Element	Enzyme	Function
Iron	Cytochromes	Electron transfer
	Succinate dehydrogenase	Aerobic oxidation of carbohydrates
	Catalase	Protection against H_2O_2
COPPER	Cytochrome oxidase	Principal terminal oxidase
	Lysyl oxidase	Lysine oxidation
	Tyrosinase	Skin pigmentation
	Ceruloplasmin	Iron utilization
	Cytocuprein	Superoxide (O_2^-) dismutation
ZINC	Carbonic anhydrase	CO_2 formation; regulation of acidity
	Carboxypeptidase	Protein metabolism
	Alcohol dehydrogenase	Alcohol metabolism
	Cytocuprein	Superoxide (O_2^-) dismutation
MANGANESE	Arginase	Urea formation
	Pyruvate carboxylase	Pyruvate metabolism
MOLYBDENUM	Xanthine oxidase	Purine metabolism
	Sulphite oxidase	Sulphite oxidation
SELENIUM	Glutathione peroxidase	Protection against haemoglobin oxidation

THE COMPLEXITY OF TRACE ELEMENT FUNCTIONS

The trace elements as a group are not functionally related in the animal. Even such physico-chemically similar elements as Ni and Co, Zn and Cd, or Se and Te are quite dissimilar physiologically. The manifestations of dietary deficiency would therefore be expected to differ greatly from element to element as an expression of breakdown or impairment of the different physiological functions in which the elements are involved. With severe deficiencies of some trace elements, characteristic clinical and pathological changes in the animal do arise. However, the situation is much more complex than this simple statement would suggest, as the following examples will make clear.

Inappetence and impaired growth occur with deficiencies of all the trace elements and thus have limited diagnostic value. They are much more obvious and dominant expressions of deficiency with some elements, notably Zn and Co, than they are with others, such as Cu and Mn. The occurrence of anaemia, of itself, gives little clue to the element involved or the physiological or biochemical function primarily impaired, because anaemia is a characteristic of Fe, Cu and Co deficiencies and of Zn, Mo and Se toxicities. The functional and structural disorders apparent to the clinician or the pathologist are merely the final expressions of defects arising at different points in a chain of metabolic events or even in a quite different chain of events. Trace element x can be vital at one point and trace element y at another. A simple or conditioned deficiency of either element would in this way lead to the same end result in the animal, although the cause would obviously be different.

The physiological role of a trace element most noticeably affected by a deficiency of that element, as evidenced by the pathological changes that become apparent, varies markedly with the intensity and duration of the dietary deficiency and with the species, age and sex of the animal. For example, with sheep in certain areas, specific lesions in the wool may be the only clinical evidence of Cu deficiency, while with rats at particular levels of Cu intake, depigmentation of hair is similarly the only sign (Mills & Murray 1960). Under these conditions it seems that all the other physiological processes requiring Cu for their proper functioning take precedence over keratinization and pigmentation and are able to secure sufficient Cu from the limited supply for their particular needs. Such differences among the cells and tissues of the body in priority of demand for trace elements and other nutrients are little understood. It seems that the *relative intensity of demand* at particular stages in the growth and productive life of the animal determines the physiological process which first fails in the competition for limited supplies of the element.

The importance of the timing of a deficiency to the nature of the biochemical effect most apparent is strikingly illustrated by the incidence of neonatal ataxia that can arise in the offspring of Cu-deficient pregnant ewes and Mn-deficient pregnant rats and guinea-pigs. The imposition of even severe Cu deficiency upon lambs that have developed normally to birth does not induce ataxia. If this deficiency is imposed upon the ewes in late pregnancy, i.e. late in the intra-uterine life of the lamb when myelin synthesis in the motor neurones of the brain and spinal cord is at its height, a neurodysgenesis develops and the lambs may exhibit ataxia at birth.

The defect responsible for the ataxia of Mn deficiency, namely impaired vestibular function arising from an anomalous development of the bony labyrinth of the inner ear, is even more critically related to the timing of the deficiency. Hurley *et al.* (1958) have pinpointed the vital period to between the 14th and the 18th day of gestation in the rat. The maintenance of Mn deficiency up to only the 14th day of gestation permits the birth of young free of ataxia. If the deficiency is continued up to and beyond the 18th day, 100% of the live young may be ataxic at birth (Table 2). It appears that there is a critical demand for Mn at this stage of uterine growth for the osseous development of a particular structure, the inner ear. The bone abnormalities that arise as a consequence of Mn deficiency are known to result from a specific effect of Mn upon cartilage mucopolysaccharide synthesis (Leach *et al.* 1969), but the reason for the special sensitivity of the skull, and particularly of the otoliths, is unknown.

TABLE 2

*Effect of Mn supplementation at various times during gestation in the rat**

Initiation of supplementation (day of gestation)†	No. of litters	Young born		Survival to 28 days	
		Total	Per litter	Live young (%)	Ataxic (%)
7-12	14	105	7.5	53	0
14	6	42	7.0	87	0
15	8	60	7.5	36	48ϕ
16	8	54	6.8	44	46ϕ
18	8	65	8.1	26	100

*From Hurley *et al.* (1958)
†Day of finding sperm considered first day of gestation
ϕMild

The importance of the severity, duration, and timing of a trace element deficiency upon the physiological function primarily affected is illustrated by numerous studies with that most protean of all trace elements, Zn. When Hurley & Swenerton (1966) fed a severely Zn-deficient diet to female rats from weaning to maturity, growth was drastically impaired and there was severe disruption of the oestrous cycles. No mating took place and the rats were completely infertile. Similar rats maintained from weaning to maturity on a *marginally* Zn-deficient diet showed no disruption of the oestrous cycles and mating took place normally. When these rats were transferred to a severely Zn-deficient diet throughout pregnancy, less than half had living young at birth and 98% showed gross congenital malformation. Subsequently it was shown that similar deleterious effects on the offspring appeared even when the Zn-deficient diet was fed only from days 6 to 14 of pregnancy (Hurley & Mutch 1973). These abnormalities apparently arose from a direct effect of lack of Zn in the foetal tissues, especially the skeletal foetal tissues, since bone deformities were obvious.

Studies carried out by Apgar (1968) have shown that severe Zn deficiency imposed upon normal adult female rats from the first day of gestation markedly affects the mothers as well as the foetuses. Seventy-five per cent of the young were found to die at birth and remarkable differences in maternal behaviour before and after parturition were observed. The Zn-deficient females delivered their litters with difficulty, suffered excessive bleeding and failed either to consume afterbirths or to prepare a nest site. The speed with which this profound effect of Zn deficiency upon maternal behaviour arose should be particularly noted. Administration of supplementary Zn as late as day 19 of pregnancy resulted in essentially normal parturition (Apgar 1973). No such behavioural disturbances arise when deficiencies of Cu and Mn are similarly imposed upon pregnant rats (Apgar 1970).

The precise mode of action of Zn in influencing maternal behaviour is unknown, but a specific effect of Zn upon behavioural processes is apparent from other studies. Thus Caldwell & Oberleas (1969) observed a significantly inferior learning ability, as measured by water maze and platform avoidance conditioning tests, in the surviving offspring of mildly Zn-deficient rats, compared with that of similar young rats from Zn-supplemented mothers. These effects of mild Zn deficiency on behaviour have since been confirmed and extended (Caldwell *et al.* 1973). Lokken *et al.* (1973) also showed that the maze acquisition ability of young rats subjected to Zn deficiency from birth to 21 days of age, and then adequately fed for 23 days, was significantly impaired compared with that of rats that had been nursed by pair-fed dams or by *ad libitum*-fed dams. These findings are consistent with earlier findings that Zn deficiency in suckling rats reduces DNA and protein synthesis in the brain and retards brain maturation, as indicated by reduced total lipid concentration (Sandstead *et al.* 1972). It is apparent that the role of Zn in brain growth, maturation and function becomes crucial in neonatal life in the rat, the critical period of brain growth in this species.

Before leaving this consideration of the intensity, duration and timing of a trace element deficiency and their relation to the physiological roles of the element, I would like to refer to some of our own work with Zn. Testicular atrophy and failure of spermatogenesis has been known to be a direct and conspicuous feature of Zn deficiency in rats for more than 20 years (Mawson & Fischer 1952). The availability of sufficient Zn for incorporation of high concentrations into sperm during the final stages of maturation is essential for the maintenance of spermatogenesis and for the survival of the germinal epithelium (Millar *et al.* 1960). In an experiment with ram lambs carried out by Underwood & Somers (1969) testicular growth was impaired and spermatogenesis ceased within 20-24 weeks on a severely Zn-deficient diet (2.4 ppm Zn). However, of particular importance in the present context are the results obtained from two further groups of ram lambs receiving diets containing 17.4 and 32.4 ppm Zn, each with unrestricted feeding. Body growth and food consumption were similar in the two groups, but testicular growth and sperm production were significantly greater in the animals receiving the larger Zn supplement. In other words, the physiological requirements for Zn for somatic growth and appetite were apparently being supplied adequately by amounts of Zn which were quite inadequate to meet the requirements for this element for another of the physiological roles of Zn, namely testicular development and spermatogenesis.

IODINE AND COBALT

The physiological roles of most of the trace elements are mediated through the presence of the elements in a wide range of active organic compounds of different structure and function. The animal body has normally the ability to synthesize these complex substances from simple inorganic or ionic forms of the element in adequate amounts and concentrations for the functional requirements of the tissues. This ability, of course, can only be discharged if the diet contains adequate amounts of the element in available forms and in proper balance with other elements and nutrients. With I and Co the position is distinctly different because the only known function of I is as a constituent of tetraiodothyronine (thyroxine) and triiodothyronine — the thyroid hormones — and the only known function of Co is as a constituent of the active cobalamins or vitamin B_{12}. The physiological roles of I in the animal are therefore apparently confined to those of the thyroid gland hormones and the physiological roles of Co to those of vitamin B_{12}.

I do not propose to discuss Co further in this paper, because its metabolic functions and those of vitamin B_{12} are to be considered in the paper to follow by Smith & Gawthorne, except to remind you of one unique, or almost unique, aspect of this trace element. I refer to the inability of the non-ruminant animal to synthesize simple forms of Co into its physiologically active form, i.e. vitamin B_{12}. Whereas the ruminant, through its ruminal microorganisms, can readily accomplish this vital transformation and supply the host tissues with their requirements for vitamin B_{12} (provided it has enough absorbable Co in its diet), the non-ruminant must obtain its supplies preformed in the diet. With these monogastric species it is thus not the Co content of the diet that matters, it is the vitamin B_{12} content that is crucial.

Since the entire functional significance of I, as was said earlier, can be accounted for through its presence in the thyroid hormones, the manifestations of I deficiency are those of a deficient supply of these hormones. However, the reverse is not necessarily true. In other words, all the signs of thyroid insufficiency can appear in the presence of adequate dietary I because of an inability of the gland either to accumulate I or to convert it into thyroactive compounds. This may arise either from the effects of goitrogenic substances or from a constitutional disability in thyroid hormonogenesis. These facts again testify to the difficulties inherent in identifying a trace element deficiency on the basis of pathological changes alone.

The thyroid hormones, and therefore I, are involved in a variety of metabolic activities, most of which derive from their fundamental effect on the rate of cellular oxidation. The mechanism of this action at the cellular or tissue level remains largely obscure. Many enzymes in various tissues are affected by variations in thyroid activity or by treatment of the animal with thyroactive substances (Hamolsky & Freedberg 1960), but direct involvement of thyroxine or triiodothyronine, and therefore I, in an enzyme system needed for the energy-transforming processes of the cells has not been demonstrated.

CHROMIUM

The essentiality of Cr for normal glucose metabolism in rats was originally demonstrated by Schwarz & Mertz (1959).

Subsequently this element was found to be necessary for growth and longevity and to be involved in lipid and protein metabolism, as well as carbohydrate metabolism (Mertz 1969).

The role of Cr in serum cholesterol homeostasis is of great interest, but is little understood. The addition of Cr to a low-Cr diet has been shown to suppress serum cholesterol levels in rats and to inhibit the tendency of these levels to rise with age. A lower incidence of spontaneous plaques and smaller amounts of stainable lipids in the aortas of the Cr-fed animals were also observed (Schroeder *et al.* 1962*b*; Schroeder & Balassa 1965; Schroeder 1969). Further evidence of Cr involvement in lipid metabolism comes from experiments demonstrating a significant increase in glucose uptake and incorporation of glucose carbon into epididymal fat with Cr plus insulin in Cr-deficient rats (Mertz *et al.* 1965).

The mode of action of Cr in glucose metabolism is rather better understood than the mechanism of the action of this element in lipid metabolism. It has been clearly shown that the effect of Cr on glucose metabolism depends on the presence of insulin; that the Cr-deficient tissue has a diminished response to insulin; and that this pattern of metal-hormone interaction occurs in all insulin-responsive systems studied, *in vitro* as well as *in vivo* (Mertz 1969). These findings have led Mertz (1969) to the hypothesis that Cr acts as a co-factor in the initial reaction of insulin with receptor sites of insulin-sensitive cell membranes.

Two further aspects of the nutritional physiology of Cr warrant brief consideration. The first is that Cr occurs in foods to a widely varying extent in combination with a small organic molecule, not yet fully characterized, which has been designated the glucose-tolerance factor or GTF. This compound is far more active in glucose metabolism than any simple Cr salt and, since both man the rat have only a limited rate of capacity to synthesize this physiologically active form of Cr from simpler forms of the element, the Cr status of a food or diet must reflect the chemical form of the Cr present, i.e. its GTF content, as well as its total Cr content (Toepfer *et al.* 1973). For example, the Cr in beer is four times as active in regulating glucose metabolism in rats, per microgram of Cr, than is the Cr in lettuce or spinach or egg-yolk (Gortner 1972).

The second aspect of Cr which I wish to mention is its practical importance in human health and nutrition. So far there is no evidence that Cr has any practical significance in the nutrition of farm animals. This may be because so little research on Cr has been done with them and because it is only within the last two years that really satisfactory methods of analysis have been developed and the ease of contamination with Cr, e.g. from stainless steel knives and needles, has been recognized. Despite these problems there is already considerable evidence that Cr intakes are insufficient in some sections of human populations. The severe impairment of glucose tolerance in malnourished (kwashiorkor or protein-calorie malnutrition) children in Jordan and Nigeria (Hopkins *et al.* 1968) and in Turkey (Gurson & Saner 1971) can be dramatically cured overnight by one oral dose of 250 μg $CrCl_3$. In similar malnourished children in Egypt, where Cr supplementation had no such effect, the tissue and food Cr levels were found to be within the normal range (Carter *et al.* 1968). In the United States, tissue Cr levels have been found to decline with age, except in the lungs (Schroeder, Balassa & Tipton 1962), in women with each pregnancy (Hambidge & Rodgerson 1969) and in children with diabetes (Hambidge *et al.* 1968). Many middle-aged and old people in the community lose their ability to metabolize sugar normally and develop what has been called 'maturity-onset diabetes'. Careful studies have shown that many of these people, but not all, can have their impaired glucose tolerance significantly improved by supplementation of their diets with Cr at the rate of 150 μg of the element per day (Levine *et al.* 1968; Hopkins & Price 1968), thus raising their total intakes from an apparently inadequate 50 μg Cr/day to 200 μg Cr/day. It is obvious that the diets consumed by such people in the U.S.A. do not always supply sufficient Cr to maintain tissue body stores or to permit satisfactory performance of the main physiological role of Cr in glucose metabolism. The extent to which Cr-responsive conditions of this nature occur in other countries, including Australia, with broadly similar dietary habits, is unknown.

SILICON

For many years Si was regarded as a rather inert element biologically, with little physiological potentiality. Research carried out during the last few years has necessitated a marked revision of this assumption. Evidence has been obtained pointing to a function for Si in the initiation of the mineralization process in bone (Carlisle 1970). Subsequently, the same investigator showed chicks raised in a low-Si environment and fed diets containing less than 0.8 ppm Si exhibited growth retardation, abnormal skeletal development and impaired feathering. When the diet was supplemented with 30 ppm Si, the bone and feather abnormalities were completely prevented and body growth increased by 35% (Carlisle 1972). Similar observations were independently made by Schwarz & Milne (1972*b*) with rats raised in plastic isolators to prevent or greatly reduce contamination from airborne dust and fed specially purified diets containing less than 5 ppm Si. The addition of 500 ppm Si to this diet increased growth by a highly significant 25.2% in one experiment and by 33.8% in another. In addition, the impairment of tooth pigmentation and the bone abnormalities in the skull observed on the low-Si diet were prevented by the supplementary Si. It was also reported that lower levels of supplementary Si gave lesser statistically insignificant responses, indicating that, relative to other trace elements, the amounts of Si needed by the rat, and to an apparently smaller extent the chick, are quite large.

In an attempt to define more precisely the physiological role of Si in animals, Schwarz (1973) produced new and exciting evidence that this element is an integral and essential component of certain mucopolysaccharides, notably hyaluronic acid and chondroitin-4-sulphate, which are important constituents of connective tissues such as cartilage and skin. Silicon, it seems, functions as a biological cross-linking agent contributing to the architecture and resilience of connective tissue. Enzymatic hydrolysis of hyaluronic acid does *not* liberate silicic acid, but leads to products of low molecular weight still containing Si in bound form. Schwarz has concluded that the Si is present as silanolate, i.e. an ether or ester-like derivative of silicic acid and that $R_1-O-Si-O-R_2$ or $R_1-O-Si-O-Si-O-R_2$ bridges play a role in the structural organization of glycosaminoglycans and polyuronides. This worker has more recently shown (Schwarz, personal communication) the high and variable concentrations of Si that occur in collagens from connective tissues. Several collagens were prepared under conditions to eliminate Si derived from water, chemicals or glass and analysed for Si by an accurate modern colorimetric method. Under these conditions, three times reprecipitated salt-soluble collagen from rat skin showed 897, acid-soluble rat skin 1997 and rat-tail tendon 1108 μg Si/g, indicating the presence of at least 3-6 atoms of Si per each α-protein chain in the collagen molecule.

FLUORINE

Biological interest in F in the past has centred largely upon its "Jekyll-and-Hyde" physiological and toxicological roles. As Dr. Jekyll it can significantly reduce the incidence of dental caries and partially prevent osteoporosis — as Mr. Hyde it can, at higher levels of intake, induce mottled enamel and other more sinister signs of fluorosis in animals and man. All attempts to demonstrate directly an essential role for F in mammalian nutrition were either negative or inconclusive (see Underwood 1971) until the indefatigible Schwarz in 1972, using his purified-diet, plastic-isolator technique, provided proof that F is an essential element for growth in the rat. Supplements to basal diets containing less than 0.04-0.46 ppm F, of 1, 2.5 and 7.5 ppm F as sodium fluoride, enhanced growth rates by 17, 30.8 and 27.9% (Schwarz & Milne 1972*a*). No other clinical signs of F deficiency were observed in the rats on the basal ration, other than some loss of pigmentation in the incisors. Subsequently Messer *et al.* (1973) showed that reproduction in mice was influenced by the level of dietary F intake. Female mice fed a diet containing 0.1-0.3 ppm F displayed a progressive development of infertility over two generations and a delayed onset of sexual maturity. No such effects were apparent when this diet was supplemented with 50 ppm F as sodium fluoride in the drinking water.

The experiments just described show that F is essential for body growth and normal reproductive performance in rats and mice, but its specific metabolic function or functions in the tissues remain obscure. Physiological concentrations of fluoride have been shown to activate several enzyme systems *in vitro* (see Schwarz & Milne 1972*a*), but biochemical defects that can be specifically related to lack of F in the living animal have not yet been revealed.

TIN, NICKEL AND VANADIUM

Little is yet known of the physiological roles of Sn, Ni and V, although it is clear that they must function in exceptionally low concentrations. Tin has been shown to be required for growth in the rat with no clinical signs or pathological changes in the animal specific to Sn (Schwarz *et al.* 1970). With Ni the position is different. Growth impairment has not been demonstrated, but chicks maintained in an all-plastic controlled environment system and consuming a diet containing less than 0.04 ppm Ni developed after four weeks an increase in skin pigmentation of the legs, swelling of the hocks and thickening of the legs near the joint area. None of these manifestations was apparent in chicks receiving the same diet supplemented with Ni at the rate of 3-5 ppm (Nielsen 1971). Preliminary studies by the same worker have revealed a relation between Ni and two hormones, thyroxine and epinephrine, but further studies are necessary to define a physiological role for Ni related to the functioning of these hormones (Nielsen 1972).

A physiological role for V in animals has long been suspected in the light of its action in certain microorganisms, its effects on phospholipid oxidation *in vitro* and its inhibition of cholesterol biosynthesis *in vitro* and *in vivo* at high doses. Not until 1970 was evidence obtained that V is necessary for the growth of feathers in chicks (Hopkins & Mohr 1971). Chicks fed a diet containing less than 0.01 ppm V showed a significant impairment in the growth of their feathers compared with control chicks on the same diet but receiving 2 ppm V in the drinking water. Subsequently, Schwarz & Milne (1971) reported a growth-stimulating effect of trace amounts of V in the diet of rats fed purified amino acid diets and kept in the isolator environment. The addition of the remarkably small amount of V of 0.1 ppm as vanadate produced a near optimal effect (Table 3). As with Sn and Ni, precise biochemical or physiological roles for V in the animal remain to be determined.

TABLE 3

*Growth response of rats to varying levels of V supplements (sodium orthovanadate, Na_3VO_4)**

Dose level (μg V/g)	Unsupplemented controls		Supplemented animals		Increase (%)	P value
	Number of rats	Average daily weight gain (g/d)	Number of rats	Average daily weight gain (g/d)		
0.01	7	1.05 ± 0.08†	7	1.27 ± 0.10	21	n.s.
0.05	16	1.04 ± 0.08	16	1.38 ± 0.08	33	< 0.01
0.10	6	1.02 ± 0.14	7	1.38 ± 0.07	35	< 0.05
0.25	14	0.87 ± 0.10	14	1.21 ± 0.09	41	< 0.02
0.50	6	1.07 ± 0.08	7	1.49 ± 0.12	39	< 0.02

*Pooled results of four successive experiments
†Mean ± standard error

REFERENCES

APGAR, J. (1968) - *Am. J. Physiol. 215* : 160

APGAR, J. (1970) - *J. Nutr. 100* : 470

APGAR, J. (1973) - *J. Nutr. 103* : 973

CALDWELL, D.F. & OBERLEAS, D. (1969) - Pan American Health Organization Science Publication No. 185

CALDWELL, D.F., OBERLEAS, D. & PRASAD, A.S. (1973) - *Nutr. Repts International 8* : 309

CARLISLE, E.M. (1970) - *Science 167* : 279

CARLISLE, E.M. (1972) - *Fedn Proc. 31* : Abstr.
CARTER, J.P., KATTAB, A., ABD-EL-HADIK, DAVIS, J.T., CHOLMY, A. & PALWARDRAN, V.N. (1968) - *Am. J. clin. Nutr. 21* : 195
GORTNER, W.A. (1972) - *Fd Technol. Aust. 24* : 504
GREEN, D.E. (1941) - *Adv. Enzymol. 1* : 177
GURSON, C.T. & SANER, G. (1971) - *Am. J. clin. Nutr. 24* : 1313
HAMBIDGE, K.M. & RODGERSON, D.O. (1969) - *Am. J. Obstet. Gynec. 103* : 320
HAMBIDGE, K.M., RODGERSON, D.O. & O'BRIEN, D. (1968) - *Diabetes 17* : 517
HAMOLSKY, M.W. & FREEDBERG, A.S. (1960) - *New Engl. J. Med. 262* : 23, 70, 129
HOPKINS, L.L. & MOHR, H.E. (1971) - *In* "Newer Trace Elements in Nutrition" (W. Mertz & W.E. Cornatzer, eds.), p. 195. Dekker : New York
HOPKINS, L.L. & PRICE, M.G. (1968) - West. Hemisphere Nutr. Congr. Puerto Rico Abstr., Vol. II, p. 40
HOPKINS, L.L., RANSOME-KURTI AND MAJAJ, A.J. (1968) - *Am. J. clin. Nutr. 21* : 203
HURLEY, L.S., EVERSON, G.J. & GEIGER, J.F. (1958) - *J. Nutr. 66* : 309
HURLEY, L.S. & MUTCH, P.B. (1973) - *J. Nutr. 103* : 649
HURLEY, L.S. & SWENERTON, H. (1966) - *Proc. Soc. exp. Biol. Med. 123* : 692
LEACH, R.M., MUENSTER, A.M. & WIEN, E.M. (1969) - *Archs Biochem. Biophys. 133* : 22
LEVINE, R.A., STREETEN, D.P.H. & DOISY, R.J. (1968) - *Metabolism 17* : 114
LOKKEN, P.M., HALAS, E.S. & SANDSTEAD, H.H. (1973) - *Proc. Soc. exp. Biol. Med. 144* : 680
MAWSON, C.A. & FISCHER, M.I. (1952) - *Can. J. med. Sci. 30* : 336
MAWSON, C.A. & FISCHER, M.I. (1952) - *Archs Biochem. Biophys. 36* : 485
MERTZ, W. (1969) - *Physiol. Rev. 49* : 163
MERTZ, W., ROGINSKI, E.E. & SCHROEDER, H.A. (1965) - *J. Nutr. 86* : 107
MESSER, H.H., ARMSTRONG, W.D. & SINGER, L. (1973) - *J. Nutr. 103* : 1319
MILLAR, M.J., FISCHER, M.I., ELCOATE, P.V. & MAWSON, C.A. (1960) - *Can. J. Biochem. Physiol. 38* : 1457
MILLS, C.F. & MURRAY, G. (1960) - *J. Sci. Fd Agric. 9* : 547
NIELSEN, F.H. (1971) - *In* "Newer Trace Elements in Nutrition" (W. Mertz & W.E. Cornatzer, eds.), p. 215. Dekker : New York
NIELSEN, F.H. (1972) - *Fedn Proc. 31* : 700 Abstr.
SANDSTEAD, H.H., GILLESPIE, D.D. & BRADY, R.N. (1972) -

Pediat. Res. 6 : 119

SCHROEDER, H.A. (1969) - *J. Nutr. 97* : 237

SCHROEDER, H.A. & BALASSA, J.J. (1965) - *Am J. Physiol. 209* : 433

SCHROEDER, H.A., BALASSA, J.J. & TIPTON, I.H. (1962) - *J. chron. Dis. 15* : 941

SCHROEDER, H.A., VINTON, W.H. & BALASSA, J.J. (1962) - *Proc. Soc. exp. Biol. Med. 109* : 859

SCHWARZ, K. (1973) - *Proc. natn. Acad. Sci. U.S.A. 70* : 1608

SCHWARZ, K. & MERTZ, W. (1959) - *Archs Biochem. Biophys. 85* : 292

SCHWARZ, K. & MILNE, D.B. (1971) - *Science 174* : 426

SCHWARZ, K. & MILNE, D.B. (1972*a*) - *Bioinorgan. Chem. 1* : 331

SCHWARZ, K. & MILNE, D.B. (1972*b*) - *Nature, Lond. 239* : 333

SCHWARZ, K., MILNE, D.B. & VINYARD, E. (1970) - *Biochem. biophys. Res. Commun. 40* : 22

STIEFEL, E.J. (1973) - *Proc. natn. Acad. Sci. U.S.A. 70* : 988

TOEPFER, E.W., MERTZ, W., ROGINSKI, E.E. & POLANSKY, M.M. (1973) - *J. agric. Fd Chem. 21* : 69

UNDERWOOD, E.J. (1971) - "Trace Elements in Human and Animal Nutrition", 3rd Edn. Academic Press : New York

UNDERWOOD, E.J. & SOMERS, M. (1969) - *Aust. J. agric. Res. 20* : 889

VALLEE, B.L. (1971) - *In* "Newer Trace Elements in Nutrition" (W. Mertz & W.E. Cornatzer, eds.). Dekker : New York

THE BIOCHEMICAL BASIS OF DEFICIENCIES OF ZINC, MANGANESE, COPPER AND COBALT IN ANIMALS

R.M. Smith and J.M. Gawthorne

Division of Nutritional Biochemistry, CSIRO
Adelaide, South Australia 5000

INTRODUCTION

Almost by definition the functions of essential trace metals in animal cells are catalytic, generally if not entirely, as components or cofactors of enzymes. The amino acid sequences of enzymes are coded for in the nucleotide sequences of chromosomal DNA. This information predetermines the eventual three-dimensional form of the protein including, for metallo-enzymes, the structure of the niche into which the appropriate metal ion must fit. Some metal-activated enzymes bind their metals so tightly that they are irreversibly inactivated if the metal is removed, whereas others show only a very loose chemical association with their metal cofactors. In all cases the metal must be present in the cell at appropriate concentrations in order to produce the amount of enzymic activity that fully expresses the controlled genetic potential of the cell. Different enzymes show different affinities for the available metal and so a deficiency of the metal will affect the activity of some enzymes more than others. But because many enzymes are present at activities seemingly far in excess of normal requirements, it is not always those enzymes most depleted in measurable activity during trace metal deficiency that limit and disrupt the processes of normal metabolism.

The present review attempts to identify and discuss those enzymic activities, critical for the function of the cell, that become rate-limiting during deficiencies of Zn, Mn, Cu and Co.

ZINC

Zinc deficiency occurs in pigs (Tucker & Salmon 1955), poultry (O'Dell *et al.* 1958), cattle (Legg & Sears 1960) and humans (Prasad *et al.* 1963) and has been induced in several species, including sheep (Underwood & Somers 1969), by feeding purified diets. The most notable effects are male infertility, retarded growth and skin lesions.

Although acute Zn deficiency does not lower the Zn concentration or reduce the activity of Zn enzymes in all tissues, some tissues (bone, testes) do show a lower Zn content and some Zn enzymes (alkaline phosphatase, carbonic anhydrase) do have diminished activity in certain tissues (Reinhold *et al.* 1967; Huber & Gershoff 1973). In general, however, the profound effects of acute Zn deficiency cannot be directly attributed to malfunctions of the well-known Zn enzymes. The primary role of Zn in nutrition appears to lie in the fundamental process of cell replication and gene expression and to involve nucleic acid and protein metabolism.

An essential function of Zn in the processes leading to DNA synthesis is now well established in mammalian cells (Sandstead & Rinaldi 1969; Williams & Chesters 1970; Rubin 1972) and evidence has been obtained that DNA polymerase from several sources contain firmly bound Zn (Chang & Bollum 1970; Slater *et al.* 1971; Stavrianopoulos *et al.* 1972; Springgate *et al.* 1973).

RNA polymerase from *Escherichia coli* also contains Zn (Scrutton *et al.* 1971) and depressed RNA polymerase activity has been reported in liver nuclei from Zn-deficient rats (Terhune & Sandstead 1972). Studies of the time course of RNA, DNA and protein synthesis in phytohaemagglutinin-stimulated lymphocytes deprived of Zn indicate a direct involvement of Zn in DNA synthesis (Chesters 1972; Williams & Loeb 1973). Work with regenerating rat liver suggests that a deficiency of Zn does not depress the rate of DNA synthesis of all cells uniformly; it leads instead to a decrease in the number of cells in which normal synthesis of DNA occurs (Fujioka & Lieberman 1964).

A concept consistent with the role of Zn in growth is that the resting cell in G1 phase is stimulated to enter S phase, subsequently to divide by a signal that triggers a series of events, one of which is the uptake of a quantity of Zn essential first for synthesis of DNA polymerase but sufficient also for the metabolic needs of both daughter cells. The concept that cells require a precise "quantum" of Zn before embarking on cell division is consistent with the near-normal Zn content of most tissues even in severely Zn-deficient animals. Each cell within the tissue contains sufficient Zn to survive but those cells programmed to divide are unable to accumulate the necessary amount of Zn. The effects of Zn deficiency are known to be most pronounced in rapidly proliferating and differentiating tissues. Zinc deficiency during pregnancy, for example, leads to gross congenital malformations in the young, accompanied by a depressed rate of incorporation of thymidine into DNA (Swenerton *et al.* 1969).

Wound healing is promoted by Zn and retarded in Zn deficiency (Pories & Strain 1966). The cause is not well understood although an involvement of Zn in the DNA synthesis that accompanies the associated cell proliferation has been suggested. In cases of extensive burns, a drastic mobilization of body Zn is known to occur (Nielsen & Jemec 1968).

Zinc deficiency is accompanied by cessation of spermatogenesis (Miller *et al.* 1964; Underwood & Somers 1969). The timing of Zn uptake during the spermatogenic cycle has been studied by Wetterdal (1958) and Parizek *et al.* (1966), and the effects of Zn deficiency on this process by Orgebin-Crist *et al.* (1971). A substantial incorporation of Zn takes place late in the cycle during maturation of spermatids to form spermatazoa. This is also the first stage to be affected by Zn deficiency; the effect is severe and specific. Monesi (1965) showed that RNA synthesis ceased after the second meiotic division very early in spermiogenesis and all the cell divisions associated with production of spermatozoa are complete prior to spermatid maturation. Thus the critical effect of Zn deficiency on male fertility may not be associated with synthesis of DNA or RNA but specifically related to the incorporation of Zn during maturation. The location of Zn in normal mature sea-urchin spermatazoa is predominantly in the metabolically active mid-piece containing the mitochondria, but also in the microfibrils of the flagellar apparatus (Morisawa & Mohri 1972). The function of this Zn is unknown.

Zinc, at physiological concentrations, inhibits ribonuclease activity (Ohtaka *et al.* 1963). Somers & Underwood (1969) found increased ribonuclease activity, accompanied by low concentrations of RNA and protein, in the testes of Zn-deficient rats. Prasad *et al.* (1973) confirmed the higher ribonuclease activity in several tissues of Zn-deficient rats, but the effect was not always accompanied by decreased concentrations of RNA, and ribonuclease activity was not shown to be related to Zn concentration in the tissues. The relationship between the substantial uptake of Zn by maturing spermatids, the prevailing activity of ribonuclease and the rate of protein synthesis remains to be studied.

Two of the gross effects of Zn deficiency, the leg deformity of chicks (Nielsen *et al.* 1967) and parakeratosis in pigs

(Dahmer *et al.* 1972), have been shown to respond to additional dietary histidine. The effect of histidine appears to arise through formation of histamine, but Zn, which also prevents the symptoms, does not affect the histamine content of the tissues (Reimann *et al.* 1971). The mechanism of the effect of histamine is not understood, but may be pharmacological. Fell *et al.* (1973) found that parakeratosis of the oesophageal epithelium of Zn-deficient rats was accompanied by an increased frequency of mitosis which decreased when Zn was given. Shortage of Zn does not appear to inhibit cell proliferation in this tissue, at least.

MANGANESE

Manganese deficiency has been found in poultry, pigs and sheep (Underwood 1971; Egan 1972). The effects of acute deficiency are similar in different species and include retarded growth, skeletal abnormalities, ataxia of the newborn and reproductive failure.

Unlike Zn, which is a functional component of many enzymes, Mn has so far been detected in only a few. At relatively high and non-physiological concentrations it stimulates many others, often by substituting for Mg. Even when the symptoms of Mn deficiency clearly indicate the site of the biochemical lesion, the isolated enzyme frequently does not contain bound Mn and in these cases a relatively specific activation by physiological concentrations of the metal *in vitro* must be accepted as evidence of a similar activation *in vivo*.

The clearest critical function of Mn is in the synthesis of glycoproteins, especially those involved in bone formation. Formation of the oligosaccharide moieties of glycoproteins is mediated by glycosyltransferases that catalyse the transfer of monosaccharides from UDP-monosaccharide donors to protein or glycoprotein acceptors with a high degree of specificity for both donor and acceptor. Chondroitin sulphate forms in cartilage as the matrix for bone formation and is synthesized through an ordered sequential action of glycosyltransferases for xylose, galactose, glucuronic acid and N-acetylglucosamine (Leach 1971). Sulphation of the oligosaccharide side-chain follows. Leach & Muenster (1962) demonstrated a reduced formation of chondroitin sulphate in the cartilage of Mn-deficient chicks and later work has shown widespread failure of glycoprotein synthesis during Mn deficiency (Tsai & Everson 1967). Manganese has been implicated in the activity *in vitro* of several glycosyltransferases (Robinson *et al.* 1966; Bosmann & Eylar 1968*a*; Spiro & Spiro 1968; Wagner & Cynkin

1969; Leach *et al.* 1969), but in most of these studies the concentrations of Mn required (5 mM-20 mM) have been well outside the physiological range of Mn concentrations in tissue water (0.01-0.1 mM). Only the galactosyltransferases have shown a high specificity for Mn at relatively low concentrations (Bosmann & Eylar 1968*b*; O'Brien & Muellenberg 1968; Khatra *et al.* 1974) and this, together with the low galactose:xylose ratio found in cartilage from Mn-deficient chicks (Leach *et al.* 1969), accompanied by low activity of the enzyme, constitutes the argument that lowered activity of galactosyltransferase is responsible for bone abnormalities in deficient animals. The neonatal ataxia that appears in young from several species when subjected to Mn deficiency during pregnancy may also be attributed to a lowered rate of glycoprotein synthesis (Schrader *et al.* 1973). The ataxia results from a congenital defect in the formation of the otoliths of the inner ear.

Deficiency of Mn results in testicular degeneration and impaired spermatogenesis in the male, defective ovulation in the female and high mortality in the offspring (Underwood 1971). No specific biochemical causes can be assigned to any of these effects although the involvement of glycoproteins and mucopolysaccharides such as hyaluronic acid in many aspects of the reproductive process (McClean & Rowlands 1942) suggests that impaired synthesis of these may be responsible for some of the effects. Reduced concentrations of hyaluronic acid have been found in the tissues of newborn guinea-pigs from Mn-deficient mothers (Tsai & Everson 1967).

Bell & Hurley (1973) have observed marked effects of Mn deficiency on the morphology of cell membranes in the tissues of mice and have recalled earlier associations in the biological effects of Mn and choline (Jukes 1940). Manganese at low concentration (50 µM) has been found to be essential for elaboration of the cell processes involved in adhesion of cultured cells to glass surfaces (Rabinovitch & de Stefano 1973).

The Mn-responsive abnormalities in glucose metabolism found in animals deficient in Mn (Everson & Schrader 1968) are still best explained as secondary to the effects of the deficiency on the beta cells of the islets of Langerhans in the pancreas (Schrader & Everson 1968) although the basis of these changes is unknown. Manganese is a component of the gluconeogenic enzyme pyruvate carboxylase (Scrutton *et al.* 1966) and may activate pyruvate kinase (Fung *et al.* 1973), but these reactions do not appear to become limiting in Mn deficiency. Similarly, although the urea-cycle enzyme arginase requires Mn for its activity (Harell & Sokolovsky 1972), there is no evidence that a deficiency of Mn restricts the capacity of the animal to produce urea (Shils & McCollum 1943).

The involvement of Zn in DNA synthesis has already been discussed. Valenzuela *et al.* (1973) have suggested that many (although not all) nucleotidyltransferases may be Zn enzymes. Manganese also is required for activity of some nucleotidyltransferases. A DNA polymerase that requires Mn at physiological concentration and probably contains Zn as well has been purified from chick embryo by Stavrianopoulos *et al.* (1972). A relatively specific requirement for Mn is exhibited by a cytoplasmic DNA polymerase purified from bone marrow cells (Byrnes *et al.* 1974) and also by an RNA-dependent RNA polymerase from rabbit reticulocytes (Downey *et al.* 1973). The joint or separate involvement of both Mn and Zn at this fundamental level of cellular activity gives grounds to suppose that some of the growth effects of deficiencies of Zn and Mn may stem from these functions. While there is good evidence that this is the case for Zn, such evidence is lacking for Mn.

COPPER

Symptoms of Cu deficiency in animals show some marked species differences as well as similarities. Depigmentation of the coat, often accompanied by changes in the physical properties of the fibres (Lee 1956), occurs in most species during Cu deficiency, and anaemia occurs in many. However, the weakening of the aorta seen in Cu-deficient pigs and poultry does not occur in sheep, and the demyelination of the central nervous system common in lambs of Cu-deficient ewes is rarely seen in other species.

Severe Cu deficiency in avian species (O'Dell *et al.* 1961) and pigs (Carnes *et al.* 1961) may lead to death by aortic rupture; this is associated with a failure to form the cross-links that complete the elaboration of the insoluble rubber-like protein, elastin. The soluble precursor tropo-elastin may be isolated in substantial amounts from the aorta of Cu-deficient pigs (Sandberg *et al.* 1971) and chicks (Rucker *et al.* 1973), but little tropo-elastin may be found in the aorta of Cu-deficient sheep (Cleary & Fanning 1974).

The cross-links of elastin are formed as the stable pyridinium compounds desmosine and isodesmosine, each the result of the condensation of four of the many peptide-linked lysine residues of tropo-elastin (Piez 1968). Condensation is preceded by oxidation of the terminal amino group of the lysines in tropo-elastin to form aldehydes, catalysed by the enzyme lysine oxidase. Lysine oxidase has been partly purified from chick embryo cartilage by Siegel *et al.* (1970) and, more extensively, from chick aorta by Harris *et al.* (1974). Both enzymes are activated by Cu and the latter enzyme was shown to contain bound Cu in the native state.

Production of aortic elastin in sheep presumably follows the same course as that in pigs and birds but the activity of lysine oxidase evidently remains sufficiently high in Cu deficiency to ensure that adequate cross-linking occurs.

Bone abnormalities in Cu deficiency have been reported in many species. In Cu-deficient chicks, bone collagen is more easily solubilized and contains less aldehyde than collagen from normal bone (Rucker *et al.* 1969). Siegel & Martin (1970) have shown that lysine oxidase intervenes in the cross-linking of soluble collagen in the cartilage of embryonic chicks. The cross-links in collagen are somewhat similar to those in elastin (Traub & Piez 1971), and it seems likely that the bone defects of Cu deficiency may also be attributed to lowered activity of lysine oxidase.

The major effects of Cu deficiency in sheep are depigmentation of black-fleeced animals, a reduction in the amount and quality of the wool and a specific demyelinating disease of young lambs known as swayback, or enzootic ataxia. The agent responsible for pigmentation in wool has not been studied but is presumably the Cu enzyme tyrosinase. Tyrosinase is responsible for oxidation and polymerization of tyrosine to form the black pigment melanin in animal tissues as well as in plants (Lerner *et al.* 1950). The cause of the decreased wool production in Cu-deficient sheep (Marston & Lee 1948) is uncertain, but the effect of Cu on wool quality is associated with the formation of disulphide cross-links between polypeptide chains in the emerging fibre (Marston 1949). The ability of inorganic Cu to catalyse the oxidation of thiols to form disulphides is well known (Cavallini *et al.* 1969).

The demyelination of the spinal cord which constitutes the essential pathology of enzootic ataxia of lambs that were Cu-deficient *in utero* (Innes & Saunders 1962) is often attributed to depleted activity of the Cu-containing respiratory enzyme, cytochrome oxidase, in the mitochondria of the associated neurones (Howell & Davison 1959; Mills & Williams 1962; Barlow 1963). The activity of cytochrome oxidase in normal mitochondria, however, seems to be in substantial excess of requirements (Gallagher *et al.* 1956; Chance *et al.* 1969; Wrigglesworth *et al.* 1973), and there has been no direct demonstration that the extent of depletion of cytochrome oxidase activity in brain from ataxic lambs is sufficient to impair respiration severely enough to cause demyelination (Lumsden 1950).

Demyelination in Cu deficiency seems to be confined to those species that undergo extensive myelination prior to birth. Thus guinea-pigs born to Cu-deficient mothers suffer from delayed myelination, particularly of the cerebellum (Everson *et al.* 1968). It is possible that the condition seen in lambs is a dysmyelination rather than a demyelination. Copper in brain has been shown to be associated with myelin-forming glial cells (Barden 1971), but its function is obscure.

The feeding of Mo and sulphate depletes liver Cu in sheep and leads to effects on wool crimp and pigmentation characteristic of Cu deficiency (Dick 1969). Matrone (1970) has suggested that formation of a Cu-Mo-S complex renders the Cu unavailable, but the precise nature of any such interaction is unknown.

The anaemia of Cu deficiency has been traced to a lack of Fe for synthesis of haemoglobin and this occurs despite elevated stores of Fe in the liver (Marston & Allen 1967; Lee *et al.* 1968). Iron is both stored and transported in the ferric state but is is released for transport as ferrous iron (Mazur *et al.* 1958). Ceruloplasmin, a Cu-protein found in plasma, has been shown to catalyse the oxidation of Fe^{++} by molecular oxygen and has also been identified as the essential plasma factor necessary for release of Fe from perfused liver into combination with the Fe-transporting protein of plasma, transferrin (Osaki *et al.* 1971). The inference that the essential action of Cu in making stored Fe available for haemopoiesis lies in the ferroxidase activity of ceruloplasmin is consistent with these observations and with the oxidative or electron transfer function of all other known Cu enzymes.

COBALT

Cobalt deficiency occurs only in ruminants (Underwood 1971) and, as shown by workers at Cornell, it responds completely to the Co-containing anti-pernicious anaemia factor, vitamin B_{12} (Smith *et al.* 1951). The surprisingly high requirement of sheep for Co to serve this purpose arises partly from the low efficiency of production of vitamin B_{12} from Co by the rumen microorganisms and partly from the low efficiency of absorption of vitamin B_{12} (Smith & Marston 1970*a*). Acute deficiency of vitamin B_{12} occurs in humans as well as in ruminants but the symptoms in the two species are markedly different. In human patients the disease of pernicious anaemia is characterized by a megaloblastic condition of bone marrow cells and by demyelination of the spinal cord. In ruminants these effects do not occur. Acute deficiency of vitamin B_{12} in sheep leads to severe and progressive loss of appetite and ultimately to death from inanition. Superimposed upon this effect is a moderate degree of metabolic inefficiency shown as a more rapid rate of body weight loss in the vitamin B_{12}-deficient animal than in its pair-fed control receiving vitamin B_{12} (Smith & Marston 1970*b*).

Most of the effects of vitamin B_{12} deficiency in sheep may be attributed to critically depressed activity of two vitamin B_{12}-containing enzymes. These are methylmalonyl-CoA mutase, which catalyses the reaction

$$\text{L-methylmalonyl-CoA} \rightleftharpoons \text{succinyl-CoA} \quad ,$$

and 5-methyl-tetrahydrofolate : homocysteine methyltransferase, which catalyses the reformation of methionine from homocysteine. The latter enzyme permits recycling of methionine following loss of its labile methyl group. It catalyses the reaction

$$\text{5-methyl-tetrahydrofolate + homocysteine} \rightleftharpoons \text{methionine + tetrahydrofolate} \ .$$

The former enzyme is responsible for one of five sequential steps in the conversion of propionate into succinate (Cannata *et al.* 1965). This pathway is of particular importance in ruminants, which derive part of their energy and much of their glucose from propionate produced in the rumen. Deficiency of vitamin B_{12} causes lowered activity of the mutase in the livers of rats (Smith & Monty 1959; Gurnani *et al.* 1960) and sheep (Marston *et al.* 1961). In sheep this is accompanied by accumulation of methylmalonic acid in the liver (Smith *et al.* 1969) and by a greatly retarded rate of clearance of injected propionate from the bloodstream (Smith & Marston 1971). The elevated concentration of propionate that appears in the blood of deficient sheep after feeding is accompanied by even higher concentrations of acetate, presumably resulting from the marked inhibitory effect of propionate on acetate metabolism in deficient sheep (Marston *et al.* 1972). Both acetate and propionate are known to depress appetite in ruminants (for review see Baile & Forbes 1974). Figure 1 shows that the level of voluntary food intake in deficient animals is inversely related to the degree of impairment of propionate metabolism and supports the view that the loss of appetite results from the known constriction in the intermediary metabolism of propionate.

Activity of the methyltransferase is also depressed in the livers of vitamin B_{12}-deficient sheep and this is accompanied

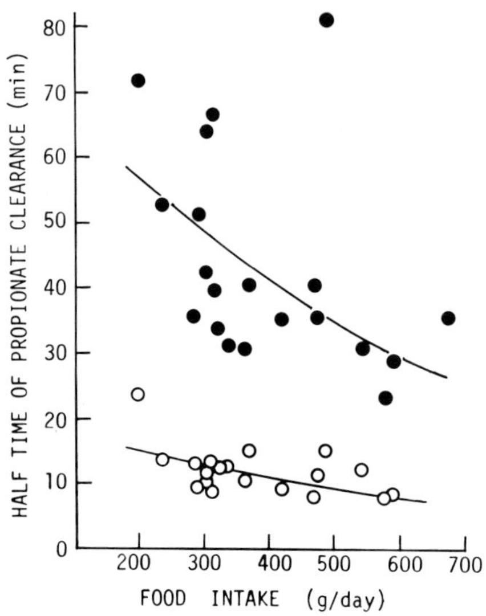

Fig. 1. The relationship between voluntary food intake and propionate clearance in vitamin B_{12}-deficient sheep. The relationship between voluntary food intake and the half-time for clearance of injected propionate is shown by closed symbols for vitamin B_{12}-deficient sheep; open symbols show results for pair-fed animals that received vitamin B_{12}. Increased impairment of propionate metabolism in deficient animals was accompanied by progressive loss of appetite. [From Marston et al. *(1972) and reproduced with the permission of the British Journal of Nutrition]*

by subnormal concentrations of S-adenosyl methionine (Gawthorne & Smith 1974) and by a fatty liver (Smith *et al.* 1974). Both defects are corrected by injection of extra methionine. The rate of turnover of the methyl group of methionine in liver is rapid (Aguilar *et al.* 1974) and restriction of methyltransferase activity evidently leads to a deficiency of available methionine. Methionine is recognized as the first-limiting amino acid for both wool growth (Reis & Schinckel 1963) and body growth (Bird & Moir 1972), even in normal sheep, and this suggests a possible basis for the impaired N retention found in vitamin B_{12}-deficient sheep (Smith & Marston 1970*b*).

A further important consequence of vitamin B_{12} deficiency is the marked depletion of liver folate stores found in vitamin B_{12}-deficient rats (Kutzbach *et al.* 1967) and sheep (Smith & Osborne-White 1973). In humans the megaloblastic anaemia of vitamin B_{12} deficiency is indistinguishable from that of dietary folate deficiency (Herbert 1973), and the response of the former to large doses of injected folate indicates a lack of available folate in the vitamin B_{12}-deficient human as well (Mollin *et al.* 1962). Despite the severe depletion of liver folates in vitamin B_{12}-deficient sheep, however, this species does not appear to develop a megaloblastic anaemia (Gawthorne *et al.* 1966).

Work with vitamin B_{12}-deficient rats (Gawthorne & Stokstad 1971) and sheep (Smith *et al.* 1974) has shown that treatment of these animals with L-methionine restores liver folates.

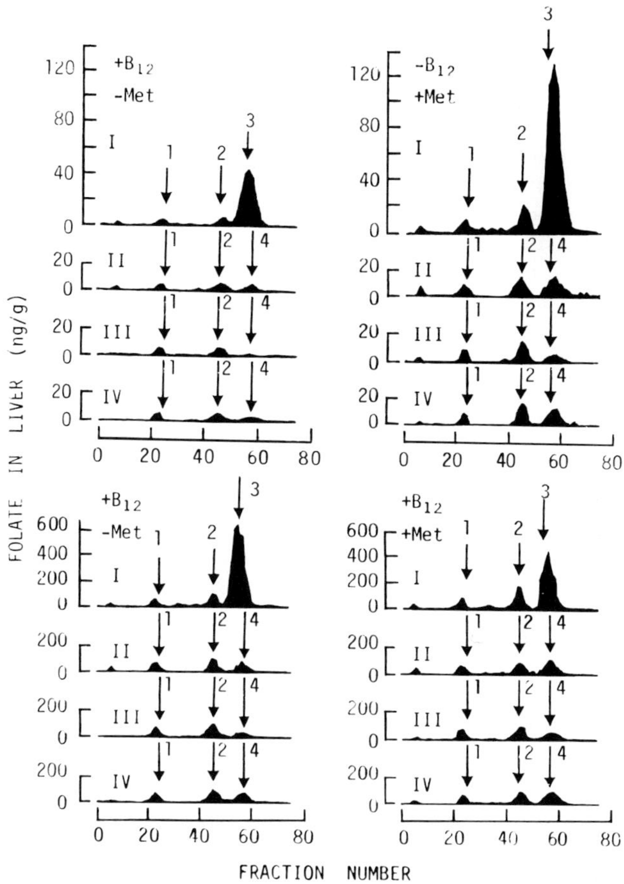

Fig. 2. Effects of vitamin B_{12} and methionine on the composition of sheep liver folates. Extracts of liver were treated with hog-kidney conjugase to convert the folates into monoglutamates before chromatography. The figure shows microbiological activity for Lactobacillus casei *(I),* Streptococcus faecalis *R. (II),* Pediococcus cerevisiae *at pH 6.3 (III) and* P. cerevisiae *at pH 5.3 (IV). The arrows indicate the elution positions of standard 10-formyl-tetrahydrofolate (1), 5-formyl-tetrahydrofolate (2), 5-methyl-tetrahydrofolate (3) and tetrahydrofolate (4). Results for animals with and without vitamin B_{12} are plotted on different scales. The results demonstrate the heavy depletion of all forms of folate suffered by vitamin B_{12}-deficient animals and the partial restoration of all these folates following treatment of the animals for 28 days with methionine. [From Smith* et al. *(1974) and reproduced with the permission of the Biochemical Journal.]*

Figure 2 shows that methionine, like vitamin B_{12}, increases the concentrations of all the major classes of folate in the livers of vitamin B_{12}-deficient sheep. The results suggest that the effect of vitamin B_{12} on folate metabolism, as well as its effect on mobilization of liver lipids, is exerted *via* methionine. However, the amounts of methionine needed for complete restoration of liver folates are considerably greater than those needed for the lipotropic effect. Evidence has been obtained that methionine exerts its effects on folate retention by promoting transport of folate into the vitamin B_{12}-deficient liver cell (Gawthorne & Smith 1974).

The megaloblastic changes in haemopoietic and epithelial cells of patients with pernicious anaemia (Whitehead *et al.* 1973) are attributed to impaired DNA synthesis and are thought to result from a lack of 5,10-methylene tetrahydrofolate, needed for production of the DNA precursor, thymidylate. The finding of Haurani (1973) of a severely depressed activity of thymidylate synthetase itself in phytohaemagglutinin-stimulated lymphocytes from patients with pernicious anaemia is therefore surprising and requires confirmation.

The demyelination in the spinal cord of patients with pernicious anaemia is also unexplained, although it has been the subject of considerable speculation. Thus, although many of the effects of vitamin B_{12} deficiency in the ruminant may be traced to specific enzymic inadequacies, the metabolic effects of human vitamin B_{12} deficiency are not so readily explained.

REFERENCES

AGUILAR, T.S., BENEVENGA, N.J. & HARPER, A.E. (1974) - *J. Nutr. 104* : 761

BAILE, C.A. & FORBES, J.M. (1974) - *Physiol. Rev. 54* : 160

BARDEN, H. (1971) - *J. Neuropath. exp. Neurol. 30* : 650

BARLOW, R.M. (1963) - *J. comp. Path. 73* : 61

BELL, L.T. & HURLEY, L.S. (1973) - *Lab. Invest. 29* : 723

BIRD, P.R. & MOIR, R.J. (1972) - *Aust. J. biol. Sci. 25* : 835

BOSMAN, H.B. & EYLAR, E.H. (1968*a*) - *Biochem. Biophys. Res. Commun. 30* : 89

BOSMAN, H.B. & EYLAR, E.H. (1968*b*) - *Biochem. biophys. Res. Commun. 33* : 340

BYRNES, J.J., DOWNEY, K.M. & SO, G.A. (1974)- *Proc. natn. Acad. Sci. U.S.A. 71* : 205

CANATA, J.J.B., FOCESI, A., MAZUMDER, R., WARNER, R.C. & OCHOA, S. (1965) - *J. biol. Chem. 240* : 3249

CARNES, W.H., SHIELDS, G.S., CARTWRIGHT, G.E. & WINTROBE, M.M.

(1961) - *Fedn Proc. 20* : 118
CAVALLINI, D., DE MARCO, C., DUPRÈ, S. & ROTILIO, G. (1969) - *Archs Biochem. Biophys. 130* : 354
CHANCE, B., AZZI, A., LEE, I.V., LEE, C.P. & MELA, L. (1969) - *In* "Mitochondria Structure and Function", FEBS Symposium, Vol. 17 (L. Ernster & Z. Drahota, eds.), p. 233. Academic Press : New York
CHANG, L.M.S. & BOLLUM, F.J. (1970) - *Proc. natn. Acad. Sci. U.S.A. 65* : 1041
CHESTERS, J.K. (1972) - *Biochem. J. 130* : 133
CLEARY, E.G. & FANNING, J.C. (1974) - *Proc. Aust. Soc. med. Res.* (in press)
DAHMER, E.J., COLEMAN, B.W., GRUMMER, R.H. & HOEKSTRA, W.G. (1972) - *J. Anim. Sci. 35* : 1181
DICK, A.T. (1969) - *Outl. Agric. 6* : 14
DOWNEY, K.M., BYRNES, J.J., JURMARK, B.S. & SO, A.G. (1973) - *Proc. natn. Acad. Sci. U.S.A. 70* : 3400
EGAN, A.R. (1972) - *Aust. J. exp. Agric. Anim. Husb. 12* : 131
EVERSON, G.J. & SCHRADER, R.E. (1968) - *J. Nutr. 94* : 89
EVERSON, G.J., SCHRADER, R.E. & WANG, T. (1968) - *J. Nutr. 96* : 115
FELL, B.F., LEIGH, L.C. & WILLIAMS, R.B. (1973) - *Res. vet. Sci. 14* : 317
FUJIOKA, M. & LIEBERMAN, I. (1964) - *J. biol. Chem. 239* : 1164
FUNG, C.H., MILDVAN, A.S., ALLERHAND, A., KOMOROSKI, R. & SCRUTTON, M.C. (1973) - *Biochemistry 12* : 620
GALLAGHER, C.H., JUDAH, J.D. & REES, K.R. (1956) - *Proc. R. Soc. Series B 145* : 134
GAWTHORNE, J.M. & SMITH, R.M. (1974) - *Biochem. J. 142* : 119
GAWTHORNE, J.M., SOMERS, M. & WOODLIFF, H.J. (1966) - *Aust. J. exp. Biol. med. Sci. 44* : 585
GAWTHORNE, J.M. & STOKSTAD, E.L.R. (1971) - *Proc. Soc. exp. Biol. Med. 136* : 42
GURNANI, S., MISTRY, S.P. & JOHNSON, B.C. (1960) - *Biochim. biophys. Acta 38* : 187
HARELL, D. & SOKOLOVSKY, M. (1972) - *Eur. J. Biochem. 25* : 102
HARRIS, E.D., GONNERMAN, W.A., SAVAGE, J.E. & O'DELL, B.L. (1974) - *Biochim. biophys. Acta 341* : 332
HAURANI, F.I. (1973) - *Science 182* : 78
HERBERT, V. (1973) - *In* "Modern Nutrition in Health and Disease", 5th Edn (R.S. Goodheart & M.E. Shils, eds.), p. 221. Lea & Febiger : Philadelphia
HOWELL, J. McC. & DAVISON, A.N. (1959) - *Biochem. J. 72* : 365
HUBER, A.M. & GERSHOFF, S.N. (1973) - *J. Nutr. 103* : 1175
INNES, J.R.M. & SAUNDERS, L.Z. (1962) - "Comparative Neuropathology". Academic Press : New York
JUKES, T.H. (1940) - *J. Nutr. 20* : 445

KHATRA, B.S., HERRIES, D.G. & BREW, K. (1974) - *Eur. J. Biochem. 44* : 537
KUTZBACH, C., GALLOWAY, E. & STOKSTAD, E.L.R. (1967) - *Proc. Soc. exp. Biol. Med. 124* : 801
LEACH, R.M. (1971) - *Fedn Proc. 30* : 991
LEACH, R.M. & MUENSTER, A.M. (1962) - *J. Nutr. 78* : 51
LEACH, R.M., MUENSTER, A.M. & WIEN, E.M. (1969) - *Archs Biochem. Biophys. 133* : 22
LEE, G.R., NACHT, S., LUKENS, J.N. & CARTWRIGHT, G.E. (1968) - *J. clin. Invest. 47* : 2058
LEE, H.J. (1956) - *J. agric. Sci., Camb. 47* : 218
LEGG, S.P. & SEARS, L. (1960) - *Nature, Lond. 186* : 1061
LERNER, A.B., FITZPATRICK, T.B., CALKINS, E. & SUMMERSON, W.H. (1950) - *J. biol. Chem. 187* : 793
LUMSDEN, C.E. (1950) - *J. Neurol. Neurosurg. Psychiat. 13* : 1
MARSTON, H.R. (1949) - *Proc. R. Soc. Series A 199* : 273
MARSTON, H.R. & ALLEN, S.H. (1967) - *Nature, Lond. 215* : 645
MARSTON, H.R., ALLEN, S.H. & SMITH, R.M. (1961) - *Nature, Lond. 190* : 1085
MARSTON, H.R., ALLEN, S.H. & SMITH, R.M. (1972) - *Br. J. Nutr. 27* : 147
MARSTON, H.R. & LEE, H.J. (1948) - *Aust. J. scient. Res. Series B 1* : 376
MATRONE, G. (1970) - *In* "Trace Element Metabolism in Animals" (C.F. Mills, ed.), p. 354. E. & S. Livingstone : Edinburgh
MAZUR, A., GREEN, S., SAHA, A. & CARLETON, A. (1958) - *J. clin. Invest. 37* : 1809
McCLEAN, D. & ROWLANDS, I.W. (1942) - *Nature, Lond. 150* : 627
MILLER, W.J., PITTS, W.J., CLIFTON, C.M. & SCHMITTLE, S.C. (1964) - *J. Dairy Sci. 47* : 556
MILLS, C.F. & WILLIAMS, R.B. (1962) - *Biochem. J. 85* : 629
MONESI, V. (1965) - *Expl. Cell. Res. 39* : 197
MOLLIN, D.L., WATERS, A.H. & HARRISS, E. (1962) - *In* "Vitamin B_{12} und Intrinsic Factor: 2 Europäeisches Symposion, 1961" (H.C. Heinrich, ed.), p. 737. Ferdinand Enke : Stuttgart
MORISAWA, M. & MOHRI, H. (1972) - *Expl. Cell Res. 70* : 311
NIELSEN, F.H., SUNDE, M.L. & HOEKSTRA, W.G. (1967) - *Proc. Soc. exp. Biol. Med. 124* : 1106
NIELSEN, S.P. & JEMEC, B. (1968) - *Scand. J. Plast. Reconstr. Surg. 2* : 45
O'BRIEN, P.J. & MUELLENBERG, C.G. (1968) - *Biochim. biophys. Acta 158* : 189
O'DELL, B.L., HARDWICK, B.C., REYNOLDS, G. & SAVAGE, J.E. (1961) - *Proc. Soc. exp. Biol. Med. 108* : 402
O'DELL, B.L., NEWBERNE, P.M. & SAVAGE, J.E. (1958) - *J. Nutr. 65* : 503
OHTAKA, Y., UCHIDA, K. & SAKAI, T. (1963) - *J. Biochem. 54*:322

ORGEBIN-CRIST, M.-C., FREEMAN, M. & BARNEY, G.H. (1971) - *Ann. Biol. Anim. Biochem. Biophys. 11* : 547

OSAKI, S., JOHNSON, D.A. & FRIEDEN, E. (1971) - *J. biol. Chem. 246* : 3018

PARIZEK, J., BOURSNELL, J.C., HAY, M.F., BABICKY, A. & TAYLOR, D.M. (1966) - *J. Reprod. Fert. 12* : 501

PIEZ, K.A. (1968) - *A. Rev. Biochem. 37* : 547

PORIES, W.J. & STRAIN, W.H. (1966) - *In* "Zinc Metabolism" (A.S. Prasad, ed.), p. 378. Charles G. Thomas : Springfield

PRASAD, A.S., OBERLEAS, D., KONIUCH, D. & DU MOUCHELLE, E. (1973) - *J. Lab. clin. Med. 82* : 461

PRASAD, A.S., SCHULERT, A.R., MIALE, A., FARID, Z. & SANDSTEAD, H.H. (1963) - *Am. J. clin. Nutr. 12* : 437

RABINOVITCH, M. & DE STEFANO, M.J. (1973) - *J. Cell Biol. 59* : 165

REIMANN, E.M., SUNDE, M.L. & HOEKSTRA, W.G. (1971) - *J. Nutr. 101* : 1623

REINHOLD, J.G., KFOURY, G.A. & THOMAS, T.A. (1967) - *J. Nutr. 92* : 173

REIS, P.J. & SCHINKEL, P.G. (1963) - *Aust. J. biol. Sci. 16* : 218

ROBINSON, H.C., TELSER, A. & DORFMAN, A. (1966) - *Proc. natn. Acad. Sci. U.S.A. 56* : 1859

RUBIN, H. (1972) - *Proc. natn. Acad. Sci. U.S.A. 69* : 712

RUCKER, R.B., GOETTLICH-RIEMANN, W. & TOM, K. (1973) - *Biochim. biophys. Acta 317* : 193

RUCKER, R.B., PARKER, H.E. & ROGLER, J.C. (1969) - *J. Nutr. 98* : 57

SANDBERG, L.B., WEISSMAN, N. & GRAY, W.R. (1971) - *Biochemistry 10* : 52

SANDSTEAD, H.H. & RINALDI, R.A. (1969) - *J. cell. Physiol. 73* : 81

SCHRADER, R.E., ERWAY, L.C. & HURLEY, L.S. (1973) - *Teratology 8* : 257

SCHRADER, R.E. & EVERSON, G.J. (1968) - *J. Nutr. 94* : 269

SCRUTTON, M.C., UTTER, M.F. & MILDVAN, A.S. (1966) - *J. biol. Chem. 241* : 3480

SCRUTTON, M.C., WU, C.W. & GOLDTHWAIT, D.A. (1971) - *Proc. natn. Acad. Sci. U.S.A. 68* : 2497

SHILS, M.E. & McCOLLUM, E.V. (1943) - *J. Nutr. 26* : 1

SIEGEL, R.C. & MARTIN, G.R. (1970) - *J. biol. Chem. 245* : 1653

SIEGEL, R.C., PINNELL, S.R. & MARTIN, G.R. (1970) - *Biochemistry 9* : 4486

SLATER, J.P., MILDVAN, A.S. & LOEB, L.A. (1971) - *Biochem. biophys. Res. Commun. 44* : 37

SMITH, R.M. & MARSTON, H.R. (1970*a*) - *Br. J. Nutr. 24* : 857

SMITH, R.M. & MARSTON, H.R. (1970*b*) - *Br. J. Nutr. 24* : 879

SMITH, R.M. & MARSTON, H.R. (1971) - *Br. J. Nutr. 26* : 41
SMITH, R.M. & MONTY, K.J. (1959) - *Biochem. biophys. Res. Commun. 1* : 105
SMITH, R.M. & OSBORNE-WHITE, W.S. (1973) - *Biochem. J. 136*:279
SMITH, R.M., OSBORNE-WHITE, W.S. & GAWTHORNE, J.M. (1974) - *Biochem. J. 142* : 105
SMITH, R.M., OSBORNE-WHITE, W.S. & RUSSELL, G.R. (1969) - *Biochem. J. 112* : 703
SMITH, S.E., KOCH, B.A. & TURK, K.L. (1951) - *J. Nutr. 44*:455
SOMERS, M. & UNDERWOOD, E.J. (1969) - *Aust. J. biol. Sci. 22* : 1277
SPIRO, M.J. & SPIRO, R.G. (1968) - *J. biol. Chem. 243* : 6529
SPRINGGATE, C.F., MILDVAN, A.S., ABRAMSON, R., ENGLE, J.L. & LOEB, L.A. (1973) - *J. biol. Chem. 248* : 5987
STAVRIANOPOULOS, J.G., KARKAS, J.D. & CHARGAFF, E. (1972) - *Proc. natn. Acad. Sci. U.S.A. 69* : 1781
SWENERTON, H., SCHRADER, R. & HURLEY, L.S. (1969) - *Science 166* : 1014
TERHUNE, M.W. & SANDSTEAD, H.H. (1972) - *Science 177* : 68
TRAUB, W. & PIEZ, K.A. (1971) - *Adv. Protein Chem. 25* : 243
TUCKER, H.F. & SALMON, W.D. (1955) - *Proc. Soc. exp. Biol. Med. 88* : 613
TSAI, H.C. & EVERSON, G.J. (1967) - *J. Nutr. 91* : 447
UNDERWOOD, E.J. (1971) - *In* "Trace Elements in Human and Animal Nutrition", 3rd Edn. Academic Press : New York
UNDERWOOD, E.J. & SOMERS, M. (1969) - *Aust. J. agric. Res. 20* : 889
VALENZUELA, P., MORRIS, R.W., FARAS, A., LEVINSON, W. & RUTTER, W.J. (1973) - *Biochem. biophys. Res. Commun. 53* : 1036
WAGNER, R.R. & CYNKIN, M.A. (1969) - *Archs Biochem. Biophys. 129* : 242
WETTERDAL, B. (1958) - *Acta radiol. Suppl. 156* : 1
WHITEHEAD, N., REYNER, F. & LINDENBAUM, J. (1973) - *J. Am. med. Ass. 226* : 1421
WILLIAMS, R.B. & CHESTERS, J.K. (1970) - *Br. J. Nutr. 24* : 1053
WILLIAMS, R.O. & LOEB, L.A. (1973) - *J. Cell Biol. 58* : 594
WRIGGLESWORTH, J.M., BAUM, H. & NICHOLS, P. (1973) - *FEBS Lett. 35* : 106

THE ROLE AND THE METABOLISM OF SELENIUM IN THE ANIMAL

K.O. Godwin

Division of Nutritional Biochemistry, CSIRO
Adelaide, South Australia 5000

INTRODUCTION

The renowned evangelist D.L. Moody told the story of one Valentine Burke, whose photograph once adorned the wall of the rogues' gallery at police headquarters in New York, but which was removed when after a change of heart Burke became a respected member of the police force. There is a recognizable parallel with the apparent transformation in the role of Se, long recognized as a toxic element and highly poisonous to stock, until its identification in 1957 as an important trace element, and its use subsequently in the virtual elimination of a world-wide disease in ruminants.

The white muscle disease story, the association of Se with Factor 3, a hepato-protective factor in rats fed certain necrogenic diets, and the subsequent demonstration of the efficacy of minute quantities of inorganic Se in protecting against the malady, has been told many times. It is not our intention to retell it; suffice it to say that Australasia, with situations such as the pumice soils of the North Island of New Zealand and the Archaean granites of Western Australia, has presented situations where the elimination of white muscle disease has been of major economic importance. Here, therefore, is an area where decisive and substantial contributions to our understanding of Se have been made, particularly where field application of the research is concerned.

THE RECOGNITION OF SELENIUM REQUIREMENTS

It was following the claim by Schwarz & Foltz, in 1957, that Se was an integral part of Factor 3, that Se was examined first in Oregon (Muth *et al.* 1958) and then in New Zealand (McLean *et al.* 1959), and found to be important in the ruminant. In New Zealand the early work of the 1960's led to the recognition of three somewhat arbitrarily separated conditions associated with a low Se intake in sheep. The first of these was frank white muscle disease in lambs and hoggets, the second was infertility in ewes, and the third was Se-responsive unthriftiness in lambs.

Affected areas were soon defined by New Zealand workers (Fig. 1) and by Australian workers (Fig. 2), and many experiments have been subsequently conducted in these and other areas.

By 1960, therefore, Se was firmly linked with a disease entity which had been recognized by veterinarians for many years. It had been studied by workers in Oregon, Cornell, Wisconsin, Montana, Nevada, Scotland, New Zealand and elsewhere. It was associated with the feeding of legumes, particularly lush clover pastures, and clearly was a complex deficiency in some way involving vitamin E and selenium.

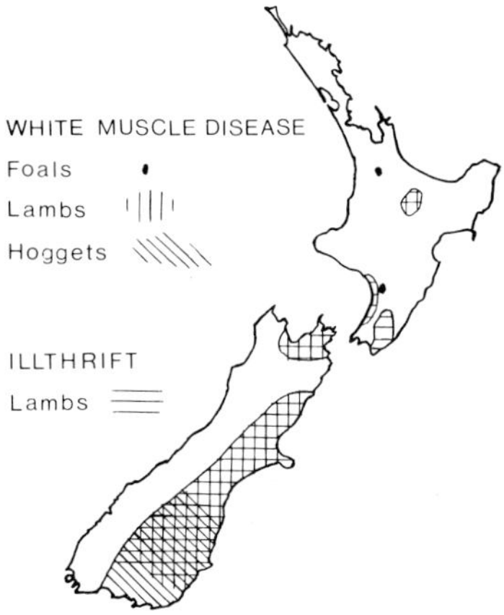

Fig. 1. Selenium-responsive areas in New Zealand. [Reprinted from Federation Proceedings 20 : 679 (1961), by W.J. Hartley & A.B. Grant]

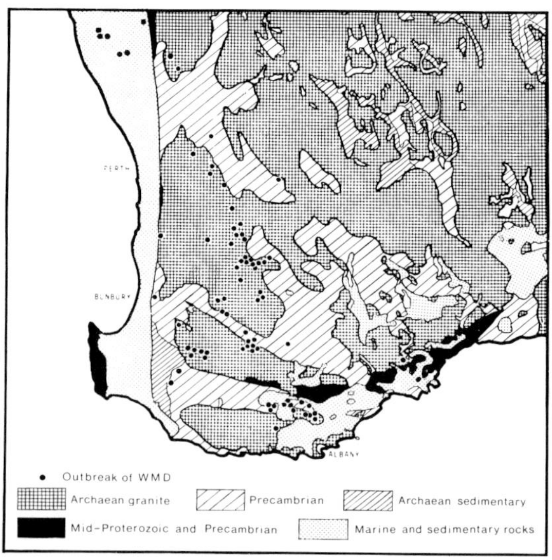

Fig. 2. Geology of south-western Western Australia showing outbreaks of white muscle disease. [Reprinted from Outlook on Agriculture 6 : 19 (1969), by M.R. Gardiner]

EXPERIMENTAL SELENIUM DEFICIENCY

Work with Se was initiated in our laboratory toward the end of 1961; the starting point was to reproduce the deficiency syndrome reported by Schwarz (1951), using Torula yeast as the sole protein source. Necrotic livers and respiratory decline in liver slices were confirmed by these studies but our main interest was in relating the syndrome in rats to muscular dystrophy in sheep, and when studies were made of respiratory behaviour in muscle slices, no characteristic abnormalities were found. Electrocardiographic records of deficient rats did, however, show a characteristic abnormality (Godwin 1965).

The studies were extended at this point to lambs. Firstly, an attempt was made, by suitable adaptation of the Torula yeast diet, to reproduce white muscle disease experimentally in lambs. Secondly the respiratory properties of liver and muscle in such animals were examined in order to compare the situation with rats. Thirdly, studies were undertaken to determine whether the ECG abnormalities reported in Se-deficient rats had any counterpart in lambs.

The diet fed to rats was modified as a milk substitute and fed in semi-liquid form to 4- to 5-week-old normal Merino lambs. The lambs were established on a pelleted form of the diet made more acceptable by the inclusion of cracked wheat, after determining that the latter had a low Se content (Godwin & Fraser 1966).

White muscle disease was produced in these lambs, usually within 8-10 weeks of their commencing on the milk substitute. In the majority of cases, electrocardiograms showed a progressive development of a characteristic abnormality; just before death the rise in amplitude of the T-wave gave way to an elevated S-T segment, similar to that seen in myocardial infarction in man. Paralleling these changes in time, a fall in blood pressure, especially marked in the extremities, was present to a greater or lesser degree. It was suggested, on the basis of these studies, that the fundamental change occurring in Se deficiency may be circulatory failure.

FIELD STUDIES

At about this time we were asked to examine the possibility that the loss of a group of lambs on a property at Myponga, South Australia, was due to Se deficiency. This led to involvement in field work which continued for some years. Trials were set up at Myponga in 1965. The sheep were part of a Corriedale stud; an incipient Se deficiency was present, with mild dystrophic conditions being occasionally found. However, the only measurable effect observed in a series of trials was in relation to wool characteristics, where in both hoggets and mature ewes staple length was increased due to Se treatment (hoggets from 7.90 to 8.94 cm, S.E. ± 0.216 $p < 0.01$; ewes from 10.22 to 11.53 cm, S.E. ± 0.249 $p < 0.001$). The significance of this observation has never been followed up.

Abnormal ECG patterns were found during the Myponga trial, but they occurred infrequently. However, their presence, associated with a mild dystrophic condition, rationalized the further extension of similar field observations. A survey was therefore conducted, with the cooperation of officers of the South Australian Department of Agriculture, in several areas, e.g. around Keith and on Kangaroo Island. In the course of this survey it was established that a higher incidence of ECG abnormalities was associated with grazing Se-low pastures than with 'normal' pastures (Godwin 1968), and that on many of these properties there was a high proportion of oestrogenic pasture, e.g. Yarloop (*Trifolium subterraneum* L. var. Yarloop). These observations led to field trials on Kangaroo Island in which the hypothesis was tested that oestrogenic infertility was associated in some situations with Se deficiency (Godwin *et al.* 1970).

THE PROBLEM OF SELENIUM ADMINISTRATION

Selenium administration to sheep had, meanwhile, during the Myponga trials in 1965, accounted for a substantial part of the handling time. Clearly a means of Se dosing, analogous to that of the Co pellet developed in this Division (Dewey *et al.* 1958) was needed. Schwarz, in discussing the effectiveness of a large number of Se compounds in protecting against dietary hepatic necrosis (1951), had rated elemental Se as 'inactive' and at least 300 times less potent than selenite. Early in 1966, groups of six rats were fed Torula yeast diets; a control group receiving no supplement had an average survival time of 38 days. One group was supplemented with 10 µg Se as sodium selenite each week, another with 1000-fold the amount of Se as elemental Se. This was offered to each rat as a small pellet mixed with food after the animals had been deprived of food for 24 hr. All supplemented animals survived and growth rates for those receiving elemental Se were similar to those receiving selenite. Warburg respirometry studies on these rats showed that respiratory decline was not present in liver slices from the Se-supplemented groups.

THE SELENIUM PELLET

A programme of testing various Se pellets in the sheep was commenced in 1967 and a successful pellet containing elemental Se was designed and manufactured (Kuchel & Buckley 1969). By examining the long-term effects of inserting four such pellets in the rumen, and by examining the distribution of radioactive Se derived from the pellets labelled with ^{75}Se (Handreck & Godwin 1970), it became apparent that it was a safe and reliable means of administering Se to the grazing animal. Work has continued in this area with the result that recently the Se pellet for sheep was accepted by the Technical Committee on Veterinary Drugs. Presently a pellet suitable for use with cattle is being tested under field conditions, and the results appear to be most satisfactory (R.E. Kuchel, unpublished observations).

FURTHER INVOLVEMENT IN FIELD WORK

The field trials on Kangaroo Island, in which the Se pellet was later used, showed that Se did have a significant effect in reducing infertility in ewes. However, there was no interaction proved between Se and oestrogenic infertility, although the Se effect was appreciable ($0.1 > P > 0.05$), as the final results showed (Godwin *et al.* 1970).

SELENIUM ANALYSIS

Involvement with field work, particularly on Kangaroo Island, led to many Se assays being required throughout all of this work. These assays have been performed by R.A. Buckley of this Division using the method of Watkinson (1966). The method has proved very sensitive and many figures have been obtained for Se levels in blood, other tissues, milk and pasture. Figures for pasture analysis in some of the locations where we experienced frank white muscle disease were of the order of 0.02-0.03 μg/g.

FEEDING LOW-SELENIUM FODDER UNDER EXPERIMENTAL CONDITIONS

In 1969, under experimental conditions, lambs were fed hay from a Se-responsive area. The hay had a Se level by analysis of 0.015 μg/g. It was hammer-milled, mixed with approximately 10% water and pelleted. The only supplement the animals received, unless they were in protected groups, was a weekly supplement of 30,000 I.U. vitamin A and 1500 μg calciferol. This has proved a very successful method of inducing white muscle disease, although the degree of muscle breakdown is seldom as severe as that generally seen with Torula yeast-type diets.

Blood Se levels fall steadily in all unsupplemented lambs and in the majority of animals dystrophic lesions are apparent within 8-10 weeks of being transferred to the Se-low feed.

EFFECTS OF EXERCISE

At this time, because of the large amounts of Se-low feed required, firstly of the synthetic diet, then latterly of the fodder, the concept was adopted that exercise might exacerbate the condition and shorten the time that the animals needed to be exposed to the dystrophogenic diets. The rationale behind this concept was reviewed briefly in 1972 (Godwin 1972). A treadmill was constructed and lambs subjected to daily exercise (0.25-0.5 km). The animals were maintained on Se-low rations. Instead of this increasing the severity of white muscle disease over a given period, the exercise decreased it. As judged by onset of lesions, severity of ECG changes and creatine phosphokinase levels, it was concluded that regular exercise had a moderating effect on the development of nutritional dystrophy in lambs (Godwin 1972). The general conclusions fitted observations made by Vignos & Watkins (1966) in studies on the effect of exercise in certain types of hereditary muscular dystrophy in man.

BIOCHEMICAL ASPECTS OF WHITE MUSCLE DISEASE

The technique for developing typical white muscle disease in lambs having been firmly established, experiments were undertaken to explain the mechanism of the dystrophic changes in muscle. The presence of calcification in muscles affected by white muscle disease had long been recognized (Muth 1955), but no systematic attempt had been made to study the retention of Ca, and certainly no suggestion had been made that Ca retention might affect the respiratory properties of the muscle tissue.

Examination of Ca levels of heart and skeletal whole muscle and of isolated mitochondria from the same tissues showed that levels were elevated even in mildly dystrophic cases. Associated with these changes in Ca level has been found a reduction in respiratory rate of the isolated and washed mitochondria (Godwin *et al.* 1974). These changes are found with only certain substrates, namely fatty acids. Currently work is in progress to follow up preliminary observations that indicate that in the intact dystrophic animal A-V differences across the dystrophic hind limb demonstrate the same impairment in substrate oxidation (I.G. Jarrett & K.O. Godwin, unpublished observations). The observed changes in respiratory properties in the skeletal muscle mitochondria are similar to changes reported by Wrogemann and others in studies on hereditary muscular dystrophy in mice and hamsters (Lin *et al.* 1970; Wrogemann *et al.* 1970). Recently Wrogemann *et al.* (1973) have suggested that the defects they observed in mitochondria from genetically dystrophic hamsters may be partly explained by the loss of pyridine nucleotides associated with the high levels of Ca in necrotic areas of the dystrophic muscle. However, although levels of NAD^+ were low in their dystrophic mitochondria, its addition did not restore respiratory control, indicating a complex abnormality. It may well be that similarities between nutritional muscular dystrophy and hereditary muscular dystrophy may continue to be a fruitful area of research where Se function is concerned.

CALCIUM RETENTION

The recognition of the early retention of Ca during the onset of white muscle disease led us to examine the excretion of Ca in the lamb fed a Se-low diet. The result proved equivocal as, although in one series of experiments there appeared to be an obvious effect of Se deficiency in reducing Ca excretion, experiments the following year failed to confirm these differences. The reason for this equivocation is unknown. However, in view of the amounts of Ca involved in the early 'calcification' of muscle compared with the amounts excreted, it seems reasonable to suggest, in retrospect, that the Ca retention by muscle is unlikely to be reflected in the figures for daily excretion. Currently experiments are being conducted using ^{45}Ca to determine turnover rates in dystrophic and supplemented lambs.

AUTORADIOGRAPHIC STUDIES USING ^{45}Ca

Using an autoradiographic technique, observations which may have a similar basis of explanation have been made using ^{45}Ca in dystrophic rats. By the inclusion of 1.5% methionine

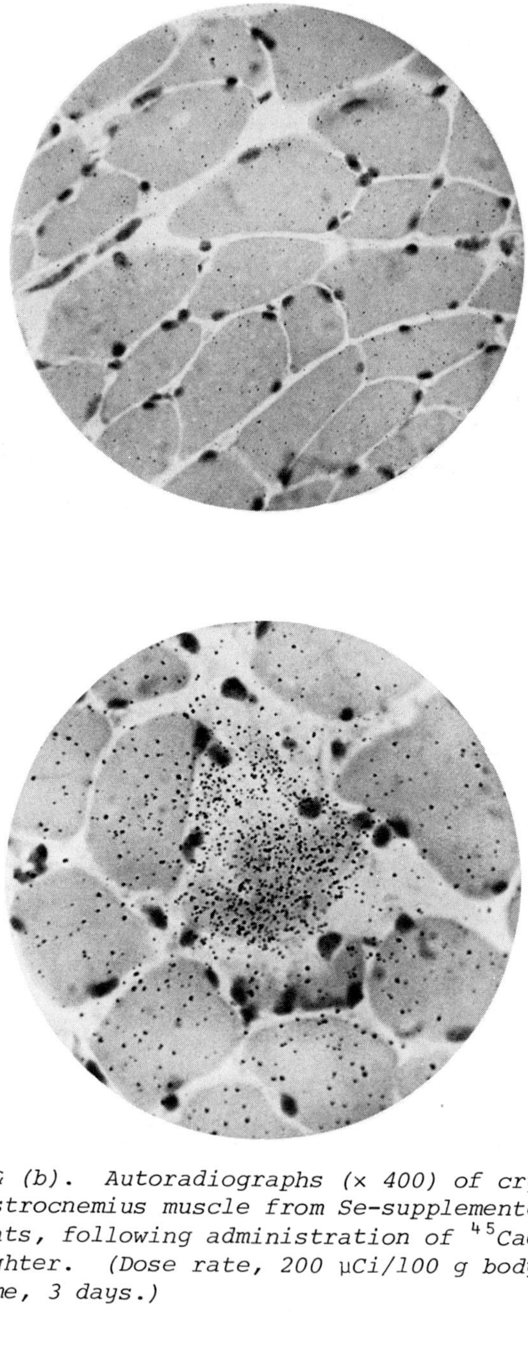

Fig. 3 (a) & (b). Autoradiographs (× 400) of cryostat sections of gastrocnemius muscle from Se-supplemented and Se-deficient rats, following administration of $^{45}CaCl_2$ one hr before slaughter. (Dose rate, 200 μCi/100 g body weight; exposure time, 3 days.)

in the Torula yeast-type diet of Schwarz (Rotruck *et al.* 1972), the dietary hepatic necrosis is prevented and typical dystrophic lesions appear in the skeletal muscle. In experiments with deficient and protected rats, we have administered ^{45}Ca as calcium chloride intraperitoneally (200 μCi/100 g body weight) 1 hr before slaughter. Muscle (gastrocnemius), kidney and liver sections have been cut with a cryostat and mounted using Appleton's technique (Appleton 1964). After 3 days' exposure, autoradiographs show that as early as 3 weeks after being fed the Se-low diet, radioactivity is concentrated in discrete fibres in muscle (Fig. 3) and discrete areas in the kidney (Fig. 4). Lesions in either of these tissues are not evident by conventional histological staining until 6-7 weeks. No unusual retention of ^{45}Ca is apparent in the livers of either the supplemented or unsupplemented animals.

INCORPORATION OF ^{75}Se INTO PROTEIN

Our interest in the next part of the Se work to be discussed was aroused quite early in the programme in view of the seeming unimportance of the form in which Se was administered. Whether given in inorganic or organic form — as selenoamino

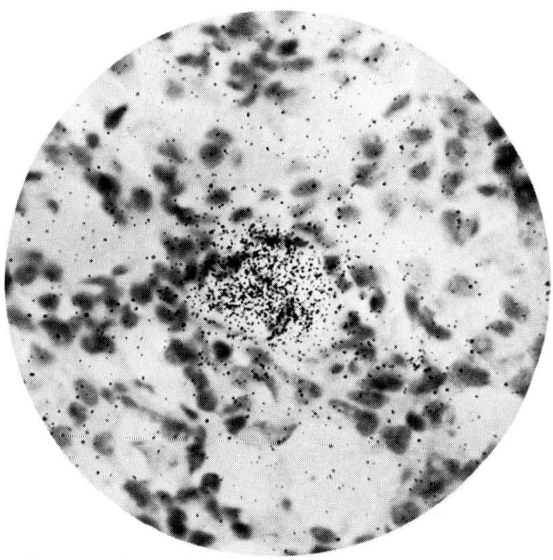

Fig. 4. Autoradiograph of kidney tubules from a Se-deficient rat. Conditions as for Figure 3.

acids — orally, intravenously, or intraperitoneally, it was apparently effective. In what form then was Se incorporated into the tissues?

Cummins & Martin (1967), working with rabbits, and Jenkins (1968), working with chicks, had claimed that inorganic Se, as selenite, was not converted to selenoamino acid in animal tissues. Working with lactating ewes and using a method of separation involving Sephadex chromatography and enzymatic digestion of protein, we were able to show the presence of ^{75}Se-selenomethionine in ewes' milk following the intraruminal administration of $Na_2{}^{75}SeO_3$. The final identification of selenomethionine was by gas-liquid chromatography, converting the labelled material to the n-propyl-N-acetyl ester and demonstrating that the effluent corresponding to the derivative contained radioactivity (Godwin *et al.* 1971). By using similar techniques, we later showed (Godwin & Fuss 1972) that a small proportion, perhaps only 1-2%, of administered ^{75}Se-selenite is converted to selenoamino acid in the rabbit.

THE SELENO-ENZYME — GLUTATHIONE PEROXIDASE

The nature of the incorporation of Se into protein now assumes more importance, since Rotruck *et al.* (1972) found that 'dietary Se' prevented autohemolysis in red cells drawn from Se-deficient rats, whereas inorganic Se added to the *in vitro* system did not.

Of even greater significance is the recent identification of glutathione peroxidase as a seleno-enzyme containing molecular proportions of Se (Rotruck *et al.* 1973; Flohe *et al.* 1973). The question becomes of even greater importance, that of how the Se is incorporated into the enzyme. Other Se-containing proteins are also being recognized (Whanger *et al.* 1973).

Using a modification of the coupled method of analysis of glutathione peroxidase of Paglia & Valentine (1967), we have examined the distribution of the enzyme in fractions of normal sheep and rat muscle. We have also found that levels of glutathione peroxidase in muscle, plasma and particularly erythrocytes from Se-low sheep are very rapidly depleted.

Recent studies have shown that under *in vitro* conditions using washed mitochondria, which in skeletal muscle contain a significant proportion of the total glutathione peroxidase of the cells, the addition of Ca ions leads to a loss of enzyme activity. The present evidence points to this being due to leakage of enzyme from the mitochondria.

CONCLUSION

Noguchi *et al.* (1973), in dicussing the role of Se and vitamin E in the chick, have suggested a general hypothesis which may have application to animal tissues in general. Essentially the hypothesis recognizes that both Se and vitamin E are incorporated into mitochondrial and microsomal membranes, and are necessary for protection of these membranes from lipid peroxidation. Glutathione peroxidase is seen as the first line of defence; when Se is low, protection of the membranes may still be afforded by vitamin E, which is within the membrane itself. The scheme is summarized in Figure 5.

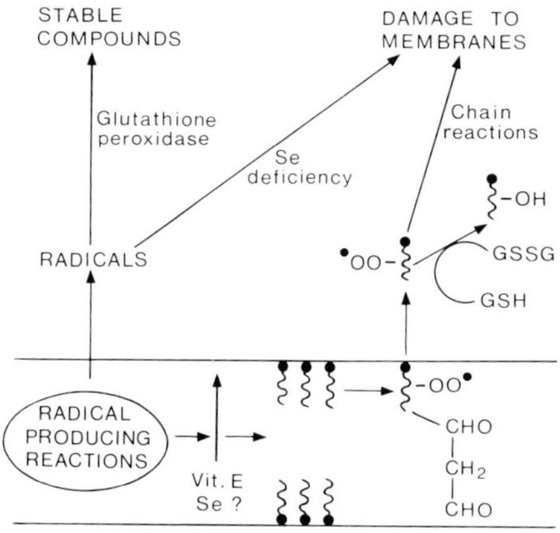

Fig. 5. Diagrammatic scheme showing the interrelationships between Se, vitamin E and mitochondrial membranes in animal tissues. [After Noguchi et al.*, J. Nutr. 103 : 1502 (1973)]*

REFERENCES

APPLETON, T.C. (1964) - *Jl R. microsc. Soc. 83* : 277

CUMMINS, L.M. & MARTIN, J.L. (1967) - *Biochemistry 6* : 3162

DEWEY, D.W., LEE, H.J. & MARSTON, H.R. (1958) - *Nature, Lond. 181* : 1367

FLOHE, L., GUNZLER, W.A. & SHOCK, H.H. (1973) - *FEBS Lett. 32* : 132

GARDINER, M.R. (1969) - *Outl. Agric. 6* : 19

GODWIN, K.O. (1965) - *Q. Jl exp. Physiol. 50* : 284

GODWIN, K.O. (1968) - *Nature, Lond. 217* : 1275

GODWIN, K.O. (1972) - *Aust. J. exp. Agric. Anim. Husb. 12* : 473

GODWIN, K.O. & FRASER, F.J. (1966) - *Q. Jl exp. Physiol. 51* : 94

GODWIN, K.O. & FUSS, C.M. (1972) - *Aust. J. biol. Sci. 25* : 865

GODWIN, K.O., HANDRECK, K.A. & FUSS, C.N. (1971) - *Aust. J. biol. Sci. 24* : 1251

GODWIN, K.O., KUCHEL, R.E. & BUCKLEY, R.A. (1970) - *Aust. J. exp. Agric. Anim. Husb. 10* : 672

GODWIN, K.O., KUCHEL, R.E. & FUSS, C.N. (1974) - *Aust. J. biol. Sci. 27* : 633

HANDRECK, K.A. & GODWIN, K.O. (1970) - *Aust. J. agric. Res. 21* : 71

HARTLEY, W.J. & GRANT, A.B. (1961) - *Fedn Proc. 20* : 679

JENKINS, K.J. (1968) - *Can. J. Biochem. 46* : 1417

KUCHEL, R.E. & BUCKLEY (1969) - *Aust. J. agric. Res. 20* : 1099

LIN, C.H., HUDSON, A.J. & STRICKLAND, K.P. (1970) - *Can. J. Biochem. 48* : 566

McLEAN, J.W., THOMSON, G.G. & CLAXTON, J.H. (1959) - *N.Z. vet. J. 7* : 47

MUTH, O.H. (1955) - *J. Am. vet. med. Ass. 126* : 355

MUTH, O.H., OLDFIELD, J.E., REMMERT, L.F. & SCHUBERT, J.R. (1958) - *Science 128* : 1090

NOGUCHI, T., CANTOR, A.H. & SCOTT, M.L. (1973) - *J. Nutr. 103* : 1502

PAGLIA, D.E. & VALENTINE, W.N. (1967) - *J. Lab. clin. Med. 70* : 158

ROTRUCK, J.T., POPE, A.L., GANTHER, H.E. & HOEKSTRA, W.G. (1972) - *J. Nutr. 102* : 689

ROTRUCK, J.T., POPE, A.L., GANTHER, H.E., SWANSON, A.B., HAFEMAN, D.G. & HOEKSTRA, W.G. (1973) - *Science 179* : 588

SCHWARZ, K. (1951) - *Proc. Soc. exp. Biol. Med. 78* : 852

SCHWARZ, K. & FOLTZ, C.M. (1957) - *J. Am. chem. Soc. 79* : 3292

VIGNOS, P.J. & WATKINS, M.P. (1966) - *J. Am. med. Ass. 197* : 843

WATKINSON, J. (1966) - *Analyt. Chem. 38* : 92

WHANGER, P.D., PEDERSEN, H.D. & WESWIG, P.H. (1973) - *Biochem. biophys. Res. Commun. 53* : 1031

WROGEMANN, K., BLANCHAER, M.C. & JACOBSON, B.E. (1970) - *Life Sciences 9* : 1167

WROGEMANN, K., JACOBSON, B.E. & BLANCHAER, M.C. (1973) - *Archs Biochem. Biophys. 159* : 267

TRACE ELEMENT INTERACTIONS IN ANIMALS

N.F. Suttle

Moredun Research Institute
Edinburgh, Scotland EH17 7JH

INTRODUCTION

The nutritional significance of a trace element is determined not only by its specific physiological and biochemical roles in animal metabolism, but also by its capacity to interact with other elements in ways which can be beneficial or harmful in terms of animal health. Many interactions between trace elements are of potential importance in the context of agricultural productivity and environmental health, and their role is likely to grow as man unwittingly increases the trace element content of his environment in exploiting reserves of minerals, once safely covered by a mantle of earth.

While the functional roles of trace elements are fairly well understood, we know little of the involvement of these nutrients in interactions with other elements and organic or inorganic ligands, particularly under conditions which are relevant to practical nutrition. The object of this review is, therefore, to stimulate investigations of the nutritional significance of trace element interactions. A theoretical scheme for investigating such interactions will be presented, together with a worked example in the form of a Cu × Mo × S interaction. Fertile areas for research in the four major categories of trace element interactions are outlined together with some appropriate experimental techniques.

ANATOMY OF A TRACE ELEMENT INTERACTION

Four interrelated areas for research can be identified in any trace element interaction. They are concerned with:

1. Identifying the dietary components involved;
2. Determining factorially the dose:response relationship for each component;
3. Locating the site(s) of interaction;
4. Determining the mechanism(s) of interaction.

The logical sequence for investigating these areas of research is that given above. The dose:response relationships are only meaningful if they have been evaluated in the presence of all components in the interaction. Furthermore, the possible mechanisms of interaction are limited once the site of interaction is known.

Few trace element interactions have been studied in depth and in many cases there are areas of research which have been totally neglected. The situation is particularly disappointing from a nutritional standpoint because it is the very area which determines the nutritional significance of an interaction that has been most neglected. A nutritional role becomes apparent if the components in any interrelationship interact at dietary concentrations which are at least occasionally found in natural diets. It has, however, been common for components to be identified in experiments involving abnormally high concentrations of the antagonist and followed by a search for a mechanism for the observed interaction. However interesting such a mechanism may be to the biochemist or physiologist, the nutritionist may find that it is totally irrelevant in the context of practical nutrition.

Reasons for the preoccupation with unnatural concentrations of potential antagonists varied. Investigations of the effects of relatively large increments in trace element concentration require lower standards of analytical precision and 'trace element hygiene', and the development of clinical symptoms of toxicity is an obvious aid to the interpretation of biochemical changes. Above all, statistically significant results are obtained more quickly from relatively small experiments when high concentrations are employed. The cost-effectiveness of such experiments should, however, be assessed in the nutritional context in terms of benefits in animal productivity or health: these benefits increase with the extent to which the biologically effective range of concentration overlaps the range of concentration found in natural diets.

The interrelationship between Cu, Mo and S provides a classic example of the complex interrelationships which can exist between trace elements: it has been extensively studied and illustrates the interplay of the four important parts in any interaction. In ruminants, intensive investigations of the dose:response relationships have established the practical significance of the interaction, but in non-ruminants the neglect of this same area of research has resulted in complete ignorance of the role of the Cu:Mo antagonism in practical nutrition.

A Cu × Mo × S INTERACTION IN RUMINANTS

Biologically Effective Concentrations of Mo and S

The outstanding work of Dick (1956) revealed that Mo could markedly impair the Cu metabolism of sheep under experimental conditions, provided that the diet contained adequate sulphate: three components to the interaction were thus identified. The antagonism was shown to have practical nutritional significance in a few localities characterized by exceptionally low (< 0.2 mg Mo/kg, Dick & Bull 1945) or high (> 7.0 mg Mo/kg, Cunningham 1950) dietary Mo concentrations, being associated with the development of Cu toxicity at the lower and Cu deficiency at the upper extreme. The effects of variations in Mo concentration within this range (0.2-7.0 mg/kg) have not been studied experimentally but in the field they are considered to be insufficient for the Cu:Mo antagonism to play an important role in the aetiology of swayback or bovine hypocuprosis (Allcroft & Lewis 1957; Hartmans 1970). In order to check this assumption the effects of varying the dietary concentrations of Mo and S within their respective normal ranges in herbage (0.5-4.5 mg Mo/kg, Whitehead 1966; Miltimore & Mason 1971; and 1.0-4.0 S/kg DM, Whitehead 1966) were investigated.

The experimental technique involved the repletion of initially hypocupraemic ewes with a Cu-supplemented diet and the use of responses in plasma Cu to assess the effects of treatments on Cu availability (Suttle 1974*a*). It was found that an increment of 4 mg/kg in dietary Mo was sufficient to reduce the availability of dietary Cu by 50% (Suttle 1974*b*), while an increment of 1 g S/kg had a similar effect. The results of one experiment in the series are presented in Table 1.

Previous investigations had been concerned solely with the involvement of inorganic S, the minor component of herbage S, in the antagonism. In view of the similar metabolic pathways of organic and inorganic S in the rumen, it was decided to check the capacity of the major organic component to potentiate the Cu:Mo antagonism. The ability of the two S sources to inhibit Cu repletion and decrease plasma Mo concentrations was found to be similar (Table 2). The third component in the antagonism should, therefore, be the total rather than the inorganic S in the diet.

The involvement of Mo and the entire S component of the diet in an antagonism which was effective over a range of concentrations frequently encountered in the field stimulated a search for the site of the antagonism.

TABLE 1

Effects of small increments in dietary Mo and inorganic S on the responses in plasma Cu (μg/ℓ) of initially hypocupraemic ewes repleted for 35 days with a Cu-supplemented diet (7 mg Cu/kg)

		Dietary Mo (mg/kg)	
		2.5	4.5
Dietary S (g/kg)	2.0	342 ± 72* (4.3)†	38 ± 42 (2.3)
	3.0	53 ± 71 (2.8)	43 ± 45 (2.4)

*S.E. of mean of 6 animals
†Availability of Cu predicted by method of Suttle (1974*a*)

TABLE 2

Effects of organic and inorganic S (3 g/kg) in the presence or absence of supplementary Mo (4 mg/kg) on Cu and Mo in plasma (μg/ℓ) of initially hypocupraemic ewes repleted for 21 days with a Cu-supplemented diet (8 mg Cu/kg)

	Dietary supplement			
	O	Mo	S	Mo + S
		Δ Plasma Cu (μg/ℓ)		
Inorganic SO_4^{2-}	345	339	216	-50
Organic S (Methionine)	219	201	146	19
		Plasma Mo (μg/ℓ)		
Inorganic SO_4^{2-}	62	1555	33	55
Organic S (Methionine)	47	1123	27	48

The Site of a Cu × Mo × S Antagonism in Sheep

The procedure used in tracing the site of interaction involved the administration of Cu and Mo by separate routes, thus by-passing suspected sites of interaction. The experiments again involved the repletion of initially hypocupraemic ewes. It was found that the administration of Mo in the diet and Cu by subcutaneous injection (Suttle & Field 1974) or continuous intravenous infusion (Suttle 1974*b*) largely circumvented the antagonism, indicating that the important site was located in the gut. In a further experiment the rumen was by-passed by administering Mo by continuous abomasal infusion while Cu was given in the diet. The mean repletion responses of three animals and of a comparable group given both Cu and Mo in the diet are shown in Figure 1. When Mo was given *via* the abomasum, it no longer inhibited the repletion response: at the same time, the urinary excretion of Mo was greatly enhanced.

It therefore appeared that the Cu × Mo × S antagonism involved a lowering of the availability of both Cu and Mo in the rumen and that inorganic and organic S could potentiate the process. These reactions pointed to a possible mechanism for the antagonism.

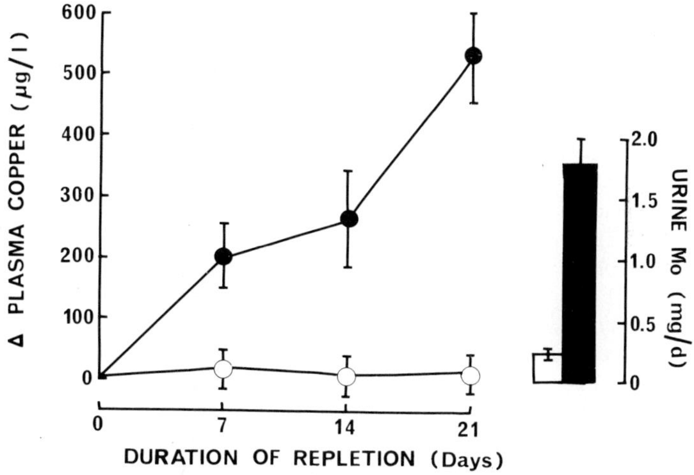

Fig. 1. Effects of administering Mo (3.2 mg/day) in the diet (o) or by continuous abomasal infusion (●) on the responses in plasma Cu (μg/ℓ) of initially hypocupraemic ewes given a Cu-supplemented diet (8 mg/kg) and also on urinary Mo excretion between 18 and 21 days. The diet was relatively rich in inorganic sulphate (3 g/kg).

A Mechanism for a Cu × Mo × S Antagonism

The hypothesis is based on two separate *in vitro* reactions. Dowdy & Matrone (1968*a,b*) have shown that the Cu^{2+} and $MoO_4{}^{2-}$ combine readily to form $CuMoO_4$ *in vitro*, and they initially claimed that this reaction provided the mechanism for the Cu:Mo antagonism in all species. In order to explain the involvement of S in the ruminant antagonism, Matrone (1970) later postulated that a bio-degradation product of S-'X' combined with the Cu:Mo complex to give a triple complex, but the nature of the complex was not given. Tridot & Bernard (1962) have found that, in a sulphide-rich alkaline medium, $MoO_4{}^{2-}$ is readily converted to thiomolybdate, $MoS_4{}^{2-}$. Since both organic and inorganic S are readily degraded to S^{2-} in the rumen, where the pH is slightly alkaline, it is conceivable that a triple complex, cupric thiomolybdate, is formed, in which the Cu is of low biological availability. The validity of the 'thiomolybdate' hypothesis is currently being investigated in the fourth stage of our research. It must be admitted, however, that it is impossible to explain all facets of the Cu × Mo × S interrelationship by a single hypothesis.

Effects of Diet and Dietary Molybdenum Concentration on the Nature of the Cu × Mo × S Antagonism in Ruminants

It is becoming clear that the site and mechanism of the Cu × Mo × S interaction changes dramatically with changes in dietary conditions which affect the sulpholytic bacteria in the rumen. Huisingh & Matrone (1972) found that the activity of sulpholytic bacteria was inhibited *in vitro* by Mo, albeit at abnormally high concentrations. Subsequent studies showed that a new and particularly active species of *Desulphovibrio* was present in the rumen of lambs reared on diets containing urea and sulphate as the predominant N and S sources (Huisingh, McNeill & Matrone 1973). Simple Cu deficiency could be readily produced by feeding such diets but a small Mo supplement (4 mg/kg), in these conditions, produced an unusual reaction in alleviating the anaemia which developed. The authors suggest that this unique effect of Mo resulted from the inhibition of sulphide formation and the decreased formation of insoluble copper sulphide (Huisingh, Gomez & Matrone 1973). Other workers, using more natural diets of lower urea content, found that 8 (N.F. Suttle, unpublished results) or > 40 mg Mo/kg (Bryden & Bray 1972) were required to inhibit S^{2-} formation.

The inhibition of sulphide formation at relatively high Mo intakes does call for additional mechanisms to explain the continuing impairment of Cu metabolism, and it is possible that systemic interactions, involving changes in plasma Cu distribution, become increasingly important as dietary Mo concentrations increase. At levels > 4 mg Mo/kg, the direct reacting Cu in plasma is increased (Suttle 1973), while 25 mg Mo/kg induces a new 'residual' Cu fraction (Suttle & Field 1968; Smith *et al.* 1968), in which the Cu may be relatively unavailable. These changes will be discussed more fully later but it is conceivable that they provide the key to the increase in urinary Cu excretion which commonly follows the use of relatively high Mo supplements in the sheep's diet and which contributes to a gradual Cu depletion (Rys *et al.* 1965; Smith *et al.* 1968; Marcilese *et al.* 1970).

The detailed studies of the responses of ruminants to relatively small Mo and S supplements have revealed not only a widespread practical nutritional significance of the Cu:Mo antagonism but also the site and possible mechanism for the interaction. The progress made would appear to support the scheme presented initially for the investigation of trace element interactions.

Cu:Mo ANTAGONISMS IN NON-RUMINANTS

The disadvantages of not following a systematic approach in investigating trace element interactions are well illustrated by reference to the Cu:Mo antagonism in non-ruminants. Following the discovery that cattle grazing pastures containing 7-100 mg Mo/kg become depleted of Cu and develop a syndrome characterized by diarrhoea, attempts were made to reproduce the syndrome in non-ruminant species. Molybdenum concentrations of 800-1000 mg/kg were required to induce molybdenosis and, not surprisingly, the many subsequent biochemical investigations in the 1950's showed the nature of the interrelationship to be quite different, involving an increase rather than a decrease in tissue Cu concentrations (Miller & Engel 1960). Many dismissed the interaction as having no practical nutritional significance and interest waned. Recent studies have shown, however, that over a similar range of Mo concentration, ruminants and non-ruminants can show a similar Cu:Mo antagonism. The guinea-pig is also depleted of Cu as Mo concentrations increase from 0.6 to 104 mg/kg and the logarithmic response curve showed that the smaller Mo supplements were the more effective in impairing Cu uptake (Suttle 1973).

Further evidence of the biological activity of low Mo intakes has recently been obtained in work with rats. Molybdenum concentrations as low as 10 mg/ℓ in the drinking water have been found to impair bone calcification (Solomons *et al.* 1973) and resistance to cold stress (Winston *et al.* 1973); an impairment of Cu metabolism may be involved in these effects and there are other indications that the Cu:Mo antagonism may be of practical nutritional significance to man and animals other than ruminants (Seelig 1972; Suttle 1974*d*). However, we know embarrassingly little of the nutritional importance of the Cu:Mo antagonism at physiological concentrations to non-ruminants, despite the intensive research of the two previous decades.

I have dealt in some detail with the Cu × Mo × S interaction in ruminants and non-ruminants because our knowledge of the nutritional significance of the antagonism in these two groups is contrastingly advanced. The contrast is related to a failure to explore fully the dose:response relationship in non-ruminants. In dealing with further trace element interactions, in the six major categories into which they fall, it should become more apparent that lack of information with regard to this particular area of research hinders the assessment of the practical nutritional significance of an interaction.

PRINCIPAL MECHANISMS FOR TRACE ELEMENT INTERACTIONS

The many known interactions involving trace elements can conveniently be grouped into six categories according to the type of mechanism involved:

1. Formation of insoluble complexes between dissimilar ions;
2. Competition for metabolic pathways between similar ions;
3. Induction of metal-binding proteins;
4. Changes in the metal component of metallo-enzymes;
5. Facilitation of trace element transport;
6. Enhancement of trace element excretion.

The categories are presented in descending order of nutritional significance at our current state of knowledge and examples will be examined in each category, giving particular emphasis to their potential nutritional roles.

Formation of Insoluble Complexes

The ability of trace elements to form insoluble complexes with anionic ligands is a major factor contributing to their generally low availability in the diet. Although the primary interaction frequently involves only one trace element, secondary interactions between elements become involved through competition for the ligand. The 'competitiveness' of an element depends, among other things, on its concentration, the relative strength of the association constant and the solubility of the product. The ligand may be inorganic (e.g. S^{2-} or PO_4^{3-}) or organic (e.g. phytate, fatty acid), and the insoluble complex may form in the diet, in the gut or in the tissues.

In the diet. Dietary phosphate lowers the availability of Zn (Vohra & Kratzer 1966) and Fe (Waddell & Sell 1964) to poultry, probably through the formation of insoluble metal phosphates. The experiments of Vohra & Kratzer indicated that phosphate formation began prior to feeding during the preparation of a mineral pre-mix. No other studies of this phenomenon have been reported but feed compounders and nutritionists would both seemingly benefit from knowing the extent of such interactions under normal and abnormal conditions of diet preparation and storage.

In the digestive tract. The processes of digestion are generally beneficial to an animal, but they can paradoxically contribute to the poor utilization of trace elements by creating conditions which are favourable for insoluble complex formation. They may, for example, permit Zn to interact with phytate and Ca, leading to the formation of insoluble zinc phytate (cf. Taylor 1965); other elements, such as Cu, Fe and Mn, may be similarly affected (Davis *et al.* 1962; Davies & Nightingale 1975). Insoluble phytate formation is unimportant in ruminant nutrition because phytate is destroyed by the rumen microflora, but these organisms are capable of generating an equally effective ligand by degrading SO_4^{2-} to S^{2-}. All heavy metal sulphides have low solubility products and, at the pH's encountered in the rumen, those metals present in the ionic form are likely to be precipitated as insoluble sulphides. A further example is provided by the interaction of trace elements with fatty acids to form insoluble soaps in the digestive tract of poultry (Dewar *et al.* 1975).

The components in interactions involving the formation of insoluble trace element complexes have generally been identified using natural or experimental diets containing an excess of the potential ligand. Under these circumstances, competition between elements for the ligand is non-existent and trace element availability is generally depressed. When the ligand is not present in excess, however, a supplement of one element may enhance the availability of others by reducing their opportunities for complex formation. Indications that such secondary interactions occur are contained in some experiments in which Cu has been added to pig rations at relatively high concentrations, as a growth stimulant; Kirchgessner & Weser (1963) observed a general increase in the availability of trace elements, while others have noted an apparent improvement in Zn availability leading to a decreased incidence of parakeratosis (Hoefer *et al.* 1960; Wallace *et al.* 1960; Ritchie *et al.* 1963).

The quantitative aspects of insoluble complex formation have received little attention. As far as phytate is concerned, we do not know the critical level of Ca to initiate the co-precipitation of other metal phytates (Byrd & Matrone 1965) or the critical level for phytates at which trace element competition becomes effective. Such information is essential if the trace element supplementation of natural or experimental rations is to be assessed rationally.

There are also doubts surrounding the nutritional role of insoluble sulphide formation in ruminants. There is evidence that most natural diets will yield an excess of S^{2-} in the rumen (Bird 1970), yet effects of dietary S (Suttle 1974*c*) and Mo (Huisingh, Gomez & Matrone 1973) supplements on Cu utilization in sheep have been explained in terms of increased or reduced CuS formation. In view of the potentially important role of S^{2-} in promoting the resistance of ruminants to heavy metal toxicities, such as those of Pb, there is need to study further the quantitative aspects of insoluble sulphide formation in the rumen and the possibility of such reactions occurring at sites beyond the rumen (Suttle 1974*c*).

It is worth noting that proof of insoluble complex formation is exceedingly difficult to obtain, since it requires selective removal of the insoluble and unreactive complex from the digesta or faeces. Evidence given for complex formation is usually empirical, being based on a fall in trace element availability when dietary concentrations of the ligand are increased and on the low availability of the preformed complex. This problem is more acute with regard to complex formation in the tissues. Alternative or contributory mechanisms to the interaction should not be ignored.

In the tissues. Although there is as yet no conclusive evidence that insoluble inorganic complexes can form in the tissues, there are two possible ways in which this type of interaction may participate in the Cu:Mo antagonism. Some years ago it was suggested that Mo inhibits the activity of sulphide oxidase in the liver, causing Cu to accumulate as insoluble CuS (Mills *et al.* 1958), and Rish (1970) has recently suggested this as the mechanism underlying the inducement of Cu deficiency by Mo in sheep. Following an observation that Cu^{2+} and $MoO_4{}^{2-}$ combined *in vitro* to form a biologically inert complex resembling cupric molybdate, Dowdy & Matrone (1968*a*) suggested that this complex might form in the tissues. The fact that molybdenotic rats possess elevated tissue Cu concentrations, yet respond to Cu therapy (Miller & Engel 1960), is certainly indicative of unavailable tissue reserves of Cu, but the formation of $CuMoO_4$ or CuS *in vivo* has not been proved.

The availability of many trace elements may, therefore, be reduced through the formation of insoluble trace element complexes, particularly in the gut. The limitation of studies to unphysiological conditions or excessive concentrations of the ligand has, however, frequently left the problem of the nutritional role of complex formation unanswered.

Interactions Between Chemically Similar Ions

Many interactions between elements have been explained on the basis of competition between chemically similar elements for a common metabolic pathway (Hill & Matrone 1970). The principle is well illustrated by the group of iso-electronic ions Cu^{2+}, Zn^{2+} and Ag^{2+}. The addition of Cd (Hill *et al.* 1963), Zn (Hill & Matrone 1962) and Ag (Hill *et al.* 1964) to the diet can induce a deficiency of Cu, and the addition of Cd can also induce a deficiency of Zn (Supplee 1961; Powell *et al.* 1964). The ability of Zn to reduce the uptake of Cu from ligated loops of intestine led Van Campen & Scaife (1967) to conclude that the gut was an important site of interaction. Studies of the gut wall led to the isolation of a Cu-rich protein of small molecular weight from the duodenum of the chick: the Cu content of this fraction was reduced by feeding Zn, Cd and Ag but not by the dissimilar group of elements, Hg, Mn and Mo. It was concluded that the interaction between Cu and the former group of elements involved competition for the carrier protein responsible for Cu absorption (Starcher 1969).

The importance of these interactions in terms of practical nutrition seemed at first to be small, since many of the antagonists were used at concentrations never encountered in natural diets: Cd for example was generally added at 100 mg/kg. The extension of these investigations to include more natural concentrations has, however, shown that as little as 3.5 and 1.5 mg Cd/kg can impair the Cu metabolism of sheep (Mills & Dalgarno 1972) and rats (Campbell & Mills 1974) respectively. In addition to showing the potential nutritional significance of the Cd × Cu interaction to man and his livestock, this work also throws some light on the mechanism of interaction. In the study with rats, 6 mg Cd/kg was as effective in inhibiting ceruloplasmin synthesis as 1000 mg

Zn/kg. The superior capacity of Cd as an antagonist may reflect a relatively strong affinity for the carrier protein or a totally different mechanism involving possibly the selective accumulation of Cd in the tissues.

A further example of competitive inhibition may be provided by an interaction between sulphate and molybdate in non-ruminants. The fact that these two tetrahedral anions have similar electronic configurations led Huisingh and his colleagues (1973) to study the effect of SO_4^{2-} on MoO_4^{2-} absorption from loops of chick intestine. The absorption of MoO_4^{2-} was effectively suppressed and this mechanism may explain the ability of SO_4^{2-} to protect the non-ruminant from molybdenosis.

An antagonism between a macro- and a micro-nutrient, such as that between SO_4^{2-} and MoO_4^{2-}, is potentially a one-sided affair, unless the concentrations of available anions at the sites of absorption or the efficiency of the absorptive mechanism is considerably more favourable to Mo than the SO_4:MoO_4 ratio in the diet. The high availability of Mo to the non-ruminant (Bell *et al.* 1964; Kinnamon 1966) must bring into question the general biological effectiveness of such an interaction. Our limited nutritional understanding of interactions is well illustrated by the fact that the study of Mo antagonists has been confined to tungstate at highly unphysiological concentrations (Leach *et al.* 1962), while the potentially important sulphate has been studied only at unphysiological concentrations of Mo (Van Reen & Williams 1956; Siegel & Monty 1961)! Has dietary sulphate a role to play in induced Mo deficiency in man (cf. Seelig 1972)? We can only speculate.

Systemic Formation of Metal-Binding Proteins

It is becoming apparent that various animal species react systemically to heavy metal loading by synthesizing proteins in the plasma and tissues. The proteins are not specific for the toxic element and may participate in trace element interactions by binding other elements. The addition of the heavy metals Cd (Shaikh *et al.* 1973; Cousins 1974) and Zn (Bremner & Marshall 1974) to the diet can, for example, induce the formation in the kidney or liver of a soluble cysteine-rich protein or metallothionein which also binds Hg and Cu. The carrier protein isolated by Starcher (1969) is probably the same as those induced by Zn and Cd and as the metallothionein isolated by Evans *et al.* (1970) from the bovine duodenum. The nutritional role of metallothioneins is thus likely to be more complex than that indicated by the early experiments, for a displacement of Cu from binding sites by Zn or by

Cd may be partly compensated by an increase in the number of sites available following an increase in metallothionein synthesis.

Molybdenum may induce the synthesis of novel plasma proteins, for the plasma of Mo-supplemented sheep contains Cu not accounted for by the familiar ceruloplasmin- and albumin-bound components (Suttle & Field 1968; Smith *et al.* 1968). The 'residual' fraction contains Cu, Mo and protein in firm apposition, since it is precipitated by TCA without losing its trace element content (Smith & Wright 1973). Acid hydrolysis studies of plasma from Mo-supplemented guinea-pigs indicate that the Cu and Mo are bound to similar sites in the protein-containing 'residual' Cu fraction (B.S.W. Smith & H. Wright, unpublished results). The induction by Mo of a protein which contains Cu in an unavailable form would explain the puzzling elevation in tissue Cu concentration found in the molybdenotic rat.

Metalloproteins can thus be intimately involved in trace element interactions but their full nutritional and physiological significance has yet to be elucidated.

Interactions Involving Metallo-enzymes

Although it has been known for some time that the essentiality of many trace elements is related to their presence in important enzymes, it has only recently been appreciated that the essential element can be replaced by other dissimilar elements. Cobalt, for example, can replace zinc in the carboxypeptidase enzyme *in vitro*, causing an increase in peptidase activity (Wacker 1970). The physiological significance of this substitution is not known but another metallo-enzyme, xanthine oxidase, shows changes in its trace element content in response to nutritional conditions. Several iso-enzymes of differing Cu and Mo contents are found in normal cattle plasma and the proportions of Cu-rich : Mo-rich iso-enzymes are positively correlated with the Cu:Mo ratio in the diet (Kovalsky *et al.* 1973,1974).

The extension of these studies to other metallo-enzymes may reveal a large reservoir of potential trace element interactions. One fruitful area of research may be the ubiquitous Cu-protein, cytocuprein: this has been found to contain Zn as well as Cu (Carrico & Deutsch 1969; Bannister *et al.* 1971) and to possess superoxide dismutase activity (McCord & Fridovich 1969). It is possible that the Zn and Cu could substitute for each other with subsequent alterations in enzyme activity. Interactions involving trace element substitutions in metallo-enzymes represent yet another field where what remains to be known far exceeds that which is known.

SPECIFIC INTERACTIONS INVOLVING THE TRANSPORT AND EXCRETION OF TRACE ELEMENTS

These two final mechanisms of trace element interaction will be dealt with briefly because they appear at present to be involved only in specific interrelationships. The requirement of Cu for normal haematopoiesis is explained by the ability of the plasma Cu-protein, ceruloplasmin, to facilitate the transport of Fe. Ceruloplasmin acts as a ferroxidase, catalysing the conversion of Fe^{2+} to Fe^{3+}, thus enabling Fe stored in the liver and reticulo-endothelial system to be transported as Fe^{3+} bound to transferrin in the plasma (Frieden 1971).

The enhancement of urinary Cu excretion in sheep given relatively high intakes of Mo, which was discussed earlier, may result from alterations in plasma Cu distribution. It conveniently emphasizes the widespread influence interactions can have on trace element metabolism, for we have seen that they can take effect from the moment a diet is mixed to the final step in metabolism, namely excretion from the body.

THE INTERDEPENDENCE OF DIFFERENT INTERACTIONS

Although the various mechanisms of interaction have been considered separately, many of them will be affected by other interactions occurring concurrently in the same animal. For example, the involvement of a trace element in the formation of insoluble complexes will obviously limit the capacity of that element to interfere with the absorption and metabolism of other trace elements. The interdependence of interactions has, however, been frequently ignored in the design of experiments. Absorption studies with isolated segments of the gut have usually been conducted under conditions which bear little resemblance to those prevailing in the gut. In particular, the absence of potential ligands in the digesta will lead to unnaturally high concentrations of the 'free' antagonist. It is, however, difficult to criticize such an approach while we know so little of the form in which trace elements are present in and absorbed from the digesta that truly physiological conditions cannot be defined.

The interdependence of interactions has also been ignored in the interpretation of experimental results. In attempting to rationalize the Cu × Mo × S antagonism in ruminants and non-ruminants, Huisingh, Gomez & Matrone (1973) have considered three separate interactions, the formation of unavailable $CuMoO_4$ and CuS and the inhibition of $MoO_4{}^{2-}$ uptake by $SO_4{}^{2-}$.

These mechanisms are obviously dependent, since the formation of either $CuMoO_4$ from Cu^{2+} and $MoO_4{}^{2-}$ or S^{2-} from $SO_4{}^{2-}$ in the gut would limit the capacity of $MoO_4{}^{2-}$ and $SO_4{}^{2-}$ to participate in the $SO_4{}^{2-} \times MoO_4{}^{2-}$ interaction.

The interdependence of trace element interactions adds yet another dimension to the complex problem of explaining 'whole animal' antagonisms in terms of specific mechanisms of interaction.

TECHNIQUES FOR INVESTIGATING NUTRITIONAL INTERACTIONS

Having said that many of the experimental techniques used in the past to investigate trace element interactions have been unsuitable from a nutritional viewpoint, it may be helpful to specify some suitable techniques. Until we have more information with regard to the fundamental processes of trace element metabolism, techniques involving isolated cell or organ systems will be of limited value and the only alternative is to use the whole animal. Conventional balance techniques are, however, unsuitable for measuring the efficiency of trace element utilization because of the generally low level of availability and the small increments in concentration that require study.

An alternative approach that is being increasingly used involves the use of systemic responses reflecting the amount of an element being absorbed and retained. A number of techniques are available: these include the measurement of accumulation of an element in a tissue storage pool (e.g. Cu in liver, Dick 1956), repletion of a plasma pool in initially depleted animals (Cu in plasma, Suttle 1974*a*) and depletion of a plasma pool in marginally depleted animals (Cu in plasma, Campbell & Mills 1974). The latter techniques have the advantages of uniform initial trace element status and a rapid response: they may also have disadvantages, however, in that unnatural diets may be required for depletion purposes and the depleted animal may not exhibit a normal trace element metabolism.

With regard to the trace element inputs which are investigated, there is still a need to maintain conditions as close to those prevailing in natural diets as possible. Inorganic supplements may be an inadequate substitute for the naturally occurring element. The use of isotopically labelled tracers can involve further assumptions with regard to the dynamics of trace element metabolism and they frequently yield results which complicate rather than clarify the apparent nature of an interaction.

CONCLUSIONS

I have deliberately looked at trace element interactions from a nutritional viewpoint because this seems to have been the aspect most neglected in the past. In taking a detailed look at the important components of the Cu × Mo × S interaction in ruminants and a brief look at other types of interaction, there are four points which emerge as crucial to any investigation.

Firstly, it is important to study trace element interactions at concentrations of the element or ligand which have some nutritional significance. Such studies will often require the use of sensitive methods for assessing the relative availability of the element under examination, but the use of marginally deficient animals holds great promise in this field.

Secondly, there would appear to be much more scope for the use of varied routes of administration in establishing the involvement of a particular site of antagonism.

Thirdly, in view of the many ways in which a single element can interact with others, we must think in terms of the relative contributions of concurrent interactions to the 'whole animal' interaction rather than single mechanisms.

Fourthly, the nutritional significance of an antagonism must be studied under physiological conditions and there is a strong case for employing the 'whole animal' approach in this field.

Present knowledge of the significance of trace element interactions in the nutrition of man and his domestic animals must inevitably represent only the proverbial 'tip of the iceberg'. The full extent of the problem will only be revealed by logical and physiological investigation of the four important areas for research into any interaction: these involve (a) identifying the components, (b) defining dose: response relationships, (c) locating the site and (d) the mechanism of interaction. The 'environmental pollution' lobby will have served at least one useful purpose if it stimulates such investigations of the many interactions which have a potential bearing upon the very health and survival of man.

ACKNOWLEDGEMENT

I am grateful to Dr. A.C. Field for his criticisms and suggestions during the preparation of this manuscript.

REFERENCES

ALLCROFT, R. & LEWIS, G. (1957) - *J. Sci. Fd Agric. 8* : S96

BANNISTER, J.V., BANNISTER, W.H. & WOOD, E.J. (1971) - *Eur. J. Biochem. 18* : 178

BELL, M.C., DIGGS, B.G., LOWREY, R.S. & WRIGHT, P.L. (1964) - *J. Nutr. 84* : 367

BIRD, P.R. (1970) - *Proc. Aust. Soc. Anim. Prod. 8* : 212

BREMNER, I. & MARSHALL, R.B. (1974) - *Br. J. Nutr. 32* : 283, 293

BRYDEN, J.McG. & BRAY, A.C. (1972) - *Proc. Aust. Soc. Anim. Prod. 9* : 335

BYRD, C.A. & MATRONE, G. (1965) - *Proc. Soc. exp. Biol. Med. 119* : 347

CAMPBELL, J.K. & MILLS, C.F. (1974) - *Proc. Nutr. Soc. 33* : 15A

CARRICO, R.J. & DEUTSCH, H.F. (1969) - *J. biol. Chem. 244* : 6087

COUSINS, R.J. (1974) - *In* "Trace Element Metabolism in Animals - II" (W.G. Hoekstra, J.W. Suttie, H.E. Ganther & W. Mertz, eds.), p. 503. University Park Press : Baltimore

CUNNINGHAM, J.J. (1950) - *In* "Copper Metabolism" (W.D. McElroy & B. Glass, eds.), p. 246. Johns Hopkins Press : Baltimore

DAVIES, N.T. & NIGHTINGALE, R. (1975) - *Proc. Nutr. Soc. 34* : 5A

DAVIS, P.N., MORRIS, L.C. & KRATZER, F.H. (1962) - *J. Nutr. 77* : 217

DEWAR, W.A., WHITEHEAD, C.C., DOWNIE, J.N. & POTTER, E. (1975) - *Proc. Nutr. Soc. 34* : 8A

DICK, A.T. (1956) - *In* "Inorganic Nitrogen Metabolism" (W.B. McElroy & B. Glass, eds.), p. 445. Johns Hopkins Press : Baltimore

DICK, A.T. & BULL, L.B. (1945) - *Aust. vet. J. 21* : 70

DOWDY, R.P. & MATRONE, G. (1968*a*) - *J. Nutr. 95* : 191

DOWDY, R.P. & MATRONE, G. (1968*b*) - *J. Nutr. 95* : 197

EVANS, G.W., MAJORS, P.F. & CORNATZER, W.E. (1970) - *Biochem. biophys. Res. Commun. 40* : 1142

FRIEDEN, E. (1971) - *Adv. Chem. 14* : 292

HARTMANS, J. (1970) - *In* "Trace Element Metabolism in Animals" (C.F. Mills, ed.), p. 441. E. & S. Livingstone : Edinburgh

HILL, C.H. & MATRONE, G. (1962) - *Proc. 12th World Poultry Congr.*, p. 219

HILL, C.H. & MATRONE, G. (1970) - *Fedn Proc. 29* : 1474

HILL, C.H., MATRONE, G., PAYNE, W.L. & BARKER, C.W. (1963) - *J. Nutr. 80* : 227

HILL, C.H., STARCHER, B.C. & MATRONE, G. (1964) - *J. Nutr.* *83* : 107

HOEFER, J.A., MILLER, E.R., ULLREY, D.E., RITCHIE, H.D. & LUECKE, R.W. (1960) - *J. Anim. Sci. 19* : 249

HUISINGH, J., GOMEZ, G.G. & MATRONE, G. (1973) - *Fedn Proc.* *32* : 1921

HUISINGH, J. & MATRONE, G. (1972) - *Proc. Soc. exp. Biol. Med. 139* : 518

HUISINGH, J., McNEILL, J.J. & MATRONE, G. (1973) - *Fedn Proc. 32* : 900

KINNAMON, K.E. (1966) - *J. Nutr. 89* : 365

KIRCHGESSNER, M. & WESER, U. (1963) - *Z. Tierphysiol. Tierernähr. Futtermittelk. 18* : 181

KOVALSKY, V.V., VOROTNITSKAYA, I.E. & TSOI, G.G. (1973) - *In* "Trace Substances in Environmental Health - VII" (D.D. Hemphill, ed.), p. 277. University of Missouri : Columbia

KOVALSKY, V.V., VOROTNITSKAYA, I.E. & TSOI, G.G. (1974) - *In* "Trace Element Metabolism in Animals - II" (W.G. Hoekstra, J.W. Suttie, H.E. Ganther & W. Mertz, eds.), p. 161. University Park Press : Baltimore

LEACH, R.M., TURK, D.E., ZEIGLER, T.R. & NORRIS, L.C. (1962) - *Poult. Sci. 41* : 300

MARCILESE, N.A., AMMERMAN, C.B., VALSECCHI, R.M., DUNAVANT, B.G. & DAVIS, G.K. (1970) - *J. Nutr. 100* : 1399

MATRONE, G. (1970) - *In* "Trace Element Metabolism in Animals" (C.F. Mills, ed.), p. 354. E. & S. Livingstone : Edinburgh

McCORD, J.M. & FRIDOVICH, I. (1969) - *J. biol. Chem.* *244* : 6049

MILLER, R.F. & ENGEL, R.W. (1960) - *Fedn Proc. 19* : 666

MILLS, C.F. & DALGARNO, A.C. (1972) - *Nature, Lond. 239* : 171

MILLS, C.F., MONTY, K.J., ICHIHARA, A. & PEARSON, P.B. (1958) - *J. Nutr. 65* : 129

MILTIMORE, L.E. & MASON, J.L. (1971) - *Can. J. Anim. Sci.* *51* : 193

POWELL, G.W., MILLER, W.J., MORTON, J.D. & CLIFTON, C.M. (1964) - *J. Nutr. 84* : 205

RISH, M.A. (1970) - *In* "Trace Element Metabolism in Animals" (C.F. Mills, ed.), p. 452. E. & S. Livingstone : Edinburgh

RITCHIE, H.D., LUECKE, R.W., BALTZER, B.V., MILLER, E.R., ULLREY, D.E. & HOEFER, J.A. (1963) - *J. Nutr. 79* : 117

RYS, R.M., KUKLEWICZ, M. & SOKOL, J. (1963) - *Rocnik. Nauk. Rolnii 83* : 145

SEELIG, M.S. (1972) - *Am. J. clin. Nutr. 25* : 1022

SHAIKH, Z.A., COLEMAN, R.L. & LUCIS, O.J. (1973) - *In* "Trace Substances in Environmental Health - VII" (D.D. Hemphill, ed.), p. 313. University of Missouri : Columbia

SIEGEL, L.M. & MONTY, K.J. (1961) - *J. Nutr. 74* : 167

SMITH, B.S.W., FIELD, A.C. & SUTTLE, N.F. (1968) - *J. comp. Path.* *78* : 449
SMITH, B.S.W. & WRIGHT, H. (1973) - *Proc. Nutr. Soc.* *32* : 25A
SOLOMONS, C.C., ERNISSE, D.J. & HANDRICH, E.M. (1973) - *In* "Trace Substances in Environmental Health - VII" (D.D. Hemphill, ed.), p. 233. University of Missouri : Columbia
STARCHER, B.C. (1969) - *J. Nutr.* *97* : 321
SUPPLEE, W.C. (1961) - *Poult. Sci.* *40* : 827
SUTTLE, N.F. (1973) - *In* "Trace Substances in Environmental Health - VII" (D.D. Hemphill, ed.), p. 245. University of Missouri : Columbia
SUTTLE, N.F. (1974*a*) - *Br. J. Nutr.* *32* : 395
SUTTLE, N.F. (1974*b*) - *In* "Trace Element Metabolism in Animals - II" (W.G. Hoekstra, J.W. Suttie, H.E. Ganther & W. Mertz, eds.), p. 612. University Park Press : Baltimore
SUTTLE, N.F. (1974*c*) - *Br. J. Nutr.* *32* : 559
SUTTLE, N.F. (1974*d*) - *Proc. Nutr. Soc.* *33* : 299
SUTTLE, N.F. & FIELD, A.C. (1968) - *J. comp. Path.* *78* : 351
SUTTLE, N.F. & FIELD, A.C. (1974) - *Vet. Rec.* *95* : 166
TAYLOR, T.G. (1965) - *Proc. Nutr. Soc.* *24* : 105
TRIDOT, G. & BERNARD, J.C. (1962) - *Acta chim. hung.* *34* : 186
VAN CAMPEN, D. & SCAIFE, P.U. (1967) - *J. Nutr.* *91* : 473
VAN REEN, A. & WILLIAMS, M. (1956) - *Archs Biochem. Biophys.* *63* : 1
VOHRA, P. & KRATZER, F.H. (1966) - *J. Nutr.* *89* : 106
WACKER, W.E.C. (1970) - *Fedn Proc.* *29* : 1462
WADDELL, D.G. & SELL, J.L. (1964) - *Poult. Sci.* *43* : 1249
WALLACE, H.D., McCALL, J.T., BASS, B. & COMBS, G.E. (1960) - *J. Anim. Sci.* *19* : 1153
WHITEHEAD, D.C. (1966) - *In* "Nutrient Minerals in Grassland Herbage", p. 42. Commonwealth Bureau of Pastures & Field Crops : Hurley
WINSTON, P.W., HOFFMAN, L. & SMITH, W. (1973) - *In* "Trace Substances in Environmental Health - VII" (D.D. Hemphill, ed.), p. 241. University of Missouri : Columbia

THE RECOGNITION AND CORRECTION OF TRACE ELEMENT DEFICIENCIES

D.J. Reuter

South Australian Department of Agriculture
Adelaide, South Australia 5000

INTRODUCTION

The basis for successful agriculture over large tracts of southern Australia was established by the now classical research undertaken between 1930 and 1950. These investigations emphasized the need for superphosphate, legume-based pasture and trace elements, notably Co, Cu, Zn, Mo and Mn. Adoption of the research findings by farmers was rapid, as responses on many of our low-fertility soils were dramatic.

The purpose of this paper is to review ways of recognizing trace element deficiencies in the field and to evaluate procedures for their correction. The topic is too broad to review comprehensively here, and so attention is directed to the underlying principles and examples given to illustrate the various aspects being discussed. The trace elements considered are Co, Cu, Fe, Mn, Se, Mo and Zn.

THE RECOGNITION OF TRACE ELEMENT DEFICIENCIES

Trace element deficiencies may occur over broad geographic areas, or they may be localized even on the scale of small isolated patches of poorer growth in an otherwise healthy crop. The disorders are often associated with particular soils or soil properties, but frequently climatic and managerial factors are also involved.

Soil, Climatic and Managerial Factors

Soil. During the process of soil formation, trace elements in the parent material undergo chemical and physical changes that influence their concentration and distribution within the developing soil profile (Hodgson 1963; Mitchell 1964,1971; Lakin & Davidson 1967; Krauskopf 1972).

Comparatively little information has been published on the total and extractable content of trace elements in Australian soil profiles. However, the trace element status of certain soil groups in Australia has been described. Blackburn & Giles (1963) showed that the pattern of trace element deficiencies observed by Tiver (1955) in south-eastern South Australia was associated with the low status of the soils and parent materials and with the modifying effects of drainage and topography. McKenzie (1959) showed that the trace element status of morphologically similar soils varied with the nature and composition of the parent material. However, Tiller (1963) concluded that the nature of the parent material is less important for many old, highly weathered and relic soils in Australia.

The distribution of trace element-deficient soils in Australia has been described by Stephens & Donald (1958), Anderson (1970) and Williams & Andrew (1970). Single or multiple deficiencies of trace elements occur in a wide variety of soils in all Australian states. In many areas the extent of deficiency is not accurately known, because the disorders are incipient, marginal or occur infrequently.

The type of deficiency that can be expected in one area can often be predicted from experience with soils of similar morphology elsewhere. Examples are: Zn deficiency of black earths and grey and brown soils of heavy texture; Mn, Cu and Co deficiencies on coastal calcareous soils; Cu and Zn deficiencies on siliceous sand-over-clay and gravelly soils. Correlations become more reliable when the morphological description of soil is based on detailed classification schemes such as the Factual Key (Northcote 1971), and where chemical and biological features that affect the availability of trace elements are also considered (see Table 7).

Climatic factors. Climate limits the potential yield and the magnitude of yield response to trace element additions. Where climate varies markedly between seasons, large between-seasons differences in nutrient uptake are observed.

The weather can affect the supply of trace elements to plants in two main ways. Firstly, by increasing soil water content and/or reducing soil aeration, the availability of Fe, Mn, Co and Mo can be increased (Lucas & Knezek 1972), whereas the availability of Se can be reduced (Allaway 1972). Also, the prevalence of Zn deficiency in cool, wet seasons has been associated with a decreased solubility of soil Zn (Bauer &

Lindsay 1965). Secondly, in seasons favourable for rapid plant growth, trace elements translocated to plant tops are rapidly diluted within the aerial tissues and this may result in deficiencies, particularly where the availability of trace elements is marginal.

The following three examples show the effect of climate on the incidence of trace element deficiencies in Australia.

In Western Australia, Gardiner & Gorman (1963) showed that the Se concentration of herbage sampled in spring (August-November) was related to annual rainfall (Fig. 1): low concentrations frequently occurred in high rainfall areas, where white muscle disease in lambs has also been reported. The relationship was modified by soil texture: higher levels tended to be found in herbage grown on "heavier" soils.

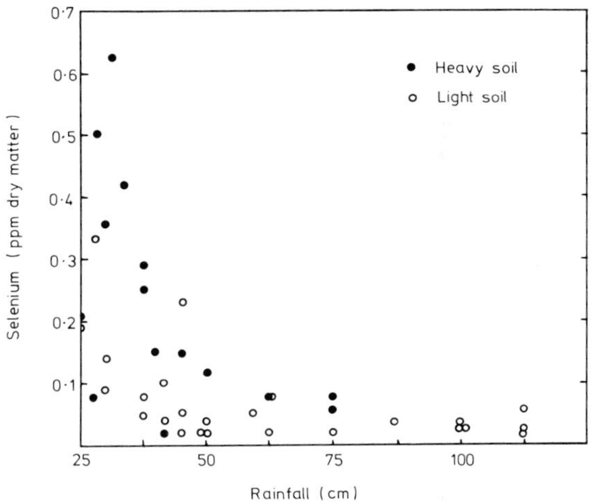

Fig. 1. The relation between Se content of plants and mean annual rainfall (Gardiner & Gorman 1963). [Reprinted by permission of the Australian Journal of Experimental Agriculture and Animal Husbandry.]

In southern Australia, Cu deficiency in sheep is more prevalent in favourable seasons with lush pasture growth in spring (Lee 1951; Bennetts 1955). In marginally deficient areas, Cu deficiency may be slight or absent in some years but definite in other seasons. In cereals, on the other hand, Cu deficiency symptoms such as wilting are more obvious in drier seasons, and especially where spring rains are below average (Riceman & Anderson 1943). This is because the extent of the root system of Cu-deficient plants is reduced (Riceman & Donald 1938).

Finally, in South Australia, Lee (1952) demonstrated that the incidence of Co deficiency in sheep varied greatly from year to year. In 5 years at Glenroy, flocks were unaffected, whereas in another 8 years, symptoms of variable severity were encountered. The cause of these between-seasons differences may be associated with changes in botanical composition of the pasture, or may be due to variations in Co intake from ingested soil, as has been suggested by New Zealand workers (Andrews *et al.* 1958; Healy 1972).

Management factors. Certain activities of man affect the incidence and severity of trace element disorders. By increasing soil pH, lime applications affect the solubility of trace elements in soils for considerable periods. Australian investigations on this subject have been reviewed by Anderson (1970). Certain impurities in commercial fertilizers, for example Zn in superphosphate, provide trace elements (Ozanne, Shaw & Kirton 1965). If purified fertilizers are used, this source is lost. In Western Europe, the intensive use of nitrogenous fertilizer on pasture has resulted in highly productive, grass-dominant swards containing lower concentrations of Fe, Cu, Co, Mo and Mn. As a result, Cu deficiency in cattle has increased (Hartmans 1973). Copper deficiency in cereals is also more severe when nitrogenous fertilizers are applied, and also where *Brassicae* precede cereal crops in the rotation (Davies *et al.* 1971). Other trace element interactions have been reviewed by Olsen (1972) and Lucas & Knezek (1972).

Exposure of the subsoil by levelling land for flood irrigation has increased the incidence of Zn deficiency in the U.S.A. (Lindsay 1972) and in Australia (Kleinig & Loveday 1962). On some soils, increasing the soil water content by irrigation raises the availability of certain trace elements (Allaway 1968), whereas draining wet soils often reduces their availability (Mitchell *et al.* 1957).

Species and cultivars often vary in their susceptibility to soil deficiencies of trace elements (Brown *et al.* 1972). Changes in the botanical composition of pastures can affect animal response where marginal deficiencies exist. The concentrations of Cu, Zn, Mo and Co are generally higher in legumes than in grasses (Beck 1962; Williams 1963; Whitehead 1966; Gladstones & Loneragan 1967,1970), whereas the reverse is true for Se (Davies & Watkinson 1966) and, with the exception of lupins (Gladstones & Loneragan 1970), for Mn. Various herbs such as capeweed [*Arctotheca calendula* (L) Levyns] may contain higher concentrations of certain trace elements than legumes (Williams 1963; Gladstones & Loneragan 1970), perhaps due to their greater depth of rooting (Ozanne, Asher & Kirton 1965). Exceptions to these general rules are found. For example, in New Zealand, Andrews (1966) found that legumes accumulated more Co than grasses on Co-sufficient soils, but the differences were negligible on Co-deficient soils. Similar observations have been made in the U.S.A. (Kubota & Allaway 1972).

Besides differences between pasture species in trace element accumulation, variation in the availability to sheep of trace elements from different grass species has recently been reported by Patil & Jones (1970). Their studies demonstrated that certain grass species when fed to lambs produced Cu and Co deficiencies whereas other species did not, although digestibilities and mineral contents of the species tested were similar.

Deficiency Symptoms

Production losses have occurred before deficiency symptoms are recognized. Symptoms are nevertheless important in diagnosis. Symptoms of trace element deficiencies in a variety of plant species have been reviewed in the monographs edited by Sprague (1961), Wallace (1964) and Chapman (1966).

Symptoms cannot be generalized completely, but in many species the symptoms are so characteristic that positive identification can be made in the field. Examples are: interveinal chlorosis (Fe), chlorosis and leaf necrosis in cereals and "marsh spot" in peas (Mn), and the "little leaf" syndrome in citrus and legumes (Zn). Sometimes the symptoms are manifest during a particular phase of plant development, for example the wilting symptoms and melanism in cereals (Cu), and this also aids identification. Specific symptoms can also vary with the severity of the deficiency. Species that show characteristic symptoms of deficiency and which are susceptible to deficiency are sometimes grown to identify marginal deficiencies.

Diagnosis cannot always be made from visible symptoms in plants. The symptoms may be slight or temporary, masked by other nutritional deficiencies, or masked by unrelated factors such as disease and weather.

Symptoms of simple and induced deficiencies in livestock have been described by Marston (1952) and Underwood (1971). The physiological role of each trace element determines which production factor (meat, wool, fertility, reproductive performance) is most impaired by deficiency. Deficiencies of Cu, Co, Se and I in ruminants have been reported in Australia. Dual deficiencies such as "coast disease" (Cu and Co) are commonly observed. Manganese and zinc deficiencies are rare in ruminants but have occurred in South Australia (Egan 1972). Disorders are often manifested at specific times in the season, such as during the spring "flush", or during stress periods, or at certain ages, as when Co deficiency occurs in lambs (Lee & Marston 1969). Furthermore, considerable genetic variability exists within and between breeds in susceptibility to Cu deficiency (Wiener 1966; Wiener & Field 1969).

Acute deficiencies in livestock are often readily identified. For example, Cu deficiency causes a defect in wool known as "steely" or "stringy" wool, a symptom which has been used to map the distribution of Cu deficiency in parts of southern Australia (Lee 1946,1951; Bennetts 1955). Subclinical deficiencies, although less obvious, can result in important economic losses. For example, symptoms of less acute Co deficiency (debility and starvation) are non-specific (Marston 1952) and can be confused with other "ill-thrift" syndromes. Similarly, subclinical Cu deficiency can lower the wool production (Marston & Lee 1948) of otherwise healthy sheep (Lee 1951). Other tests are needed to confirm these "grey" areas of marginal deficiency.

Response Experiments

Response experiments conducted in the field are used both to detect and to measure the magnitude of production losses caused by deficiencies. On the evidence of likely gains in production, farmers decide whether remedial action is necessary. In Australia, field responses and symptoms together have largely defined the geographic distribution of trace element deficiencies.

Response experiments are expensive and time-consuming, as they need to be replicated on similar soils at several locations within a region. Fertilizer experiments that compare a range of application rates of one or more trace elements over several seasons not only estimate the biological and economic significance of the deficiencies and their interactions, but can also estimate residual effects, a factor frequently overlooked when fertilizer recommendations are formulated. Another approach often used for identifying deficiencies is to apply trace elements as foliar sprays to affected crops and to observe the recovery of plant vigour or colour.

Direct administration of trace elements to grazing livestock and measurements of improvements in productivity and health have been widely used for defining deficiencies. The measurements recorded need to be related to the physiological role of the element being tested. In New Zealand the distribution of Co and Se deficiencies were ascertained from several hundred short-duration, simple livestock response experiments (Andrews 1956; Andrews *et al.* 1968). In relating responses to soil type and pasture content, Andrews (1956) concluded:

> *The most conclusive method of diagnosing cobalt-deficiency disease is to assess the response of animals to supplements of cobalt administered by mouth.... If cobalt dosed animals make significantly better weight*

gains than untreated animals, cobalt deficiency is confirmed.

A feature of field experimentation is the marked variation in response from season to season. Clearly, field research workers need to collect additional relevant information which can help to explain the reasons for this variation.

Glasshouse experiments are used to screen the fertility status of a large number of soils under uniform growing conditions. The advantages of this approach lie in the economy of time and labour, the number of treatments that can be imposed, and continuity of observation. However, the utility of these investigations is questionable for trace elements, as the fertility of the subsoil is rarely assessed, and the intensity of factors such as soil moisture, temperature, light and rooting density that greatly affect the availability and utilization of trace elements by plants, are unlikely to be comparable with field conditions.

Diagnostic Tests

Additional diagnostic tests are used either to examine suspected occurrences of deficiencies in the field or to identify the cause of unknown disorders. Such tests need to be appropriate, reproducible and rapid. The introduction of atomic absorption and direct-reading emission spectrometers has simplified analysis for many trace elements.

The objective of diagnostic testing is to separate deficient and non-deficient situations. Because application of trace elements often costs little in relation to the production benefits realized, there is little need to use the tests to predict the quantity of trace element needed to remedy the deficiency. The three stages of testing are: (1) selection of the appropriate test, (2) calibration of the test against response and estimation of a "critical level" which delineates deficiency from adequacy, and (3) collection of representative samples. Sample collection has been discussed for soils by Oertel (1959) and Cox & Kamprath (1972), and for plants by Jones (1972).

Soil tests. Soil analysis can indicate the need to apply trace elements at sowing time. Except where the total concentration of trace elements in soil is very low or very high, total levels are poorly correlated with plant requirements. Extraction and equilibration of soil with chemical reagents has been widely used for predicting the availability of trace elements to plants (Cox & Kamprath 1972). The procedures tend to be empirical due to an incomplete understanding of the soil/soil solution/plant system. Generally it is recognized (Lindsay 1972) that soil tests should reflect both the intensity factor (concentration) and the capacity factor (ability to replenish), and most tests quantify the concentration of water-soluble, exchangeable, complexed and/or reducible fractions within the soil. It may sometimes be desirable to simultaneously estimate more than one element in the extracting solution.

The utility of a soil test is generally gauged by its correlation with plant growth and concentration or uptake of nutrients by plants grown on soils supplying from low to high amounts of trace element. Ideally, the test must cater for wide variation in soil properties. Tests are too often calibrated in glasshouse experiments where conditions may differ from those in the field. Calibrations in the field are rarely reported. Certainly, poorer correlations can be expected in the field, as plant yield and nutrient uptake depend on many factors unrelated to trace element supply. The predictive value of soil tests for trace element supply has often been improved by considering soil properties such as pH, cation exchange capacity, organic matter, bulk density and lime content (McKenzie 1966; van Luit & Henkens 1967; Cox 1968; Adams *et al.* 1969), either in conjunction with the test or as part of a multivariate statistical treatment (Dolar & Keeney 1971). At the very least it can be assumed that soil tests are likely to be more reliable in regions where similar soils occur.

Soil tests are of little value for predicting deficiencies in tree crops (Chapman 1966). Soil tests for Fe and Mn seem unwarranted as symptoms are readily seen. With careful interpretation, soil tests have been used to define trace element disorders in grazing livestock, particularly in conjunction with herbage analysis (Mitchell *et al.* 1957; Kubota 1964; Kubota, Lazar, Simonson & Hill 1967). Sampling stream sediments has also proved useful (Thornton & Webb 1969).

The "critical" soil concentration, below which deficiency can be expected, is estimated either by regression analysis of per cent yield response on soil test concentration (Henkens 1961; Dawson & Bhella 1972) or by separating deficient and non-deficient soils on the basis of yield response and symptom rating (Fig. 2) and plant analysis (Mitchell *et al.* 1957; Hoff & Mederski 1958; van Luit & Henkens 1967). The utility of the "critical"value is determined by the variance of the data used for calibration. On their own, soil tests separate extremely deficient soils from adequately supplied soils, and leave a "grey" area between these extremes.

Plant analysis. Foliar analysis is widely used as an aid for diagnosing trace element deficiencies in tree crops

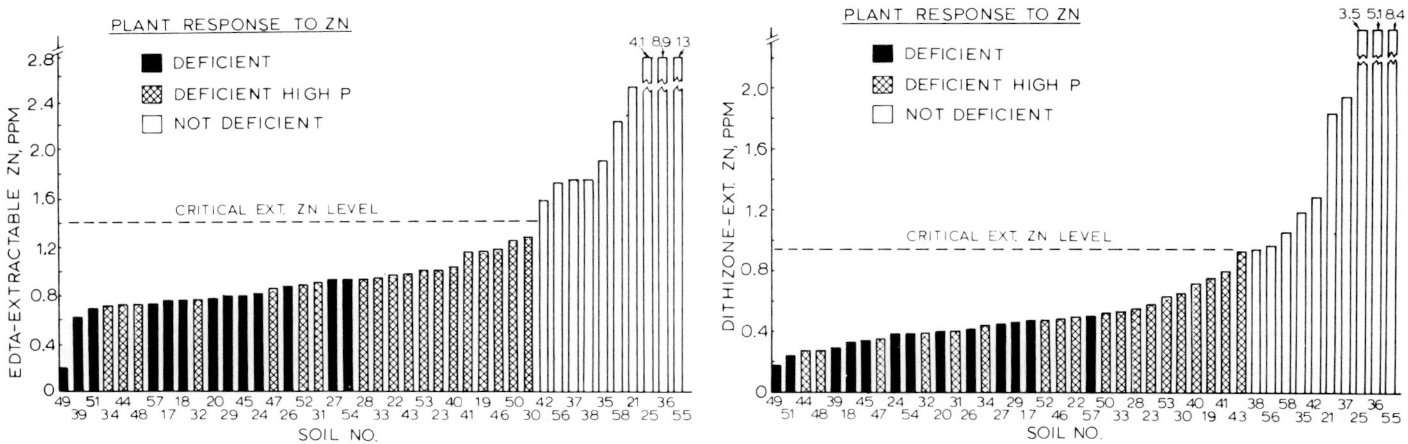

Fig. 2. Calibration and derivation of the critical concentration of two soil tests for predicting Zn deficiency in maize (Trierweiler & Lindsay 1969). [Reprinted by permission of the Soil Science Society of America Incorporated]

(Chapman 1966; Raupach 1975), and is being used increasingly for annual field crops and pasture. An advantage of plant analysis is that it reflects the "available" nutrient status of soils, as integrated by the plant. Used in conjunction with soil analysis and plant symptoms, a more definitive assessment of nutritional status is often possible.

As with soil analysis, problems of calibration and interpretation arise, and these have been reviewed by Bates (1971) and Jones (1972). The problems include: the considerable variation in the accumulation of trace elements by different plant species, the change in concentration in top growth with plant age (Fig. 3a), the change in concentration of one element as affected by the level of other nutrients or environmental factors, and the unequal distribution of trace elements within plant tissues (Williams & Moore 1952). Of particular importance is the relation between concentration and yield, as the concentration of nutrients in the tops of deficient plants can be similar to that of healthy plants (Fig. 3b).

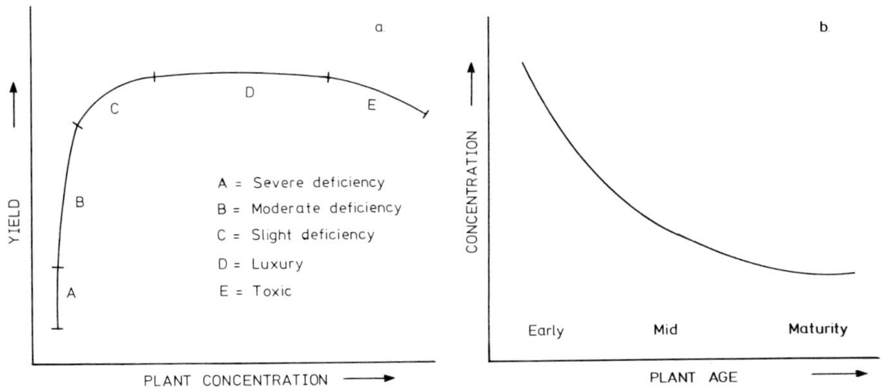

Fig. 3a. Functional relationship between plant yield and nutrient concentration.

Fig. 3b. The decrease in plant nutrient concentration with increasing plant age.

Plant analysis needs to be related to specific stages of plant development and to specified plant organs or tissues (Jones 1972). For example, Bates (1971) emphasized the importance of sampling when deficiencies first appear, and of sampling tissues of similar physiological age, such as the youngest fully expanded leaf.

Plant analysis has been used successfully for assessing deficiencies in ungrazed pastures, especially if legumes are used (Mitchell *et al.* 1957; McNaught 1970). For grazed pasture, interpretation is more complex, but herbage analysis has been used successfully for establishing the cause of trace element disorders in ruminants (Fig. 4) on a specific or regional basis (Cunningham 1960; Kubota *et al.* 1961; Kubota 1964; Kubota, Allaway, Carter, Cary & Lazar 1967; Kubota, Lazar, Simonson & Hill 1967; Hogan *et al.* 1971) and for estimating the trace element supply in conserved feeds.

Critical concentrations in plants, below which deficiencies can be expected in crops and livestock are given by Jones (1972) and Underwood (1971) respectively.

Blood and organ analysis. The use of blood and organ analysis for diagnosing trace element deficiencies in livestock has been reviewed by Underwood (1971). The concentrations of Cu and vitamin B_{12} in liver, although less convenient to sample are sensitive to low intakes and are considered to be more reliable than blood analysis. Variations between animals in liver biopsy concentrations can be relatively high. Tentative diagnostic standards for vitamin B_{12} concentration in sheep (Andrews *et al.* 1960; Andrews & Hart 1962; Andrews 1965; Marston 1970) and cattle (Skerman *et al.* 1959) have been reported. The concentration of Se in the liver and kidney cortex is also relatively sensitive to Se status, and diagnostic criteria are available (Andrews *et al.* 1968).

The concentration of Se in whole blood is sensitive to the level of intake and is therefore a useful screening index of Se status (Judson & Obst 1975). Concentrations of Cu in blood

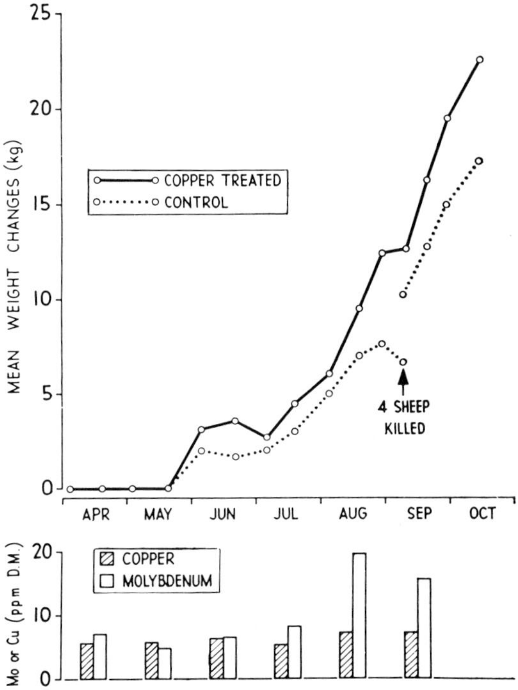

Fig. 4. Seasonal variation in Cu and Mo concentrations in pastures in relation to liveweight responses in sheep to Cu therapy (Hogan et al. *1971). [Reprinted by permission of the New Zealand Journal of Agricultural Research*]

consistently below 10 µmol/ℓ are indicative of simple or induced deficiency (Underwood 1971), and mean serum vitamin B_{12} concentrations less than 0.2 µg/ℓ often indicate deficiency in sheep, although considerable flock and age variability occurs (Andrews & Stephenson 1966). Serum vitamin B_{12} analysis is not used for diagnosing Co deficiency in cattle (H.J. Lee, personal communication).

Enzyme activity and metabolite accumulation. As the biochemical functions of trace elements are often related to specific enzymes, more reliable diagnosis of deficiencies may be achieved either by estimating enzyme activity or by measuring the accumulation of metabolites caused by lack of enzymic activity. Such tests depend less on comparison of standards established by correlation with yield response, but need to be adaptable to routine laboratory analysis. Examples are given by Randall (1970), who demonstrated that the activity of nitrate reductase is substantially reduced in Mo-deficient plants, and by Kessler (1961), who showed that ribonuclease activity in apple leaves increases sharply with Zn concentrations below 15 ppm. The ceruloplasmin oxidase activity in plasma has been used to screen for Cu deficiency in cattle (Todd 1969) and may be reliable for assessing Mo-induced Cu deficiency (Bingley & Anderson 1972). The estimation of methyl-malonic acid and formiminoglutamic acid in urine has also been suggested as test for diagnosing Co deficiency in young sheep (Gawthorne 1968; Andrews *et al.* 1970; Hogan *et al.* 1973), and the high activity of aspartate aminotransferase and creatine phosphokinase in serum has been associated with white muscle disease (Oksanen 1967; Judson & Obst 1975).

CORRECTION OF TRACE ELEMENT DEFICIENCIES

Soil Applications

Trace element compounds. A wide variety of compounds are used to correct trace element deficiencies (Table 1). Australian manufacturers chiefly use oxides, sulphates and carbonates. Insoluble compounds can sometimes be used successfully if incorporated into the soil in a finely divided state (Steenbjerg 1951). In South Australia, 24 standard fertilizers containing trace elements are marketed and many special mixes are also available.

Incorporation with macronutrient carriers. Trace elements are usually applied with a macronutrient carrier. The most difficult problem facing manufacturers is to produce fertilizers of uniform composition. Until recently in Aus-

TABLE 1

Trace element compounds used in fertilizers

	% Element (approx.)
COPPER	
*Copper sulphate† - $CuSO_4 \cdot 5H_2O$	25
*Cupric oxide - CuO	75
*Copper carbonate ore	6-25
Copper oxychloride†	52
COBALT	
*Cobalt sulphate† - $CoSO_4 \cdot 7H_2O$	21
IRON	
*Ferrous sulphate† - $FeSO_4 \cdot 7H_2O$	18
Iron chelates of various types†	vary
MANGANESE	
*Commercial manganese sulphate	25
*Manganous sulphate† - $MnSO_4 \cdot H_2O$	32
Manganous oxide† - MnO	41-68
MOLYBDENUM	
*Sodium molybdate† - $Na_2MoO_4 \cdot 2H_2O$	38
*Molybdenum trioxide - MoO_3	58
*Ammonium molybdate - $(NH_4)_6Mo_7O_{24} \cdot 4H_2O$	54
ZINC	
*Zinc sulphate† - $ZnSO_4 \cdot H_2O$	35
*Zinc sulphate† - $ZnSO_4 \cdot 7H_2O$	23
*Zinc oxide† - ZnO	65-71
*Zinc carbonate ore	20-40
*Dross	80

*Compounds used in Australia (Anon. 1970)
†Compounds also used in foliar sprays

tralia, fertilizers containing trace elements were prepared by mixing finely divided trace element salts with cured granulated macronutrient carriers. Because of differences in particle size, shape and density, the trace element salts segregated from the carrier during storage and application. This resulted in uneven distribution of nutrients to the soil (Silverberg *et al.* 1972), uneven crop growth, and inefficient use of applied nutrients (Reuter *et al.* 1972).

Three manufacturing techniques have been introduced to overcome this problem: (1) blending similar-sized granules of both trace element salts and macronutrient carriers, (2) coating powdered trace element salts onto the surface of carriers with binders, (3) incorporating trace elements into macronutrient carriers before ammoniation and/or granulation.

The advantages and limitations of the first two techniques have been discussed by Silverberg *et al.* (1972). The third method has been introduced more recently into Australian manufacturing processes (e.g. Reuter *et al.* 1972*b*). Chemical reactions between the trace element salt and the carrier usually result under the temperature and pH conditions existing during granulation. Reaction products and the conditions under which they form have been described by Lehr (1972).

Adding a water-soluble trace element salt does not necessarily mean that the element will be water soluble in the final product; for instance, Mortvedt (1968) showed that the effectiveness of zinc sulphate, incorporated into a N-P-K fertilizer or concentrated superphosphate, decreased as the level of water-soluble Zn was reduced by ammoniation. This was presumably due to the formation of insoluble $ZnNH_4PO_4$. Incorporating high levels of the hygroscopic salt $ZnSO_4 \cdot 7H_2O$ with superphosphate during granulation produces physically unsatisfactory products. On the other hand, when some insoluble trace element sources, such as ZnO, CuO and metallic Cu and Zn, are incorporated with certain carriers, they become soluble (Lehr 1972). Since it is generally agreed that at least part of the trace element in granular fertilizers needs to be in a water-soluble form (Silverberg *et al.* 1972; Giordano & Mortvedt 1972), the availability of trace elements to plants contained in fertilizer reaction products needs assessing. Giordano & Mortvedt (1972) show that the availability to plants of trace elements in granular fertilizers depends on the movement from the granule into the soil, the type of macronutrient carrier, fertilizer and soil pH, and the concentration of the trace element in the carrier.

Rate and method of application. Recommended rates of application of trace elements are established by local field testing. The rates depend both on soil and crop characteristics, which affect the severity of the deficiency, and on the profitability of the yield response per unit of trace element applied.

Table 2 indicates the rates at which trace elements are commonly applied to soils in Australia to correct deficiencies. These rates tend to be lower than those tabulated by Sauchelli (1969), Mortvedt & Cunningham (1971) and Murphy & Walsh (1972) for other countries.

The quantity of trace elements needed to correct deficiencies on mineral soils is usually much less than that needed on organic soils, where elements such as Cu are strongly

TABLE 2

Rates of trace elements applied to soil in Australia

	Rate of element (kg ha^{-1})	Reference†
COBALT		
Pastures	0.09-0.24	a,b
COPPER		
Mineral soils		
Cereals and pastures	0.7-3	a,b
Organic soils		
Pastures	2-4	c
MANGANESE		
Cereals	4-6	a,b,d
Pastures	6	e
Vegetables	17-28	f
MOLYBDENUM		
Cereals and pastures	0.04-0.10	a,b,g
ZINC		
Light soils		
Cereals and pastures	0.6-2.4	a,b
Heavy soils		
Cereals	1.8-29	b,h
Pastures	1.8-3.6	b
Linseed	7-29	b,i

†a. Gartrell & Glencross (1969); b. Tiver (1955); c. Anderson (1946); d. Reuter *et al.* (1972*b*); e. Higgs & Burton (1955); f. Skene & Kefford (1954); g. Anderson (1956); h. Duncan (1967*a*); i. Duncan (1967*b*).

bound (Ennis & Brogan 1961). When coastal peats are developed in the eastern U.S.A., 60 kg ha^{-1} Cu is applied initially, followed by annual maintenance dressings of 16 kg ha^{-1} Cu (Murphy & Walsh 1972).

With the exception of selenate and possibly molybdate (Jones & Belling 1967), the mobility of trace elements in most soils is small. Transference of trace elements to plant roots is primarily *via* diffusion over short distances (Barber *et al.* 1966; Halstead *et al.* 1968; Wilkinson *et al.* 1968). Thus the efficiency with which plant roots recover applied trace elements largely depends on the distance between roots and fertilizer. For example, Giordano & Mortvedt (1966) demonstrated that yield response of maize to Zn was related to the number of fertilizer granules applied; further response was not obtained beyond a critical number of granules. Viets (1966) pointed out that the percentage of the soil volume effectively contacted by roots varies from 1 to 5%, values that are consistent with the percentage recovery, by a single crop, of trace elements mixed with the soil.

It is therefore not surprising that broadcast applications are usually less effective than banding or mixing trace elements in soil; where broadcasting is practised, higher application rates are needed (Murphy & Walsh 1972). Soil applications are usually not recommended for tree crops. On the other hand, banding may concentrate toxic quantities of trace elements near the seed in sandy soils (Anon. 1968). Mixing trace elements within the soil, as by broadcasting followed by cultivation, has often proved to be the most efficient method of application, presumably because it increases the contact between plant roots and applied fertilizer.

Residual effects. Fertilizer applications to soil may not only increase the yield of the crop to which they are applied, but may also improve the yield of succeeding crops. Repeated applications over a number of years gradually reduce the level of response to a nutrient, and may even induce toxicities. The latter have been observed on sandy soils following the repeated use of fungicides containing Cu (van Alphen 1957). Similarly, the disposal of swine effluent containing high concentrations of Cu on mineral soils may also cause toxicity (Baker 1974). In general terms the residual value of soil-applied trace elements has received relatively little attention.

The magnitude of the residual response depends on the soil, crop, nutrient, quantity and type of fertilizer, and the method of application. The residual effects of Fe and Mn are negligible, as their salts are rapidly immobilized by soil. In contrast, applications of Cu, Co, Zn, Mo and Se may last for considerable periods (Table 3). An example of the long-term residual value of Cu and Zn is provided by the siliceous sands of the Coonalpyn Downs, South Australia. Following initial applications of 2 kg ha^{-1} Cu and 1.8 kg ha^{-1} Zn to establish improved pastures, only superphosphate was needed thereafter (Riceman 1961). Riceman (personal communication) based this conclusion on observations that: (1) pasture and livestock production was satisfactory for at least 7 years after the initial application; (2) applied Cu and Zn are not leached to any great extent (Jones *et al.* 1957; Jones & Belling 1967); and (3) the Cu and Zn removed in farm produce (Table 4) approximately equalled their addition as impurities in superphosphate.

Gartrell & Glencross (1969) also conclude that a single application of Cu, Zn or Mo suffices to correct deficiencies on many mineral soils of the cereal-pasture areas of Western Australia. On other soils the residual effect of Mo and Co may last for only a few years (Swaine 1959; Mannetje *et al.* 1963; Adams *et al.* 1969). Repeated large applications of Cu in particular, and also other trace elements, are required on peat and muck soils.

Few field experiments have measured long-term residual effects of trace elements. Yet such information can be important where farm incomes are marginal, where trace element

TABLE 3

Duration of residual effects of trace element applications to soil

	Crop	Elemental rate ($kg\ ha^{-1}$)	Residual effect (years)	Ref.[†]
COBALT	Pasture	0.06-0.12	⩾ 2	a
	Pasture	0.24	5	b
	Pasture	0.06	⩾ 2	c
	Pasture	0.24	1 to ⩾ 2	d
COPPER				
Mineral soils	Cereals/pastures	0.7-3.0	Many	e,f
	Cereals	2.8-5.6	⩾ 8	g
	Cereals	23	⩾ 10	h
Organic soils	Wheat	2-16	⩾ 2	i
	Various	9.4	⩾ 10	j
MOLYBDENUM	Pasture	0.14	⩾ 6	k
	Pasture	0.09	⩽ 3	l
	Pasture	0.04	5-6	m
SELENIUM	Pasture	0.5-0.9	Several	n
	Pasture	0.03-0.06	1-2	o
ZINC	Cereals/pastures	0.6-2.4	Many	e,f
	Maize	5.6-28.0	⩾ 4	p

†a. Rossiter *et al.* (1948); b. Lee (1951); c. Bennetts (1955); d. Adams *et al.* (1969); e. Gartrell & Glencross (1969); f. Riceman (1961); g. Reith (1968); h. Davies *et al.* (1971); i. Fleming & Delaney (1961); j. Pizer *et al.* (1966); k. Anderson (1952); l. Swaine (1959); m. Scott (1963); n. National Research Council (1971); o. Davies & Watkinson (1966); p. Brown *et al.* (1962).

TABLE 4

Quantity of trace elements removed in farm products

	Removal (g ha^{-1})		
	Cu	Zn	Mn
Wheat cv. Halberd*			
grain yield (910-4100 kg ha^{-1})	10-25	10-55	35-125
grain and straw (2250-12050 kg ha^{-1})	35-110	15-115	135-510
Barley cv. Clipper*			
grain yield (1400-4000 kg ha^{-1})	10-40	25-80	25-70
grain and straw (3850-9625 kg ha^{-1})	30-105	40-140	105-330
Wool†	0.3	5	0.06

*Unpublished data supplied by J.E. Schultz and R.J. French for cereals grown on nine non-deficient sites in South Australia during 1972 and 1973.

†Unpublished data of D.J. Reuter, R.J. Hannam, G.J. Judson and M.E. Dodson, from wool of 30 non-deficient wethers, stocked at 12 sheep/hectare, mean for two years.

inputs are relatively costly, or where nutrient imbalance results from overuse of fertilizers. Other approaches have been used to estimate maintenance requirements. For example, the need to reapply Cu is ascertained in southern Australia by the appearance of "steely wool" in flocks and by the loss of pigmentation in the wool of black sheep (Lee 1951; Bennetts 1955). Others have monitored changes in trace element status with time using soil and plant tests (Boawn *et al.* 1960; Brown *et al.* 1962,1964; Follett & Lindsay 1971). An appreciation of residual value can also be gained from balance studies as used for Cu and Zn by Riceman (1961) and for Se by Allaway *et al.* (1967).

Australian usage. The quantities of trace elements purchased as fertilizers in Australia are shown in Figure 5A for the years 1959 to 1972. The major trace element inputs are Cu and Zn. With the exception of the 1961/62 season, usage remained relatively constant until 1969/70, when it decreased sharply. The rural recession during the latter period restricted fertilizer usage generally and reduced the amount of new land development. In South Australia, farmers questioned whether repeated trace element applications were still necessary, and usage declined accordingly (Fig. 5B). The figure also indicates annual variation in usage of Cu and Zn in South Australia.

Western Australia is by far the largest user of trace elements (Table 5) among the Australian states. The southern states — Western Australia, South Australia and Victoria — account for nearly all the Cu, Zn and Mn, approximately 80% of the Co and Fe, and about 60% of the Mo used in Australia. Use is made of significant quantities of Mo in New South Wales and of Co in Tasmania.

Foliar Sprays

The application of trace elements in foliar sprays is a recognized means of correcting deficiencies in field crops (Caldwell 1955; Wittwer *et al.* 1963; Mortvedt & Cunningham 1971; Murphy & Walsh 1972).

Generally trace element deficiencies in tree crops are more efficiently and economically corrected by foliar appli-

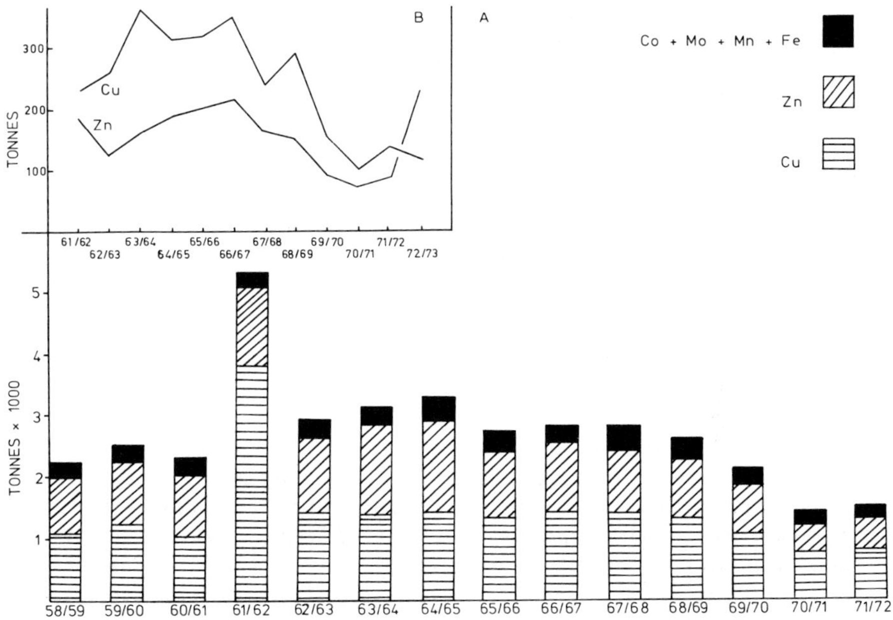

Fig. 5. A. Trace Elements (as elements) purchased in fertilizers in Australia between 1959 and 1972. B. Trends in the use of Cu and Zn in South Australia between 1962 and 1973. The data for 1959-1970 were derived from estimates published by the Department of Primary Industry (Anon. 1964,1970) and for the period 1971-1972 from data kindly supplied by major fertilizer manufacturers. Usage is expressed on an elemental basis. The data are estimates only and for some years the information is incomplete.

TABLE 5

Estimated consumption of trace elements by Australian states [expressed as a % of the total Australian usage over six seasons (1962-1968)]

	Co	Cu	Fe	Mn*	Mo	Zn
Western Australia	46	61	9	43	34	70
South Australia	17	23	37	54	8	14
Victoria	16	13	34	3	15	15
Tasmania	21	2	14	-	7	-
New South Wales	-	-	-	-	27	-
Queensland	-	1	6	-	9	1

*Based on 5 years figures.

cation than by soil application. Trunk injections, sprays during dormancy and swabbing of pruning cuts have also been used. Deficiencies of Mn and Fe are almost always corrected by foliar sprays, as soil applications are rapidly immobilized. Deficiencies of Cu, Zn, Mo and Co may also be corrected by foliar sprays, but in many cases soil applications are preferred, primarily because these applications may have good residual effects, but also because sprays require additional farm operations and equipment. Foliar sprays are also used to correct deficiencies that develop during crop growth or which cannot be prevented by soil applications. "Marsh spot" (Mn deficiency) in peas in prevented by spraying Mn at flowering (Henkens & Jongman 1965). Foliar sprays may also be used in conjunction with soil applications where severe deficiencies occur (Reuter *et al.* 1972*a*).

Low application rates used in foliar sprays (Table 6) usually have a similar effect on crop yield as do relatively higher rates applied to the soil. For example, McNall (1967) indicated that Zn sprays were 12 times and Fe sprays 25-100 times more efficient than soil applications. Foliage scorches if the concentration of spray material is too high (Caldwell 1955) and therefore tolerance concentrations are usually recommended.

Nutrients enter the foliage *via* epidermal cells and cells within stomates (Wittwer & Teubner 1959; Franke 1967) in two stages — an initial rapid non-metabolic penetration followed by slower metabolic accumulation. Uptake varies with cuticle thickness and structure, stomate number per unit area, and leaf age. The time required for 50% absorption varies from 1-2 days for Zn and Mn to 10-20 days for Fe and Mo (Bukovac & Wittwer 1957).

Environmental conditions at and after application influence the effectiveness of foliar sprays. Moderately high temperatures (21°C) and conditions such as high humidity which prevent drying of the spray deposit enhance absorption of Mn by soyabean leaves (Mederski & Hoff 1958). Durkee (1967) concluded that early morning was the best time to spray; humidity is high, dew is often present, and stomates are open. Heavy rains after spraying reduce the effectiveness of sprays (Caldwell 1955). Spray efficiency is also influenced by factors such as droplet size, spray volume, crop interception, wetting agents, spray retention and tenacity (Somers 1958).

TABLE 6

Compounds and application rates used in foliar sprays to correct trace element deficiencies

	Compound	Elemental rate (kg ha^{-1})	Spray volume (ha^{-1})	Ref.*
COPPER				
Cereals	$CuSO_4 \cdot 5H_2O$	0.2	560-1120	a
		0.1	500	b
	Cu-oxychloride	0.6-1.2	224	a
Citrus	$CuSO_4 \cdot 5H_2O$	0.09	100	c
IRON				
Cereals		0.5	500	b
Corn, sorghum, etc.	$FeSO_4 \cdot 7H_2O$	1.0-1.5	224-336	d
Peas		1.0	112-224	e
MANGANESE				
Cereals	$MnSO_4 \cdot H_2O$	0.9-1.8	56-224	f
Peas	$MnSO_4 \cdot H_2O$	1.6	112-224	e
Vegetables	$MnSO_4 \cdot 4H_2O$	0.7-2.9	1122	g
MOLYBDENUM				
Cereals	$Na_2MoO_4 \cdot 2H_2O$	0.01	500	b
ZINC				
Cereals, linseed	$ZnSO_4 \cdot 7H_2O$	0.25	112	h,i
Corn, sorghum, etc.	$ZnSO_4 \cdot H_2O$	0.3-0.5	224-336	d
Citrus	$ZnSO_4 \cdot H_2O/ZnO$	0.27	100	c
Pinus radiata	$ZnSO_4 \cdot 7H_2O$	2.5% solution		j

*a. Davies *et al.* (1971); b. King & Alston (1975); c. Leonard (1967); d. Reuss & Lindsay (1963); e. Seeliger & Moss (1971); f. Reuter *et al.* (1972*a*); g. Skene & Kefford (1954); h. Duncan (1967*a*); i. Duncan (1967*b*); j. Raupach (1967).

Frequently deficiency symptoms of Fe and Mn reappear in new growth after an initial foliar application and spray applications need to be repeated (Reuss & Lindsay 1963; Henkens & Jongman 1965; Seeliger & Moss 1971; Reuter *et al.* 1972*a*,*b*). This effect is clearly related to the "partial mobility" of foliar-applied trace elements within the plant, the order of mobility being Zn > Cu > Mn > Fe > Mo (Bukovac & Wittwer 1957). Although foliar-applied Fe (Brown *et al.* 1965) and Mn (Single 1958; Vose 1963*a*; Henkens & Jongman 1965) is redistributed primarily to the younger tissues, the amount translocated is often inadequate to meet the requirements of new growth.

Prevention in Livestock

In Australia, soil applications have generally corrected trace element deficiencies in both herbage plants and livestock. In some situations, however, deficiencies or nutrient imbalances in livestock are only evident during certain times of the year (Lee 1951; Hogan *et al.* 1971) and with deficiencies such as "molybdenosis" (Cu deficiency induced by high Mo intake) remedial therapy can only be accomplished by direct administration to livestock.

Feeding supplementary fodders and grain may deliberately or inadvertently correct deficiencies; for example, in the U.S.A., grain from Se-adequate areas has been used to prevent Se deficiency in livestock in other areas (Kubota, Allaway, Carter, Cary & Lazar 1967). Other preventive methods are now discussed.

Injections and drenches. Subcutaneous and intramuscular injections of Cu compounds (Harvey & Sutherland 1953; Cunningham 1959; Alexander *et al.* 1967; Camargo *et al.* 1968), subcutaneous injections of Se compounds (Blaxter 1963; Andrews *et al.* 1968) and intramuscular injections of vitamin B_{12} (Marston 1970; Hogan *et al.* 1973) have proved effective for preventing deficiencies in the foetus and in young and mature ruminants. Selenium injected in stages does not accumulate in edible tissues (Allaway *et al.* 1967). The doses are transported to storage organs such as the liver and kidney (Camargo *et al.* 1962; Hartley 1967; Marston 1970), and have a relatively long therapeutic effect.

Injections are more effective than oral drenches, which are not absorbed efficiently from the alimentary tract. Drenches therefore need to be given more frequently and this requires more frequent mustering of stock. Nevertheless, drenches containing Cu and Co (Lee 1951; Bennetts 1955) and Se (Andrews *et al.* 1968; Godwin *et al.* 1970) are used at strategic times during the season.

Intra-ruminal "bullets". The development of the intra-ruminal Co bullet by Dewey *et al.* (1958) provided a major breakthrough in the prevention of Co deficiency in ruminants in Australia. The lodging of the bullet and "grinder" [used to abrade calcium phosphate deposits from the bullet surface (Dewey *et al.* 1969)] into the reticulum or the rumen ensured a small but continual supply of Co to rumen microflora for the synthesis of vitamin B_{12}. Rejection of the bullets is low in Merino sheep (Dewey *et al.* 1958), but a 20% rejection rate in Romney sheep found in New Zealand by Millar & Andrews (1964) has favoured the use of cobaltized superphosphate in that country (Andrews 1970). The Co requirement for young rapidly growing lambs (0.2 mg Co/day) is greater than that of mature sheep (0.08 mg Co/day) (Lee & Marston 1969). Providing ewes with bullets usually but not always ensures an adequate Co status in unweaned lambs (Andrews & Stephenson 1966). Sometimes additional Co sprays are applied to pastures from lambing to weaning.

Selenium bullets have increased the concentration of Se in blood and tissue of mature Merino sheep for periods of up to 12 months (Kuchel & Buckley 1969). Excessive quantities of Se did not accumulate in edible tissues. More recently Andrews *et al.* (1974) administered a different type of Se bullet to Romney lambs grazing a Se-deficient pasture in New Zealand and significantly increased their liveweight and Se status. However, over 12 months about 15% of the bullets were rejected and, as with Co bullets, acquired a coating of calcium phosphate.

Addition of trace elements to licks and stock water. Licks are used in drier climates and are popular where marginal or spasmodic deficiencies occur in livestock (Lee 1951). However, it is impossible to control intake by individual animals (Harvey 1952).

Copper and cobalt salts have sometimes been added to stock water, but precipitation and seasonal differences in the quantity of water consumed are disadvantages (Lee 1951; Bennetts 1955).

Other Corrective Measures

Frits, in which metallic salts are incorporated into glass matrices, release a slow continual supply of trace elements to soil. The rate of release is controlled by particle size and matrix composition. Molybdenum (Henkens & Smilde 1967) and selenium frits (Watkinson & Davies 1967) appear to be promising materials. Frits containing Fe are unsatisfactory (Rhoads *et al.* 1956), and Cu and Mn frits have been shown to be inferior to other trace element compounds (Henkens &

Smilde 1966,1967). The effectiveness of Zn frits depends on the nature of the soil (Hoeft & Walsh 1971) as well as on particle size (Holden & Brown 1965).

Synthetic chelates containing trace elements are effective under some conditions, but their high cost restricts their use to high-value crops, lawns and ornamentals (Nelson 1965). The stability of these compounds in soil is related to pH (Norvell & Lindsay 1969); for example, Fe-EDTA is unstable in some soils above pH 7, as soil Ca replaces the Fe which then reacts with soil to form insoluble compounds. Fe-EDDHA, Fe-DTPA and Fe-EDTA are normally recommended for calcareous, neutral and acid soils respectively (Lindsay *et al.* 1966). The effectiveness of Zn and Cu chelates may be partially lost by Fe^{3+} and Ca^{2+} in soil replacing the chelated metals (Lindsay & Norvell 1969; Norvell & Lindsay 1969). By a similar mechanism, applications of Mn-DTPA have corrected both Fe and Mn deficiencies (Boxma & de Groot 1971).

In glasshouse experiments, fusions of S with Mn (Ludwick *et al.* 1968) and with Zn and Cu compounds (Sharpee *et al.* 1969) effectively provided these elements to a number of successive crops. The response was related to the solubility of the compounds used and inversely related to fertilizer particle size. Iron-sulphur fusions were ineffective sources of Fe (Sharpee *et al.* 1969). These materials have not been evaluated in the field.

Coating seeds with Mo compounds in solution or slurry is sometimes more effective than applications to the soil (Reisenauer 1963). However, sodium molybdate pelletted with subterranean clover seed markedly reduces the survival of inoculated *Rhizobium* (Gartrell 1969). Deficiencies of other trace elements can be partially prevented by various seed treatments but other corrective measures are more effective and popular (Murphy & Walsh 1972).

Varietal susceptibility to trace element deficiencies is well recognized and has been reviewed by Vose (1963*b*), Gerloff (1963) and Brown *et al.* (1972). In some situations, growing tolerant species or cultivars may reduce or avoid the need for fertilizers (Brown *et al.* 1972).

SUMMARY

As agricultural systems become more intensive, trace element deficiencies and imbalances are more likely to occur. Although corrective measures are generally inexpensive compared with other farm costs, failure to prevent deficiencies can substantially lower farm incomes, particularly where marginal deficiencies occur over large areas and where high-value produce is grown.

Trace element disorders involve a complex interaction of many factors (see Table 7). The importance of these factors and their interactions needs to be elucidated for an understanding of a disorder and its occurrence. Multi-factor analysis has been used in a few studies. For example, the distribution of Co (Kubota 1967), Se (Kubota, Allaway, Carter Cary & Lazar 1967; Ehlig *et al.* 1968; Geering *et al.* 1968) and Mo-induced Cu (Kubota *et al.* 1961; Kubota, Lazar, Simonson & Hill 1967) deficiencies in parts of the U.S.A. have been defined by systematically studying pedological, soil, plant

TABLE 7

Factors contributing to trace element deficiencies

COPPER	MANGANESE
Low total Cu	High soil pH
Free $CaCO_3$	Neutral-alkaline peats
High organic matter content (e.g. peats)	Free $CaCO_3$
Low organic matter content (sands)	Unconsolidated soil
High N, P and Zn	Poor drainage
Moisture stress	High Fe
High Mo and SO_4^{2-} (livestock)	Liming
	Moisture stress
IRON	COBALT
High soil pH	Alkaline and calcareous soils
Free $CaCO_3$	High soil Fe and Mn
High HCO_3^-	Liming
High Mn	Moisture stress
Poor drainage	
Extreme moisture changes	MOLYBDENUM
Liming	Low soil pH
Plant species	High soil Fe and Al oxides
	High SO_4-S
ZINC	Leached soils
Low Zn	Low seed Mo content
Low organic matter content	
Free $CaCO_3$	SELENIUM
High pH	Low soil pH
High clay content	Waterlogging and high rainfall
High N, P	High soil Fe oxides
Liming	High SO_4-S
Low temperature	Pasture species
Land levelling	

and animal parameters. This type of research warrants further application.

Simple deficiencies can usually be recognized by their symptoms, but other factors frequently mask the diagnosis, especially when the deficiencies are incipient. The development of various diagnostic tests has assisted in the recognition of field disorders. Most tests have both strengths and weaknesses, due in part to their empirical nature and the inherent problems of calibration and interpretation. Research effort needs to be directed towards establishing the *most appropriate* sample to take, and *when* and *how* it should be taken.

Several aspects need further research to improve the efficiency of application. Possibly foremost is the need to measure or estimate the residual effects of previously applied trace elements to soil. A related problem is to define the rate of immobilization and the amount of trace element immobilized in soil, and the value of trace elements in animal excreta. The whole question of residual effects is particularly relevant to many southern Australian soils. In the virgin state they may be extremely deficient in one or more trace elements, but recent evidence shows that residual effects of previous applications may last for many years.

Because trace element uptake from applied fertilizer occurs mainly through diffusion over short distances, more needs to be known of the soil and environmental factors that influence plant root distribution in relation to fertilizer placement.

REFERENCES

ADAMS, S.N., HONEYSETT, J.L., TILLER, K.G. & NORRISH, K. (1969) - *Aust. J. Soil Res. 7* : 29

ALLAWAY, W.H. (1968) - *Adv. Agron. 20* : 235

ALLAWAY, W.H. (1972) - *Cornell Vet. 63* : 151

ALLAWAY, W.H., CARY, E.E. & EHLIG, C.F. (1967) - *In* "Symposium: Selenium in Biomedicine" (O.H. Muth, ed.), p. 273. AVI Publising Co. Inc. : Westport, Connecticut

ALEXANDER, G.I., HARVEY, J.M., LEE, H.J. & STUBBS, W.C. (1967) - *Aust. J. agric. Res. 18* : 169

ANDERSON, A.J. (1946) - *J. Coun. scient. ind. Res. Aust. 19* : 394

ANDERSON, A.J. (1952) - *J. Aust. Inst. agric. Sci. 18* : 159

ANDERSON, A.J. (1956) - *Soil Sci. 81* : 173

ANDERSON, A.J. (1970) - *J. Aust. Inst. agric. Sci. 36* : 15

ANDREWS, E.D. (1956) - *N.Z. Jl Agric. 92* : 239

ANDREWS, E.D. (1965) - *N.Z. Jl Agric. 8* : 788
ANDREWS, E.D. (1966) - *N.Z. Jl Agric. 9* : 829
ANDREWS, E.D. (1970) - *N.Z. Agric. Sci. 5* : 11
ANDREWS, E.D., GRANT, A.B. & BRUNSWICK, L.F.C. (1974) - *N.Z. vet. J. 22* : 46
ANDREWS, E.D. & HART, L.I. (1962) - *N.Z. Jl agric. Res. 5* : 403
ANDREWS, E.D., HART, L.I. & STEPHENSON, B.J. (1960) - *N.Z. Jl agric. Res. 3* : 364
ANDREWS, E.D., HARTLEY, W.J. & GRANT, A.B. (1968) - *N.Z. vet. J. 16* : 3
ANDREWS, E.D., HOGAN, K.G., STEPHENSON, B.J., WHITE, D.A. & ELLIOTT, D.C. (1970) - *N.Z. Jl agric. Res. 13* : 950
ANDREWS, E.D. & STEPHENSON, B.J. (1966) - *N.Z. Jl agric. Res. 9* : 491
ANDREWS, E.D., STEPHENSON, B.J., ANDERSON, J.P. & FAITHFUL, W.C. (1958) - *N.Z. Jl agric. Res. 1* : 125
ANON. (1964,1970) - Department of Primary Industry - Chemical Fertilizers
ANON. (1968) - *In* "Agricultural Research in the South-East of South Australia and Adjacent Areas of Victoria". CSIRO Aust. : Melbourne
BAKER, D.E. (1974) - *Fedn Proc. 33* : 1188
BARBER, S.A., HALSTEAD, E.H. & FOLLETT, R.F. (1966) - Transactions Commissions II and IV of the International Society of Soil Science, Aberdeen, p. 299
BATES, T.E. (1971) - *Soil Sci. 112* : 116
BAUER, A. & LINDSAY, W.L. (1965) - *Proc. Soil Sci. Soc. Am. 29* : 413
BECK, A.B. (1962) - *Aust. J. exp. Agric. Anim. Husb. 2* : 40
BENNETTS, H.W. (1955) - *J. Agric. West. Aust. (3rd Ser.) 4* : 42
BINGLEY, J.B. & ANDERSON, N. (1972) - *Aust. J. agric. Res. 23* : 885
BLACKBURN, G. & GILES, J.B. (1963) - CSIRO Divisional Report 2/63
BLAXTER, K.L. (1963) - *J. Nutr. 17* : 105
BOAWN, L.C., VIETS, F.G., CRAWFORD, C.L. & NELSON, J.L. (1960) - *Soil Sci. 90* : 329
BOXMA, R. & DE GROOT, A.J. (1971) - *Pl. Soil 34* : 741
BROWN, A.L., KRANTZ, B.A. & MARTIN, P.E. (1962) - *Proc. Soil Sci. Soc. Am. 26* : 167
BROWN, A.L., KRANTZ, B.A. & MARTIN, P.E. (1964) - *Proc. Soil Sci. Soc. Am. 28* : 236
BROWN, A.L., YAMAGUCHI, S. & LEAL-DIAZ, J. (1965) - *Pl. Physiol. 40* : 35
BROWN, J.C., AMBLER, J.E., CHANEY, R.L. & FOY, C.D. (1972) -

In "Micronutrients in Agriculture" (J.J. Mortvedt, P.M. Giordano & W.L. Lindsay, eds.), p. 389. Soil Science Society of America Inc. : Wisconsin
BUKOVAC, M.J. & WITTWER, S.H. (1957) - *Pl. Physiol. 32* : 428
CALDWELL, T.H. (1955) - *N.A.A.S. q. Rev. 30* : 241
CAMARGO, W.V. deA., LEE, H.J. & DEWEY, D.W. (1962) - *Proc. Aust. Soc. Anim. Prod. 4* : 12
CHAPMAN, H.D. (1966) - "Diagnostic Criteria for Plants and Soils". University of California : Riverside
COX, F.R. (1968) - *Agron. J. 60* : 521
COX, F.R. & KAMPRATH, E.J. (1972) - *In* "Micronutrients in Agriculture" (J.J. Mortvedt, P.M. Giordano & W.L. Lindsay, eds.), p. 289. Soil Science Society of America Inc. : Wisconsin
CUNNINGHAM, I.J. (1959) - *N.Z. vet. J. 7* : 15
CUNNINGHAM, I.J. (1960) - N.Z. Department of Agriculture Bulletin 378
DAVIES, D.B., HOOPER, L.J., CHARLESWORTH, R.R., LITTLE, R.C., EVANS, C. & WILKINSON, B. (1971) - N.A.A.S. Technical Bulletin 21, p. 88
DAVIES, E.B. & WATKINSON, J.H. (1966) - *N.Z. Jl agric. Res. 9* : 317
DAWSON, M.D. & BHELLA, H.S. (1972) - *Agron. J. 64* : 308
DEWEY, D.W., LEE, H.J. & MARSTON, H.R. (1958) - *Nature, Lond. 181* : 1367
DEWEY, D.W., LEE, H.J. & MARSTON, H.R. (1969) - *Aust. J. agric. Res. 20* : 1109
DOLAR, S.G. & KEENEY, D.R. (1971) - *J. Sci. Fd Agric. 22* : 273
DUNCAN, O.W. (1967*a*) - *Qd J. agric. Anim. Sci. 24* : 287
DUNCAN, O.W. (1967*b*) - *Qd J. agric. Anim. Sci. 24* : 301
DURKEE, G.E. (1967) - cited in: *Agric. Chem. 22* : 23
EGAN, A.R. (1972) - *Aust. J. exp. Agric. Anim. Husb. 12* : 131
EHLIG, C.F., ALLAWAY, W.H., CARY, E.E. & KUBOTA, J. (1968) - *Agron. J. 60* : 43
ENNIS, M.T. & BROGAN, J.C. (1961) - *Ir. J. agric. Res. 1* : 35
FLEMING, G.A. & DELANEY, J. (1961) - *Ir. J. agric. Res. 1* : 81
FOLLETT, R.H. & LINDSAY, W.L. (1971) - *Proc. Soil Sci. Soc. Am. 35* : 600
FRANKE, W. (1967) - *A. Rev. Pl. Physiol. 18* : 281
GARDINER, M.R. & GORMAN, R.C. (1963) - *Aust. J. exp. Agric. Anim. Husb. 3* : 284
GARTRELL, J.W. (1969) - *Aust. J. exp. Agric. Anim. Husb. 9* : 432
GARTRELL, J.W. & GLENCROSS, R.N. (1969) - *J. Agric. West. Aust. (4th Ser.) 9* : 517
GAWTHORNE, J.M. (1968) - *Aust. J. biol. Sci. 21* : 789
GEERING, H.R., CARY, E.E., JONES, L.H.P. & ALLAWAY, W.H.

(1968) - *Proc. Soil Sci. Soc. Am. 23* : 35
GERLOFF, G.C. (1963) - *A. Rev. Pl. Physiol. 14* : 107
GIORDANO, P.M. & MORTVEDT, J.J. (1966) - *Proc. Soil Sci. Soc. Am. 30* : 649
GIORDANO, P.M. & MORTVEDT, J.J. (1972) - *In* "Micronutrients in Agriculture" (J.J. Mortvedt, P.M. Giordano & W.L. Lindsay, eds.), p. 505. Soil Science Society of America Inc. : Wisconsin
GLADSTONES, J.S. & LONERAGAN, J.F. (1967) - *Aust. J. agric. Res. 18* : 427
GLADSTONES, J.S. & LONERAGAN, J.F. (1970) - *Proc. XI Int. Grassld Congr., Brisbane,* p. 350
GODWIN, K.O., KUCHEL, R.E. & BUCKLEY, R.A. (1970) - *Aust. J. exp. Agric. Anim. Husb. 10* : 672
HALSTEAD, E.H., BARBER, S.A., WARNCKE, D.D. & BOLE, J.B. (1968) - *Proc. Soil Sci. Soc. Am. 32* : 69
HARTLEY, W.J. (1967) - *In* "Symposium: Selenium in Biomedicine" (O.H. Muth, ed.), p. 79. AVI Publishing Co. Inc. : Westport, Connecticut
HARTMANS, J. (1973) - *Third World Conference on Animal Production 2* : 2(c) 1
HARVEY, J.M. (1952) - *Aust. vet. J. 28* : 209
HARVEY, J.M. & SUTHERLAND, A.K. (1953) - *Aust. vet. J. 29* : 261
HEALY, W.B. (1972) - *N.Z. Jl agric. Res. 15* : 289
HENKENS, Ch.H. (1961) - *Versl. landbouwk. Onderz. No. 67.10*
HENKENS, Ch.H. & JONGMAN, E. (1965) - *Neth. J. agric. Sci. 13* : 392
HENKENS, Ch.H. & SMILDE, K.W. (1966) - *Neth. J. agric. Sci. 14* : 165
HENKENS, Ch.H. & SMILDE, K.W. (1967) - *Neth. J. agric. Sci. 15* : 21
HIGGS, E.D. & BURTON, M.V. (1955) - *J. Dep. Agric. S. Aust. 59* : 114
HODGSON, J.F. (1963) - *Adv. Agron. 15* : 119
HOEFT, R.G. & WALSH, L.M. (1971) - *Agron. J. 63* : 84
HOFF, D.J. & MEDERSKI, H.J. (1958) - *Proc. Soil Sci. Soc. Am. 22* : 129
HOGAN, K.G., LORENTZ, P.P. & GIBS, F.M. (1973) - *N.Z. vet. J. 21* : 234
HOGAN, K.G., MONEY, D.F.L., WHITE, D.A. & WALKER, R. (1971) - *N.Z. Jl agric. Res. 14* : 687
HOLDEN, E.R. & BROWN, J.R. (1965) - *J. agric. Fd Chem. 13* : 180
JONES, G.B. & BELLING, G.B. (1967) - *Aust. J. agric. Res. 18* : 733
JONES, G.B., RICEMAN, D.S. & McKENZIE, J.O. (1957) - *Aust. J.*

agric. Res. 8 : 190

JONES, J.B. (1972) - *In* "Micronutrients in Agriculture" (J.J. Mortvedt, P.M. Giordano & W.L. Lindsay, eds.), p. 319. Soil Science Society of America Inc. : Wisconsin

JUDSON, G.J. & OBST, J.M. (1975) - *In* "Trace Elements in Soil-Plant-Animal Systems" (D.J.D. Nicholas & A.R. Egan, eds.). Academic Press : New York

KESSLER, B. (1961) - *In* "Plant Analysis and Fertilizer Problems", p. 314. American Institute of Biological Science : Washington, D.C.

KING, P.M. & ALSTON, A.M. (1975) - *In* "Trace Elements in Soil-Plant-Animal Systems" (D.J.D. Nicholas & A.R. Egan, eds.). Academic Press : New York

KLEINIG, C.R. & LOVEDAY, J. (1962) - *Aust. J. exp. Agric. Anim. Husb. 2* : 228

KRAUSKOPF, K.B. (1972) - *In* "Micronutrients in Agriculture" (J.J. Mortvedt, P.M. Giordano & W.L. Lindsay, eds.), p. 7. Soil Science Society of America Inc. : Wisconsin

KUBOTA, J. (1964) - *Proc. Soil Sci. Soc. Am. 24* : 246

KUBOTA, J. (1967) - *Soil Sci. 106* : 122

KUBOTA, J. & ALLAWAY, W.H. (1972) - *In* "Micronutrients in Agriculture" (J.J. Mortvedt, P.M. Giordano & W.L. Lindsay, eds.), p. 525. Soil Science Society of America Inc. : Wisconsin

KUBOTA, J., ALLAWAY, W.H., CARTER, D.L., CARY, E.E. & LAZAR, V.A. (1967) - *J. agric. Fd Chem. 15* : 448

KUBOTA, J., LAZAR, V.A., LANGAN, L.N. & BEESON, K.C. (1961) - *Proc. Soil Sci. Soc. Am. 25* : 227

KUBOTA, J., LAZAR, V.A., SIMONSON, G.H. & HILL, W.W. (1967) - *Proc. Soil Sci. Soc. Am. 31* : 667

KUCHEL, R.E. & BUCKLEY, R.A. (1969) - *Aust. J. agric. Res. 20* : 1099

LAKIN, H.W. & DAVIDSON, D.F. (1967) - *In* "Symposium: Selenium in Biomedicine" (O.H. Muth, ed.), p. 27. AVI Publising Co. Inc. : Westport, Connecticut

LEE, H.J. (1946) - M.Ag.Sc. Thesis, University of Adelaide

LEE, H.J. (1951) - *J. Dep. Agric. S. Aust. 54* : 1

LEE, H.J. (1952) - Proceedings of Specialist Conference in Agriculture, 1949, Melbourne, p. 262

LEE, H.J. & MARSTON, H.R. (1969) - *Aust. J. agric. Res. 20* : 905

LEHR, J.R. (1972) - *In* "Micronutrients in Agriculture" (J.J. Mortvedt, P.M. Giordano & W.L. Lindsay, eds.), p. 459. Soil Science Society of America Inc. : Wisconsin

LEONARD, C.D. (1967) - *Farm Technology 23* (6)

LINDSAY, W.L. (1972) - *Adv. Agron. 24* : 147

LINDSAY, W.L., HODGSON, J.F. & NORVELL, W.A. (1966) - Transactions Commissions II and IV of the International Society of Soil Science, Aberdeen, p. 305

LINDSAY, W.L. & NORVELL, W.A. (1969) - *Proc. Soil Sci. Soc. Am. 33* : 62

LUCAS, R.E. & KNEZEK, B.D. (1972) - *In* "Micronutrients in Agriculture" (J.J. Mortvedt, P.M. Giordano & W.L. Lindsay, eds.), p. 265. Soil Science Society of America Inc. : Wisconsin

LUDWICK, A.E., SHARPEE, K.W. AND ATTOE, D.J. (1968) - *Agron. J. 60* : 232

MANNETJE, L., SHAW, N.H. & ELICH, T.W. (1963) - *Aust. J. exp. Agric. Anim. Husb. 3* : 20

MARSTON, H.R. (1952) - *Physiol. Rev. 32* : 66

MARSTON, H.R. (1970) - *Br. J. Nutr. 24* : 615

MARSTON, H.R. & LEE, H.J. (1948) - *Aust. J. scient. Res. (Ser. B) 1* : 376

McKENZIE, R.M. (1959) - *Aust. J. agric. Res. 10* : 52

McKENZIE, R.M. (1966) - *Aust. J. exp. Agric. Anim. Husb. 6* : 170

McNALL, L.R. (1967) - *Fertilizer Solutions 11* : 8

McNAUGHT, K.J. (1970) - *Proc. XI Int. Grassld Congr., Brisbane,* p. 334

MEDERSKI, H.J. & HOFF, D.J. (1958) - *Agron. J. 50* : 175

MILLAR, K.R. & ANDREWS, E.D. (1964) - *N.Z. vet. J. 12* : 9

MITCHELL, R.L. (1964) - *In* "Chemistry of the Soil" (F.E. Bear, ed.), 2nd Edn, p. 320

MITCHELL, R.L. (1971) - *Tech. Bull. Minist. Agric. 21* : 8

MITCHELL, R.L., REITH, J.W.S. & JOHNSTON, I.M. (1957) - *J. Sci. Fd Agric. 8* : 51

MORTVEDT, J.J. (1968) - *J. agric. Fd Chem. 16* : 241

MORTVEDT, J.J. & CUNNINGHAM, H.G. (1971) - *In* "Fertilizer Technology and Use" (R.A. Olson, T.J. Army, J.J. Hanway & V.J. Kilmer, eds.), 2nd Edn, p. 413. Soil Science Society of America Inc. : Wisconsin

MURPHY, L.S. & WALSH, L.M. (1972) - *In* "Micronutrients in Agriculture" (J.J. Mortvedt, P.M. Giordano & W.L. Lindsay, eds.), p. 347. Soil Science Society of America Inc. : Wisconsin

NATIONAL RESEARCH COUNCIL (1971) - *In* "Selenium in Nutrition". National Academy of Sciences : Washington, D.C.

NELSON, L.B. (1965) - *Adv. Agron. 17* : 62

NORTHCOTE, K.H. (1971) - "A Factual Key for the Recognition of Australian Soils", 3rd Edn. Rellim Technical Publications : Glenside, S.A.

NORVELL, W.A. & LINDSAY, W.L. (1969) - *Proc. Soil Sci. Soc. Am. 33* : 86

OERTEL, A.C. (1959) - *Aust. J. agric. Res. 10* : 58

OKSANEN, H.E. (1967) - *In* "Symposium: Selenium in Biomedicine" (O.H. Muth, ed.), p. 215. AVI Publishing Co. Inc. : Westport, Connecticut

OLSEN, S.R. (1972) - *In* "Micronutrients in Agriculture" (J.J. Mortvedt, P.M. Giordano & W.L. Lindsay, eds.), p. 243. Soil Science Society of America Inc. : Wisconsin

OZANNE, P.G., ASHER, C.J. & KIRTON, D.J. (1965) - *Aust. J. agric. Res. 16* : 785

OZANNE, P.G., SHAW, T.C. & KIRTON, D.J. (1965) - *Aust. J. exp. Agric. Anim. Husb. 5* : 29

PATIL, B.D. & JONES, D.I.H. (1970) - *Proc. XI Int. Grassld Congr., Brisbane,* p. 726

PIZER, N.H., CALDWELL, T.H., BURGESS, G.R. & JONES, J.L.O. (1966) - *J. agric. Sci., Camb. 66* : 303

RANDALL, P.J. (1970) - *Proc. XI Int. Grassld Congr., Brisbane,* p. 338

RAUPACH, M. (1967) - *Adv. Agron. 19* : 307

RAUPACH, M. (1975) - *In* "Trace Elements in Soil-Plant-Animal Systems" (D.J.D. Nicholas & A.R. Egan, eds.). Academic Press : New York

REISENAUER, H.M. (1963) - *Agron. J. 55* : 459

REITH, J.W.S. (1968) - *J. agric. Sci., Camb. 70* : 39

REUSS, J.O. & LINDSAY, W.L. (1963) - Colorado Agricultural Experimental Station Pamphlet 59

REUTER, D.J., HEARD, T.G. & ALSTON, A.M. (1972*a*) - *Aust. J. exp. Agric. Anim. Husb. 13* : 434

REUTER, D.J., HEARD, T.G. & ALSTON, A.M. (1972*b*) - *Aust. J. exp. Agric. Anim. Husb. 13* : 440

RHOADS, W.A., WALLACE, A. & ROMNEY, E.M. (1956) - *Soil Sci. 81* : 359

RICEMAN, D.S. (1961) - *Wool Technol. Sheep Breed. 8* : 111

RICEMAN, D.S. & ANDERSON, A.J. (1943) - *J. Dep. Agric. S. Aust. 47* : 16

RICEMAN, D.S. & DONALD, C.M. (1938) - CSIR Pamphlet 78

ROSSITER, R.C., CURNOW, D.H. & UNDERWOOD, E.J. (1948) - *J. Aust. Inst. agric. Sci. 14* : 9

SAUCHELLI, V. (1969) - "Trace Elements in Agriculture". Van Nostrand Reinhold Co. : New York

SCOTT, R.S. (1963) - *N.Z. Jl agric. Res. 6* : 567

SEELIGER, M.T. & MOSS, D.E. (1971) - *J. Agric. S. Aust. 75* : 38

SHARPEE, K.W., LUDWICK, A.E. & ATTOE, O.J. (1969) - *Agron. J. 61* : 746

SILVERBERG, J., YOUNG, R.D. & HOFFMEISTER, G. (1972) - *In* "Micronutrients in Agriculture" (J.J. Mortvedt, P.M. Giordano & W.L. Lindsay, eds.), p. 431. Soil Science Society of America Inc. : Wisconsin

SINGLE, W.V. (1958) - *Ann. Bot. 22* : 479
SKENE, J.K.M. & KEFFORD, R.O. (1954) - *J. Dep. Agric. Vict. 52* : 505
SKERMAN, K.D., SUTHERLAND, A.K., O'HALLORAN, M.V., BOURKE, J.M. & MUNDAY, B.L. (1959) - *Am. J. vet. Res. 20* : 977
SOMERS, E. (1958) - *N.A.A.S. q. Rev. 46* : 68
SPRAGUE, H.B. (1961) - "Hunger Signs in Crops". David McKay Co. : New York
STEENBJERG, F. (1951) - *Physiologia Pl. 4* : 677
STEPHENS, C.G. & DONALD, C.M. (1958) - *Adv. Agron. 10* : 167
SWAINE, F.G. (1959) - *J. Aust. Inst. agric. Sci. 25* : 51
THORNTON, I. & WEBB, J.S. (1969) - *In* "Trace Element Metabolism in Animals" (C.F. Mills, ed.), p. 397. E. & S. Livingstone : Edinburgh
TILLER, K.G. (1963) - *Aust. J. Soil Res. 1* : 74
TIVER, N.S. (1955) - Department of Agriculture, South Australia Bulletin 441
TODD, J.R. (1969) - *In* "Trace Element Metabolism in Animals" (C.F. Mills, ed.), p. 448. E. & S. Livingstone : Edinburgh
TRIERWEILER, J.F. & LINDSAY, W.L. (1969) - *Proc. Soil Sci. Soc. Am. 33* : 49
UNDERWOOD, E.J. (1971) - "Trace Elements in Human and Animal Nutrition", 3rd Edn. Academic Press : New York and London
VAN ALPHEN, Th.G. (1957) - *Centrum voor Landbouwdocumentatie 20*
VAN LUIT, B. & HENKENS, Ch.H. (1967) - *Versl. landbouwk. Onderz. No. 695*
VIETS, F.G. (1966) - *In* "Zinc Metabolism" (A.S. Prasad, ed.), p. 90. Charles G. Thomas : Springfield
VOSE, P.B. (1963*a*) - *J. exp. Bot. 14* : 448
VOSE, P.B. (1963*b*) - *Herb. Abstr. 33* : 1
WALLACE, T. (1964) - "The Diagnosis of Mineral Deficiencies in Plants by Visual Symptoms". HMSO : London
WATKINSON, J.H. & DAVIES, E.B. (1967) - *N.Z. Jl agric. Res. 10* : 116
WHITEHEAD, D.C. (1966) - *Mimeogrd Publs Commonw. Bur. Past. Fld Crops, No. 1/1966*
WIENER, G. (1966) - *J. comp. Path. 76* : 435
WIENER, G. & FIELD, A.C. (1969) - *In* "Trace Element Metabolism in Animals" (C.F. Mills, ed.), p. 92. E. & S. Livingstone : Edinburgh
WILKINSON, H.F., LONERAGAN, J.F. & QUIRK, J.P. (1968) - *Proc. Soil Sci. Soc. Am. 32* : 831
WILLIAMS, C.H. & ANDREW, C.S. (1970) - *In* "Australian Grasslands" (R.M. Moore, ed.), p. 321. Australian National University Press : Canberra
WILLIAMS, C.H. & MOORE, C.W.E. (1952) - *Aust. J. agric. Res.*

3 : 343
WILLIAMS, R.D. (1963) - *Mimeogrd Publs Commonw. Bur. Past. Fld Crops No. 1/1959*
WITTWER, S.H., BUKOVAC, M.J. & TUKEY, H.B. (1963) - *In* "Fertilizer Technology and Usage" (M.H. McVickar, G.L. Bridger & L.B. Nelson, eds.), 1st Edn, p. 429. Soil Science Society of America Inc. : Wisconsin
WITTWER, S.H. & TEUBNER, F.G. (1959) - *A. Rev. Pl. Physiol.* *10* : 13

APPLICATION OF REGIONAL GEOCHEMICAL RECONNAISSANCE TO AGRICULTURE

A.S. Joyce

School of Applied Geology
South Australian Institute of Technology
Adelaide, South Australia 5000

INTRODUCTION

The aim of this paper is to indicate that regional geochemical reconnaissance techniques, similar to those practised in mineral exploration, can be adapted to agricultural reconnaissance and to recommend how this can be achieved.

The increasing awareness that quite subtle variations in both absolute and relative abundances of minor and trace elements in soils play a major part in determining the thriftiness of plants and animals has generated considerable interest in regional variations in soil chemistry. A major problem is posed by the fact that soils are capable of profound variation, both in profile and laterally, even when developed over homogeneous parent material. Consequently, in many areas the density of sampling required to confidently describe the chemistry of soil by direct sampling is prohibitive. The solution is to switch attention to some indirect method of sampling which will provide an average composition of the soils of a relatively large area with the maximum economy of samples.

Four basic methods appear open:

1. *Geological maps* provide some information on the gross distribution of elements in the parental materials from which soils are derived, but they are not an adequate answer. Firstly, many soils of agricultural importance are transported. Secondly, it is common for soils of different chemical and physical character to develop even over homogeneous parental rock within a small area. Finally, many rock types show considerable chemical variation and, in any case,

detailed chemical data are rarely available for map areas of significant size.

2. *Remote sensing* by aircraft or spacecraft on the basis of colour or reflectivity variation induced in plants by variations in soil chemistry can be used. This method is limited mainly to areas nurturing a very restricted variety of plants, whether natural or cultivated. Other limitations are that not all plants show significant colour or reflectivity responses to chemical characteristics which may retard their own thriftiness or the thriftiness of animals or other plants introduced to the area.

3. *Airborne scintillometer surveying* based on radiation from U, K and Th has been demonstrated as a reconnaissance technique for mapping soils in Oklahoma (Schwarzer & Adams 1973). The method probably has considerable potential for rapid assessment of the homogeneity and patterns of soil types in a survey area, but it does nothing towards assessing the abundance of elements other than U, K and Th.

4. *Stream sediment geochemistry* offers the widest scope for agricultural reconnaissance. Low density sampling of stream sediments is a widely practised and successful method of prospecting for concealed mineral deposits. The method is based on the premise that the chemistry of a sediment sample approximates to the average composition of the soils within the catchment area upstream of the sampling point. Numerous mineral discoveries attest qualitatively to the correctness of this assumption but few quantitative assessments have been attempted or recorded.

One attempt by CRA Exploration Pty. Ltd. recorded that an area of 20 km^2 in the Upper Kaverong catchment on Bougainville has an average soil content of 660 ppm Cu expressed as 670 ppm Cu in stream sediment at the mouth of the catchment basin (D.H. Mackenzie, personal communication). The average Cu content of the soil was estimated on closely spaced sampling, which indicated 18 km^2 of soil containing 400 ppm Cu developed on volcanic ash, 1.2 km^2 of weathered mineralization containing 5000 ppm Cu, and 0.8 km^2 of soils containing 200 ppm Cu developed on volcanic ash.

AGRICULTURAL APPLICATIONS

A marked contrast between the application of stream sediment geochemistry to mineral exploration and its application to agricultural reconnaissance is that elements sought as indicators of ore deposits are commonly enriched by many orders of magnitude above their abundances in common earth materials, whereas some elements of interest to agriculture require detection at very low concentrations. This presents more of a practical analytical problem than a theoretical limitation and it can be solved by refined techniques which are available to lower detection limits and to improve precision.

Proof that stream sediment geochemistry can be adapted for agricultural purposes is afforded by publications emanating from the Applied Geochemistry Research Group at Imperial College since 1964 (Webb *et al.* 1971; Thornton 1972). The reconnaissance method applied in the United Kingdom was initiated primarily for mineral exploration. It is based on sampling fine sediment from active channels at a density of about one sample per 2.6 km^2. Each sample is sieved and the fraction finer than 80 mesh (210 μ) is analysed for Ag, Bi, Co, Cr, Cu, Fe, Ga, Mn, Mo, Ni, Pb, Sn, Ti, V and Zn spectrographically. As, Se, Cu, Mn and Zn are analysed by colorimetric or atomic absorption procedures in selected samples.

Perhaps the most impressive published examples of the application of this type of data to agriculture concern Cu and Mo (Webb & Atkinson 1965; Thornton *et al.* 1969; Thornton & Webb 1970; Webb & Thornton 1971; Webb *et al.* 1971; Thornton 1972; Thornton *et al.* 1972). The earliest studies were conducted in part of County Limerick, Eire, where a strong correlation was established between anomalously high Mo in stream sediments and high Mo in soils and herbage. Previously suspected Mo-induced Cu deficiency in cattle within the anomalous area was confirmed by blood tests.

Other stream sediment surveys covering about 2500 km^2 in Derbyshire and North Staffordshire, U.K., detected about 150 km^2, within which Mo was anomalously high. Follow-up studies confirmed that rocks, soils and herbage within the area contained more than common abundances of Mo. Molybdenum-induced Cu deficiency in animals had previously been reported in only about 40 km^2 of the anomalous area. Blood Cu tests carried out on animals within parts of the area delineated as anomalous by the stream sediments, but not previously recognized on clinical symptoms, demonstrated widespread existence of low blood Cu values. Subsequent supplementation trials over two years involving Cu injections at the beginning of the grazing season showed an impressive mean increase of 40% in the live weight gain of injected animals compared with untreated control animals. With few exceptions, the animals involved in the trials showed no obvious signs of Cu deficiency.

Extension of the reconnaissance has disclosed additional extensive Mo-rich areas, some of which were known to be characterized by Mo-induced Cu deficiencies in cattle, while others were not suspected previously.

In addition to areas where high Mo values have induced Cu deficiencies in livestock, some other parts of England have been shown to have basic deficiencies in Cu which can be recognized in stream sediment data and correlated with Cu-deficient plants and cattle.

Other reconnaissance results of agricultural significance have concerned Se toxicity (Webb & Atkinson 1965), Mn deficiency in cereals and cattle (Webb *et al.* 1968; Thornton & Webb 1970), Co deficiency in sheep (Webb *et al.* 1968; Thornton 1972), Pb and As toxicity in livestock and Zn toxicity in cereals (Thornton & Webb 1970).

RECOMMENDED SAMPLING PROCEDURE

It is clear that the experiments conducted in the United Kingdom have demonstrated sufficient potential for the agricultural application of stream sediment geochemistry to warrant commencement of experiments in Australia. Also it is a particularly opportune time to commence since several of the state geological surveys are preparing regional geochemical maps to supplement 1:250,000 geological map sheets for mineral exploration purposes. For example, the Queensland Geological Survey has conducted a broad-scale stream sediment sampling programme of the Rockhampton 1:250,000 sheet (Wall 1974) and the South Australian Mines Department is preparing the Barker 1:250,000 sheet (I. Faulks, personal communication).

Since a large proportion of the cost of geochemical reconnaissance is attributable to sampling costs, it is highly desirable that all interested parties cooperate at this early stage to adopt acceptable standard collection techniques, sampling densities and sample storage and retrieval facilities. Also it would be convenient if a standard suite of elements were analysed routinely by standard techniques, but this is less important than the standard collection of samples and their retention for subsequent retrieval should the need arise to perform additional analyses. Elements of interest in various mineral exploration programmes include Ti, V, Cr, Mn, Fe, Co, Ni, Cu, Zn, As, Zr, Nb, Mo, Ag, Sn, Sb, Ba, Ta, W, Au, Pb, Bi, rare earth elements, Th and U. Many of these elements are of interest to agricultural scientists, but P, Se and Cd would be worthwhile additions. Depending on the extent of geological and agricultural knowledge within specific survey areas, some of these 28 elements could be omitted.

The tried and tested sampling medium is fine silty sediment collected from active stream channels in locations not recognizably contaminated by very locally derived material such as landslides or collapsed banks.

Artificial contamination by factory or domestic discharge, roadway drains, etc., must be taken into account either by ensuring that samples are collected upstream of the source of contaminants or by recording potential contamination on the sample record at the time of collection (for some purposes it may be desirable to have a record of the degree of local pollution in streams). Suspected recent contamination of stream sediment can sometimes be confirmed by establishing a significant difference in chemistry between presently active sediment and older perched alluvium nearby.

The quantity of sample normally collected in mineral exploration is about 100 g; this is placed in a pre-numbered wet-strength kraft paper packet which can be dried conveniently by the sun or in an oven. The choice of 100 g has been conditioned by the need to minimize the bulk carried while sampling the difficult terrain often traversed in mineral exploration, at the same time maintaining adequate sampling precision. For routine geochemical mapping of the type herein proposed, individual samples of about 300-500 g are recommended in order to permit splits to be made available on request (samples of this size are desirable in prospecting for elements such as Sn and Au, in any case).

For most purposes, it matters little whether samples are collected as single "grabs" from the sediment or as composite samples, but some elements do occur in heavy resistate minerals and are dispersed mechanically, so composite sampling across the various velocity domains of the channel is desirable, where practical. Experience in the Flinders Ranges, South Australia, indicates that the channels need not be carrying water at the time of collection, provided they show evidence of having been active recently.

It has become apparent in mineral exploration that only streams with catchment areas of less than 20-50 km^2 can be sampled meaningfully. Major streams are too diluted by widespread and common materials. A basic consideration in designing an adequate sampling density is the size of individual geological formations and soil series which it is intended should be recognizable as influences on the chemistry of each sample. For example, Se contents in excess of 3 ppm in a stream sediment may be indicative of toxic quantities of Se in the soils of a given area (Webb & Atkinson 1965), but "normal" quantities of Se in sediment are less than 0.5 ppm. If Se-rich areas of 1 km^2 are sufficiently large to warrant recognition, then a sampling density of about one sample per 5 km^2 is required to avoid dilution of the Se anomaly to a level at which it would probably not be recognized.

A density of about one sample per 2 km^2 is recommended in order to permit recognition of targets comparable in size to the units generally recorded on regional soil or geological maps. This density also minimizes the risk of suppression or enhancement of true geochemical values by unsuspected local dilution or contamination. Ultimate practical controls on sampling are the nature of the drainage pattern itself, cost and, to a lesser extent, ease of access.

Precise sample locations are best recorded on air photographs at the time of collection but require conversion to map coordinates for data storage. Additional data which should be recorded at each sample site include the date, pH, width of channel, depth of water, rate of flow, degree of turbulence, type of banks, nature of stream bottom and grain size of sediment. These features aid subsequent interpretation.

Most stream sediment analyses are performed in the laboratory on material fine enough to pass through 80 mesh nylon sieving cloth (apertures nominally 204 μ). The purpose of selecting this size fraction is

1. to minimize, or in some cases eliminate, the need to grind the specimen,
2. to minimize sampling error in selecting a small analytical portion to represent the large field sample,
3. to exclude coarse quartz-rich sand and pebbles and concentrate silt- and clay-sized particles which carry the bulk of elements released during chemical weathering and subsequently scavenged by clays, Fe and Mn hydroxides, and other fine secondary minerals.

There are sound reasons for selecting a different size fraction for specific elements in specific terrains (Joyce 1974). However, for multipurpose surveys investigating a range of elements entering the drainage system variously by chemical and mechanical dispersion, the size fraction finer than 80 mesh is adequate and widely used.

In the interest of economizing on sample bulk, some geochemists recommend wet sieving in the field. Unless the sediment is dominantly very coarse, this procedure is undesirable as it may lead to selective loss of clay-sized material, which harbours much of the metal content of the sediment, especially that which is loosely bound by the sorption process and, therefore, most relevant to agriculture.

Analytical procedures vary somewhat but atomic absorption, optical emission and X-ray fluorescence spectrometry are the prime tools. Care must be taken to avoid introducing excessive sampling variance at the analytical stage (Howarth &

Lowenstein 1971). For atomic absorption procedures, a sample size of 0.5-1.0 g generally yields satisfactory precision (cf. Wall & Lam 1974). Samples for X-ray fluorescence are necessarily larger than 1 g, but samples for optical emission are small and pose problems with precision unless care is taken in fine grinding and representative sampling. It is important that variations in analytical technique be avoided within individual survey regions, because measured quantities of the elements present are not necessarily "total" quantities or accurate quantities.

Because of their interest in "available" elements, agricultural scientists are certain to raise the question of selective extraction techniques prior to analysis. For maximum interdisciplinary benefit and because there may be significant differences in "availability" of elements in soils compared with their derived sediment (where physico-chemical conditions may be radically different), it is clear that analyses for regional reconnaissance should attempt to determine total quantities of the element sought. The time for selective extraction techniques arises at follow-up stages and may be performed on splits of selected original samples, if these are available for retrieval.

To put the cost of widespread regional geochemical reconnaissance in perspective, the recommended sampling density of one sample per 2 km^2 would involve about 4 million samples to cover the whole of Australia. Since very large tracts of Australia are unsuitable for stream sediment sampling (and also unsuitable for effective agriculture), the number of samples is probably better expressed as about 7000 samples per complete 1:250,000 map sheet. Webb (1973) estimated that 8 million geochemical samples are collected annually in countries outside the U.S.S.R. and 10 million within the U.S.S.R. Thus, the number of samples involved in this recommendation can be attained within a realistic time and at a realistic cost. Great economy would result if appropriate legislation required exploration companies to lodge the residues of their stream sediment samples with the state mines departments at some suitable stage, in the same way that they are required to lodge representative drill core obtained during their exploration in some states.

PRESENTATION OF RESULTS

Maps presenting only raw analytical data on a topographic base are generally inadequate for ready recognition of regional geochemical patterns because there are too many similar numbers and too much local variance. This is overcome by using various coding and smoothing procedures. A summary, illustrations and selected annotated bibliography of these procedures are presented elsewhere (Joyce 1974).

The simplest procedures involve coding specific ranges of geochemical values as spots of graded size, spots of various colours, or lines ("worms") of graded width outlining the drainage pattern. In this fashion, data are presented for all locations sampled but they have been simplified for easier recognition of regional patterns. "Grey-level mapping" (Howarth 1971) is a simple computerized method of coding sample locations on maps using a line printer, but coding by hand is often more convenient. The selection of class limits for coding should not be arbitrary, but rather selected after careful examination of the distribution of values within the total population being examined. Any statistical approach which will permit recognition of subpopulations can be used to define meaningful threshold values separating natural subpopulations (usually there is considerable overlap between subpopulations). Histograms are often adequate but it must be appreciated that their construction involves arbitrary initial decisions on class intervals. It is important to appreciate that geochemical populations are rarely normally distributed. Lognormal distributions are more common, but other forms of distribution are also encountered. For this reason, any parametric statistical tests which are applied to geochemical data are suspect or negated. Non-parametric statistics should be used.

Comparison of histograms and cumulative frequency distribution curves plotted on log-probability graph paper is a simple practical procedure widely used in mineral exploration to establish threshold values separating mineralized samples, enriched in a given indicator element, from non-mineralized samples containing "background" abundances. The elegance of this method lies in the fact that the logarithmic and probability scales have the effect of generating straight segments on the distribution curve to represent constituent lognormally distributed populations of values. Natural thresholds for coding geochemical data are, therefore, chosen at inflexion points on the distribution curve. Mineral explorers are primarily concerned with the uppermost inflexion point, but lower inflexion points of relevance in trace element deficiency studies are equally conspicuous (Fig. 1). Where difficulty is encountered in determining limiting thresholds by inspection, procedures have been developed to search for subpopulations (Sinclair 1974).

An alternative to discrete coding of every sample point on

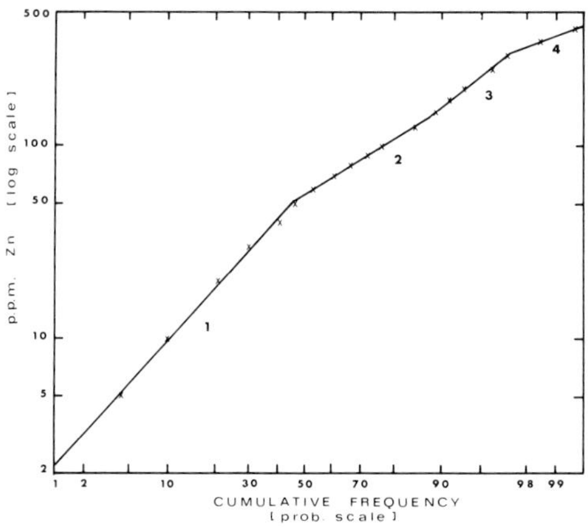

Fig. 1. Cumulative frequency distribution of Zn values in the minus 80 mesh fractions of 261 stream sediment samples from western Tasmania.

the map is to generate contour maps. Contouring of raw data is generally too complex, so trend surface analysis or moving average analysis is preferable. Trend surface analysis suffers from excessive constraint on the patterns capable of being generated because of the need to assume a variety of simple forms, then seek a best fit. These forms are unrealistic in relation to the complex patterns reasonably anticipated from the distribution of soil series and geological formations in most large regions.

A better compromise is to smooth the geochemical data using moving average ("rolling-mean") techniques and to generate contours based on natural population thresholds determined in the fashion described above for coded maps. In essence, moving average analysis is based on moving a window of suitable size randomly or regularly over a raw data map and averaging the elemental abundance of samples viewed through the window at any given location (because of the previous comments concerning lognormality in geochemical data, a geometric mean is preferable to an arithmetic mean). The choice of window size determines the degree of smoothing achieved and the permissible degree of smoothing depends on the expected natural complexity of any given region.

It is worth noting that there are two levels of pattern recognition possible on geochemical maps. The first involves recognition of gross patterns in the smoothed data, and the second involves recognition of local deviations on a scale relevant to small targets such as mineral deposits or soil series in small areas. Fine detail of this type is achieved by deviation maps showing the departure of individual samples from local values predicted by the regionally smoothed data.

All of the comments made so far on data presentation refer primarily to single-element maps. To reduce the volume of output generated in a regional multi-element survey and to aid interpretation, especially of covariation of elements, it may be desirable to process results by discriminant analysis, principal component or factor analysis, or cluster analysis prior to coding of individual sample locations or regional contouring. Such procedures condense the raw data into a smaller number of derivative variables which can be grasped and interpreted more readily.

A major investigation into methods of demonstrating the presence or absence of regional geochemical patterns is being conducted currently by the Regional Geochemical Branch of the United States Geological Survey, using Missouri as a type area.

INTERPRETATION

The first task in a regional survey is to recognize geochemical patterns, if they exist. The second task, interpreting what those patterns mean in relation to soil series and/or underlying geology, depends on understanding the interplay of environmental influences which govern the dispersion of elements both chemically and mechanically from parental rock material through one or more soil-forming cycles and into the sediment sampled. It is not a safe assumption that a high content of a given element in soil will necessarily be expressed as a high content in derived stream sediment, and *vice versa*. Variations in pH, encounters with immobile scavenger substances or mobile complexing agents, and many other factors blur the correlation between sediment, soil and geology. For example, Thornton & Webb (1970) have drawn attention to the observation that high Mn and Co analyses in sediment from North Wales are associated with acid moorland soils which are low in Mn and Co; the explanation offered is that the Mn and Co are mobilized in the soils, where Eh and pH are low, and immobilized in the stream, where Eh and pH are higher.

However, bearing in mind these warnings against oversimplified interpretation, it is fair comment that sufficient expertise exists to cope with interpretation, and refinements are forthcoming.

EXISTING DATA AVAILABLE IN AUSTRALIA

In addition to government-sponsored regional geochemical surveys of whole 1:250,000 map sheets, such as Ayr (Marshall 1967) and Rockhampton (Wall 1974) in Queensland and Barker in

South Australia, there are numerous government reports on local surveys in which sampling has been quite intense. In addition, the state mines departments maintain technical files of reports which exploration companies must lodge recording the results of work carried out within the areas covered by their exploration licences and prospecting authorities. This information is available upon appropriate request after a suitable time-lapse designed to protect the interests of the explorer.

While the usefulness of these data is limited by the explorer's original choice of elements, there is a significant bank of analytical information on elements significant to agriculture in areas scattered widely throughout Australia. The elements Cu, Pb and Zn are generally included in mineral exploration surveys. Thus, a literature survey alone would outline a number of locations where Pb toxicity or Cu or Zn deficiency might exist or arise in livestock or plants. In view of the rapid expansion in eastern Australia of the planting of pine forests, some of which have been shown to require Zn supplementation to hasten maturity, an assessment of existing stream sediment data on Zn might facilitate planning of new plantations and point up some of the existing forests which are either adequately endowed with Zn or critically deficient.

Parts of the west coast of Tasmania which are being rapidly stripped of natural forest by the wood-chip industry and are therefore requiring re-afforestation or conversion to some other agricultural pursuit have been subjected to repeated and very high density sediment sampling programmes during mineral exploration for diverse targets, including Cu, Pb, Zn, Sn, Ni, Co and platinoids. For example, Figure 1 illustrates the distribution of Zn values in the minus 80 mesh fraction of 261 sediments from an area of about 100 km^2 in western Tasmania. The main subpopulations in the data correlate with

1. Zn-poor sediment developed over quartzite conglomerate and glacigene deposits derived from quartzite conglomerate,
2. sediment of intermediate Zn content derived by weathering of mainly acid to intermediate volcanic rocks,
3. Zn-rich sediment derived mainly from fine-grained sedimentary rocks, and
4. anomalously high Zn related to sulphide mineralization.

It is noteworthy that soils developed on the quartzite conglomerate and glacigene deposits doe not support the myrtle forest flora typical of most of the area, but only tea trees and coarse grasses. Thus it is suggested that examination of Tasmanian Mines Department records might provide valuable data for rapid assessment of the agricultural potential of the areas suffering denudation.

North Queensland is another area where some agricultural assessment might be attempted on existing data. In this case the elements which might prove most interesting are Cu and Mo in relation to grazing areas. Analyses for both these elements are available for the total Ayr 1:250,000 map sheet and for smaller areas explored by companies (data filed with the Queensland Mines Department). For example, one of the areas prospected in the Connors Arch near Eungella covers an area of 2500 km^2 and was sampled for stream sediment at a density of nearly one sample per km^2. The elements Pb, Zn, Cu, Ni, Ag, Mo, As and U were determined. Sediments correlated with two geological units, one sedimentary and one granitic, have notably low Cu contents (about 6 ppm); in addition, sediment from the Cu-poor granitic unit has the highest Mo content of the whole survey area (ranging from 2 to 18 ppm). The affected area is only about 26 km^2 occurring in rugged terrain and, consequently, it is probably insignificant in relation to local agriculture. However, it does suggest the possibility of similar imbalances occurring elsewhere within the same geological setting.

CONCLUSIONS

Experiments in the United Kingdom have demonstrated that regional geochemical reconnaissance based on sampling of stream sediments can be adapted to yield information on existing or potential trace element imbalances affecting plants and animals.

Serious attempts should be made to develop adequate standard sampling procedures and to establish joint enterprises with the state mines departments and the Bureau of Mineral Resources to produce multipurpose regional geochemical maps at 1:250,000 scale.

Storage and retrieval facilities should be established for reconnaissance samples so that additional analytical or mineralogical procedures can be performed when the need arises. Exploration companies should be encouraged to ultimately make available the residue of reconnaissance samples.

The Bureau of Mineral Resources and the state mines departments have a large amount of reconnaissance analytical data on file for many parts of Australia. Examination of these data would almost certainly yield new information of agricultural significance.

REFERENCES

HOWARTH, R.J. (1971) - *Math. Geol. 3* : 95
HOWARTH, R.J. & LOWENSTEIN, P.L. (1971) - *Trans. Inst. Min. Metall., Lond. 80* : B363
JOYCE, A.S. (1974) - *Exploration Geochemistry*. Techsearch, Adelaide, 157p.
MARSHALL, N.J. (1967) - *Bur. Min. Resour. Aust. Rec. 1967/129* (unpub.)
SCHWARZER, T.F. & ADAMS, J.A.S. (1973) - *Econ. Geol. 68* : 1297
SINCLAIR, A.J. (1974) - *J. Geochem. Explor. 3* : 129
THORNTON, I. (1972) - *Billiton-Venture, Christmas 1972*, p. 22
THORNTON, I., KERSHAW, G.F. & DAVIES, M.K. (1972) - *J. agric. Sci., Camb. 78* : 151
THORNTON, I., MOON, R.N.B. & WEBB, J.S. (1969) - *Nature, Lond. 221* : 457
THORNTON, I. & WEBB, J.S. (1970) - *J. Indian Soc. Soil Sci. 18* : 357
WALL, L.N. (1974) - *Abs. Geol. Soc. Aust. Joint Specialist Gps Meeting, Brisbane*, p. 51
WALL, L.N. & LAM, J.S. (1974) - *Qd Govt Min. J. (June)* : 216
WEBB, J.S. (1973) - *Trans. Inst. Min. Metall., Lond. 82* : 33
WEBB, J.S. & ATKINSON, W.J. (1965) - *Nature, Lond. 208* : 1056
WEBB, J.S., NICHOL, I. & THORNTON, I. (1968) - *Rep. 23rd Int. geol. Congr., Prague 6* : 131
WEBB, J.S. & THORNTON, I. (1971) - *In* "Mineral Studies with Isotopes in Domestic Animals", p. 189. International Atomic Energy Agency : Vienna
WEBB, J.S., THORNTON, I. & NICHOL, I. (1971) - *Minist. Agric., Fish., Fd Tech. Bull. 21* : 1

DIAGNOSIS OF TRACE ELEMENT DEFICIENCIES IN WHEAT ON EYRE PENINSULA, SOUTH AUSTRALIA

P.M. King* and A.M. Alston†

*South Australian Department of Agriculture
Adelaide, South Australia 5000
†Waite Agricultural Research Institute
The University of Adelaide
Glen Osmond, South Australia 5064

INTRODUCTION

Deficiencies of trace elements — principally Cu, Mn and Zn — have been reported to occur in wheat on solidized solonetz and solodic soils, lateritic podzolic soils, calcareous sands and terra rossas of Eyre Peninsula, South Australia (Tiver 1956; French 1958). However, the extent of the deficiencies and the spatial relationships of the soils involved have not been defined.

Recently two areas, totalling 4600 hectares and containing representatives of the most important agricultural soils on the Peninsula, were surveyed at Stokes and Wharminda (King & Alston 1975). Land units based on geomorphology and topography were defined; the soils were mapped (1:20,000), and their position within the land units was described. This paper reports an investigation in which the occurrence of trace element deficiencies on the soils of the surveyed areas was determined by measuring the grain yield response of wheat in the field to foliar application of Cu, Zn, Mn, Fe, Mo and B. The use of soil and plant analyses to identify Cu-deficient soils is discussed. Some preliminary results of the investigation have been published (King & Alston 1974).

EXPERIMENTAL CONDITIONS AND TECHNIQUES

Twenty half-replicated 2^6 factorial experiments were conducted during 1970 and 1971. The annual rainfalls at Wharminda for the two years were 266 and 388 mm respectively. Rainfall was not recorded at Stokes, but rainfall at Port Lincoln in 1970 and 1971 was 435 and 698 mm respectively. Although Port Lincoln is 50 km from Stokes, isohyets show that rainfall at Stokes is more likely to be similar to that at Port Lincoln that at any other recording station in the area (French 1958). The mean annual rainfalls at Wharminda and Port Lincoln are 359 and 484 mm respectively. The soils at the experimental sites (2-4, 6, 7, 9, 10, 12-24), which are listed in Table 1, have been described (King & Alston 1975). No trace element fertilizers had been applied previously.

Wheat was sown at 65 kg ha^{-1} in plots 20 m long and 2 m (12 rows) wide. The cultivars used were those then recommended for the districts by the South Australian Advisory Committee on Wheat Quality, viz. Halberd at Wharminda and Pinnacle at Stokes. Basal fertilizers were sodium tripolyphosphate (40 kg ha^{-1} P) drilled with the seed and ammonium sulphate (50 kg ha^{-1} N) broadcast on the soil surface and incorporated into the soil with the back tynes of the seed drill. An additional 7 kg ha^{-1} N was applied at tillering as urea in foliar sprays at Sites 2, 3, 4, 6, 7 and 9, where symptoms of N deficiency had appeared.

The trace element fertilizer treatments were applied in foliar sprays at the two-leaf stage, at late tillering and at the boot stage. According to the treatment, the wheat received on each occasion in 500-1500 ℓ ha^{-1} deionized water: 100 g ha^{-1} Cu, 200 g ha^{-1} Zn, 500 g ha^{-1} Mn and 500 g ha^{-1} Fe as sulphates; 200 g ha^{-1} B as boric acid; and 10 g ha^{-1} Mo as sodium molybdate.

The wheat was sampled by cutting quadrats totalling 15 or 20 m of row per plot. Whole tops were taken at the boot stage, and straw at maturity. Grain yields were determined by reaping the remainder of the plots. Sub-samples of the plant material were dried at 70°C and weighed.

Ten samples of soil were collected from the A_1 horizon at each site at the beginning of the growing season. The samples were bulked, air-dried and passed through a 2 mm nylon sieve. Total Cu in the < 2 mm fraction was determined in hydrofluoric-perchloric acid digests (Hanna 1964). Extractable Cu was determined by three methods -

Ethylene diamine tetra-acetic acid (EDTA): 20 g soil were shaken for 16 hr with 40 ml 0.05 M disodium EDTA and 1 M ammonium acetate solution adjusted to pH 6.0 with ammonium hydroxide (Viro 1955).

Diethylene triamine penta-acetic acid (DTPA): 10 g soil were shaken for 1 hr with 20 ml 0.001 M DTPA in 0.01 M calcium chloride solution, buffered with triethanolamine

TABLE 1

Details of soils and occurrences of trace element deficiencies in the factorial and ancillary experiments

Soil mapping unit (a)	Soil categories (b)	Principal profile form (c)	Site number (d)	Occurrence of deficiencies (e)
1	DEEP SILICEOUS SAND	Uc2.21	2,25	
		Uc2.21	3	Cu, Fe
2	CALCAREOUS SAND	Uc1.11	4	Cu, Mn
		Uc1.11	26	Mn
		Uc1.11	27	Mn(f)
		Uc1.11	28	Mn, Zn
	SOLODIZED SOLONETZ AND SOLODIC SOILS (Wharminda)			
	A_1 horizons light brown sand: total depth of A horizons -			
5	> 45 cm	Dy5.83	6	Cu
		Dy5.83	29,30	
6	30-45 cm	Dy5.83	7	
		Dy5.83	31	Cu
7	15-30 cm	Dy5.43	9	Cu
		Dy5.43	32	
8	< 15 cm	Dy5.43	10	Cu
	A_1 horizons brown loamy sand: total depth of A horizons < 15 cm: solum depth -			
9	> 15 cm	Dy5.83	12	Cu, Mo
		Dy5.83	33	
10	< 15 cm	Dy5.83	13	Zn
		Dy5.83	34	Mn(f)
	SOLONIZED BROWN SOILS			
	A_1 horizons -			
11	low in carbonate	Gn1.83	14	Zn
12	high in carbonate	Gc1.12	35	Zn
		Gc1.12	36	Zn(f)

continued on next page

TABLE 1 *continued*

Soil mapping unit (a)	Soil categories (b)	Principal profile form (c)	Site number (d)	Occurrence of deficiencies (e)
	LATERITIC PODZOLIC SOILS			
	A_1 horizons -			
17	with concreted laterite	Dy3.82	15	Cu, Zn
		Dy3.82	16	Cu
		Dy3.82	38	Cu
18	red-brown with pisolitic laterite	Dr2.62	19	Cu, Mo, Fe(g)
		Dr2.22	39	Cu(f)
19	grey with pisolitic laterite	Dy3.82	17	
		Dy3.82	18	Cu
		Dy3.82	40	
		Dy3.82	41	Cu(f)
	SOLODIC SOILS (Stokes)			
	A_1 horizons -			
20	grey with white quartz gravel; parent material white granitic gneiss	Dy3.82	22,42	
		Dy3.82	23	Cu, Mo, Mn(g)
21	brown with pink sand grains; parent material pink granitic gneiss	Db2.22	20	Cu
		Db2.62	21	Cu
23	grey with white sand grains; parent material alluvium (gully line soils)	Dy3.82	24	Cu

(a) King & Alston (1975)

(b) The names of the great soil groups are used according to the definitions of Stace *et al*. (1968)

(c) Northcote (1971)

(d) Factorial experiments 2-24, ancillary experiments 25-42

(e) Underline = significant effects of fertilizer treatment on grain yield; no underline = deficiency symptoms observed or attributes other than grain yield affected.

(f) Grain yields not measured.

(g) Soil analysis indicates deficiencies unlikely.

at pH 7.3 (Lindsay & Norvell 1969).

Calcium nitrate: 20 g soil were shaken successively with three 70 ml volumes of 0.01 M calcium nitrate solution adjusted to pH 6.0 with ammonium hydroxide (Tiller *et al.* 1972). The first shaking was for 16 hr; the second and third were for 1 hr each. The extracts were bulked.

Cu was measured in the soil extracts, and in nitric-perchloric acid digests of plant material from the control plots, by atomic absorption spectrophotometry. The bulked calcium nitrate extracts were buffered at pH 6.0 with sodium acetate-acetic acid, and the Cu concentrated into methyl isobutyl ketone with ammonium pyrrolidine dithiocarbamate before measurement (Allan 1961).

In addition to the factorial experiments, 17 ancillary experiments (Sites 25-36, 38-42) were conducted to provide further information on the occurrence of deficiencies on the different soils. Copper, zinc and manganese were applied singly or in various combinations in foliar sprays (as in the factorial experiments) or at time of sowing as sulphates in solid form at 1 kg ha^{-1} Cu, 1.7 or 3.4 kg ha^{-1} Mn and 1.6 kg ha^{-1} Zn. The soils were classified as deficient or not deficient on the basis of either grain yield measurements or the appearance of unequivocal symptoms of deficiency in plants on the control plots.

EFFECT OF SOILS AND FERTILIZER APPLICATION ON YIELD

Yields of wheat on the soils at Wharminda (Sites 2-14) were lower than at Stokes (Table 2). This is attributed to the difference in rainfall between the two areas, and to the sandy texture of the A horizons and associated low water holding capacity and poor nutrient status of the soils at Wharminda: the lowest yields were obtained where the A horizons were deepest. At Stokes the yields on the solodic soils (Sites 20-24) were higher than those on the lateritic podzolic soils (Sites 15-19). The latter soils are shallower and more prone to waterlogging in winter and drying out early in the spring.

The yield response of wheat to application of Cu in the factorial experiments (main effects) varied greatly from soil to soil, and was proportionally greater at the higher yielding sites (Table 2). Symptoms of Cu deficiency identical to those described by Riceman (1943) were observed in 14 of the 20 factorial experiments. However, deficiency symptoms in wheat on the control plots were not always associated with increased grain yields where Cu was applied. Vegetative

TABLE 2

Dry matter yield responses of wheat to the application of Cu in the factorial experiments and concentrations of Cu in the wheat and in the soil

Site number	Date of sowing	Dry matter yield (kg ha^{-1}): Boot stage, Whole tops, O	Boot stage, Whole tops, +Cu	Maturity: Straw, O	Straw, +Cu	Grain, O	Grain, +Cu		Cu concentration in wheat (ppm): Boot stage, Whole tops	Maturity, Grain	Cu concentration in soil (ppm): Total	Extractable: EDTA	DTPA	$Ca(NO_3)_2$
2	19-5-71	-	-	-	-	650	600		1.4	-	2.0	0.29	0.07	0.007
3	8-6-71	1240	1500***	1320	1800***	230	700***	D	1.8	1.3	1.5	0.11	0.01	0.001
4	17-6-71	2610	2800	2070	2250	700	1010*	D	2.6	1.5	2.0	0.23	0.02	0.003
6	18-5-71	2070	1920	2570	2470	790	800	D	3.3	3.0	9.0	0.38	0.09	0.007
7	20-5-71	2580	2740	-	-	890	800		1.6	2.6	2.5	0.43	0.12	0.005
9	21-5-71	2050	2490	2430	2760**	760	1080***	D	0.9	2.0	2.0	0.18	0.05	0.004
10	25-5-71	4910	4940	5160	5370	2280	2210	D	2.8	2.5	2.5	0.30	0.09	0.005
12	7-6-71	4390	4350	4270	4440	1050	1160*	D	4.4	3.0	3.0	0.35	0.09	0.004
13	17-6-71	5170	5300	3510	3480	1580	1650		2.9	4.7	2.5	0.30	0.10	0.004
14	8-6-71	3560	3570	3360	3280	1030	1000		4.9	4.5	3.0	0.50	0.16	0.006
15	22-6-70	2430	4020***	2590	3080***	15	1910***	D	1.8	1.8	4.0	0.14	0.05	0.001
16	10-6-71	3450	4030**	2980	3360	950	1230*	D	1.6	1.8	4.0	0.33	0.07	0.002
17	24-6-70	4540	4510	3810	3820	1210	1250		6.9	4.0	5.0	1.33	0.34	0.018
18	10-6-71	5070	5640*	3320	3710*	2120	2370*	D	3.5	2.4	6.0	1.34	0.34	0.009
19	9-6-71	4740	5240**	4340	4380	2160	2130	D	2.9	3.4	12.0	1.18	0.33	0.008
20	23-6-70	5140	5270	5160	5420	2050	2620*	D	3.5	5.0	4.0	0.48	0.11	0.006
21	10-6-71	5140	6640***	4890	6660***	1810	3810***	D	1.8	2.0	6.5	0.59	0.15	0.004
22	22-6-70	4520	4210	3910	4030	2360	2390		4.0	4.9	4.5	0.84	0.24	0.012
23	9-6-71	6210	6510	6440	7030*	2520	3100***	D	3.4	3.8	5.5	0.65	0.13	0.005
24	9-6-71	3800	3900	3760	3970*	2260	2620***	D	1.7	1.5	3.0	0.57	0.15	0.014

* Significant at P = 0.05
** Significant at P = 0.01
*** Significant at P = 0.001

D = Deficiency symptoms present

growth was usually less affected by Cu deficiency than was grain production, although there was no effect on grain yield at Site 19, where application of Cu increased dry matter yield at the boot stage. In no instance did application of Cu significantly ($P = 0.05$) decrease grain yield.

Responses to trace elements other than Cu were much less frequent. Table 3 gives the grain yield of wheat where significant response (main effects) occurred. Application of Zn reduced grain yield at Site 4, the only site where there was a large response to application of Mn. Both Fe and Mo increased grain yield at two sites but in no case was yield increased by application of B.

TABLE 3

Effect of applying Zn, Mn, Fe and Mo on straw and grain yield of wheat in the factorial experiments

Site number	Dry matter yield (kg ha^{-1})				Deficiency symptoms
	Straw		Grain		
	O	+Zn	O	+Zn	
4	2400	1930	1210	500*	Mn, Cu
14	3150	3490***	950	1080*	Zn
	O	+Mn	O	+Mn	
4	1770	2260	560	1150**	Mn, Cu
	O	+Fe	O	+Fe	
3	1490	1630	440	490*	-
19	4340	4380	2080	2210*	-
	O	+Mo	O	+Mo	
19	4370	4350	2080	2210**	Mo
23	6970	6490	2700	2920*	Mo

* Significant at $P = 0.05$
** Significant at $P = 0.01$
*** Significant at $P = 0.001$

Interaction occurred between Cu and Zn, Cu and Mn, Zn and Fe and Mn and Mo, but the effects on grain yield were generally small. The most common interaction was Cu and Zn, which significantly affected grain yield at four sites (15, 21, 22, 23). All but Site 22 were Cu deficient, and Zn application increased the severity of the Cu deficiency and decreased yield in the absence of added Cu. At Site 15, application of Cu and Zn together produced higher yields than did Cu or Zn alone, indicating that this site was deficient in both elements. These effects are shown in Table 4, which also lists the other significant interactions and possible interpretations that could be put on them in relation to the occurrence of deficiency.

TABLE 4

Effect of interactions on the grain yield of wheat in the factorial experiments, and possible interpretation with respect to trace element deficiencies

Site number	Grain yield of wheat (kg ha^{-1})				LSD (P = 0.05)	Deficiencies
	Fertilizer treatment					
	O	Cu	Zn	Cu+Zn		
15	27	1850	4	1960	80	Cu, Zn
21	2020	3800	1600	3820	290	Cu
22	2420	2290	2290	2490	170	
23	2690	2970	2350	3220	300	Cu
	O	Cu	Mn	Cu+Mn		
3	210	780	260	620	111	Cu
	O	Zn	Fe	Zn+Fe		
9	1020	850	850	960	190	
22	2490	2280	2210	2500	170	
	O	Mn	Mo	Mn+Mo		
4	790	970	330	1320	420	Mn
23	2420	2980	3090	2740	300	Mo, Mn*

*Response to Mn was unexpected on this soil, which had a pH of 6.3 and contained 70 ppm total Mn and 3 ppm calcium nitrate-extractable Mn in the A_1 horizon.

RELATION OF DEFICIENCIES TO SOIL TYPE

The factorial and ancillary experiments indicated particular sites where trace element deficiencies occurred (Table 1). If the sites are characteristic of the soils they represent, the results of the field experiments can be used in conjunction with soil maps to delineate areas deficient or potentially deficient in trace elements.

Copper deficiency occurred on a wide range of soils: calcareous sands, siliceous sands, solodized solonetz, solodic and lateritic podzolic soils. The response of wheat at Wharminda appeared to be related to the nature of the surface soil. Deficiency occurred frequently on soils with light-brown sand A_1 horizons (Soil Mapping Units 1, 5-8), but infrequently where the A_1 horizons were brown loamy sand and sandy loam (Soil Mapping Units 9-12). The latter soils, besides having higher clay contents, also contain more organic matter. At Stokes, wheat on seven of the nine sites on lateritic podzolic soils (Soil Mapping Units 17-19) and four of the six sites in solodic soils (Soil Mapping Units 20, 21, 23) responded to Cu application and it seems reasonable to conclude that soils in this area should generally be regarded as potentially deficient in Cu.

While Cu deficiency was widespread, responses to Zn and Mn applications were confined largely to calcareous soils. Soils that could be considered potentially deficient in Zn are the solonized brown soils with calcareous loam horizons at or near the surface (Soil Mapping Units 11, 12). Mn deficiency is likely on the calcareous sands (Soil Mapping Unit 4). Both Zn and Mn deficiency occurred on shallow solodic soils (Soil Mapping Units 9, 10) where cultivation or erosion has exposed carbonate at the soil surface. Erosion is not uncommon on Eyre Peninsula, and the possible effects that it might have on the occurrence of deficiencies should not be overlooked.

A major problem in delineating areas of trace element deficiency arises because the soil mapping units are defined on the basis of easily recognizable morphological properties that may be poorly correlated with the availability of trace elements. For example, the solodized solonetz and solodic soils with light-brown sand A_1 horizons at Wharminda were divided into four mapping units (5-8) according to the total depth of sand in the A horizons, but Cu-deficient and non-deficient sites were found on units with shallow and deep A horizons (Table 1). Further assessment is required, both within and between the soil mapping units, of those soil properties that exert most control on availability.

Clearly not all the soils that might be regarded as potentially deficient on the basis of a soil survey and the results of field experiments will respond to fertilizer application. However, plant and soil analyses can be used to distinguish responsive and unresponsive sites within the potentially deficient areas.

PLANT AND SOIL ANALYSES

Data are presented only for Cu since there were relatively few yield responses to the application of other trace elements in the factorial experiments. The Cu concentration in plants from the plots to which no trace elements had been added and in the soils sampled before the wheat was sown are given in Table 2. The concentration of Cu in the tops of the wheat at the boot stage was poorly related to both dry matter and grain yield responses to Cu application. However, there was a good relationship between Cu concentration in the grain and grain yield response. The responsive and unresponsive sites are largely separated by a critical concentration of 2.5 ppm in the grain, although application of Cu increased grain yield on two solodic soils at Sites 20 and 23, where the Cu concentrations in the grain were 5.0 and 3.8 ppm (Fig. 1). These anomalies may have arisen because the experimental design used requires very uniform sites. The response to application of Cu at Site 20 was not uniform over the site and, since the control plot was located in an unresponsive area, some doubt exists about the validity of using the concentration of Cu in the wheat from that plot to represent the whole site. The critical value and the range of Cu concentrations in healthy and deficient wheat grain are similar to those reported by Caldwell (1971) for south-east England, where < 2 ppm is considered to indicate deficiency while > 3 ppm indicates sufficiency.

The relation between yield response and Cu concentration in the grain was not greatly affected by the nature of the soil. Grain analysis is therefore likely to be useful for

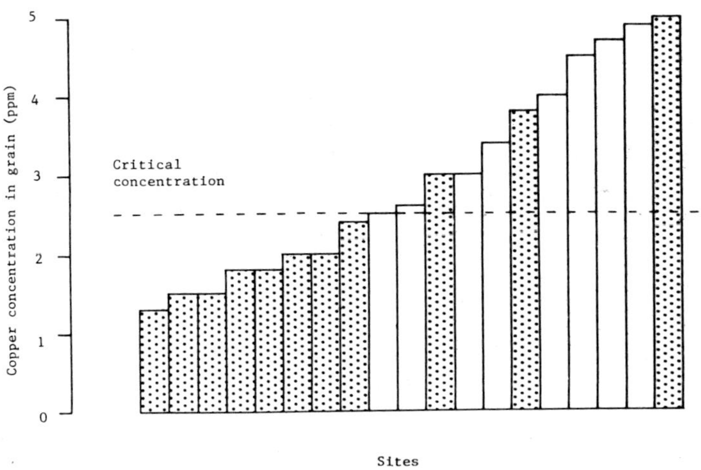

Fig. 1. Copper concentration in the grain of wheat from the control plots (no trace element fertilizers added) in the factorial experiments. Sites where grain yield was increased by application of Cu ⊡; sites where grain yield was not increased by application of Cu □.

identifying areas deficient in Cu, particularly when information on the soil is lacking.

The nature of the soil clearly affected the relationship between yield response and the amount of Cu extracted from the soil, and differences between soils or groups of soils must be recognized before soil analysis can be used effectively. This is illustrated in Figure 2 for EDTA-extractable Cu. The curves in Figure 2 are hand-fitted because there were not enough sites to test the fit statistically for each group of soils. The relation between yield response and DTPA- and calcium nitrate-extractable Cu are similar to those for EDTA. The relations shown in Figure 2 are attributed to the properties of the soils, although it could be argued that differences between soils at Stokes and Wharminda might have arisen, in part, because of differences in rainfall or because different cultivars were grown in the two areas, viz. Halberd at Wharminda and Pinnacle at Stokes. However, in a glasshouse experiment, Halberd was found to be slightly more susceptible to Cu deficiency and more responsive to application of Cu than was Pinnacle (A.M. Alston, unpublished data). If this difference occurred in the field, the effect of growing one

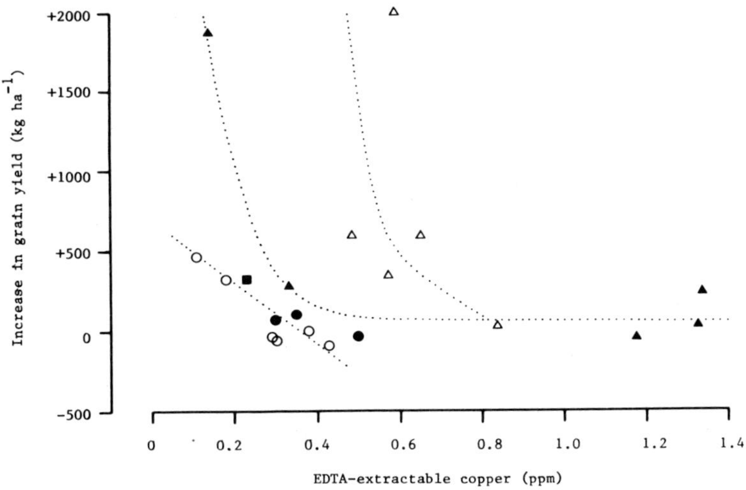

Fig. 2. Increases in grain yield due to application of Cu in the factorial experiments versus *EDTA-extractable Cu in the A_1 horizons of the soils. Wharminda: calcareous sand ■; deep siliceous sand, solodized solonetz and solodic soils with light-brown sand A_1 horizons ○; shallow solodized solonetz and solodic soils with brown loamy sand A_1 horizons ●. Stokes: lateritic podzolic soils ▲; solodic soils △.*

of the cultivars at all sites would be to accentuate the difference between the curves for soils of the two areas. Although rainfall may affect the response to application of Cu, the difference in rainfall between Stokes and Wharminda is not considered a major factor since data for 1970 (a dry year) and 1971 (a wet year) appear to fall on the same curves in Figure 2.

Tentative critical concentrations of extractable Cu are given in Table 5. A critical concentration of 0.75 ppm EDTA-extractable Cu has been reported for a wide range of soils by McKenzie (1966), Reith (1968) and Cox & Kamprath (1972). The

TABLE 5

Tentative critical concentrations of extractable Cu in three groups of soils (ppm)

Groups of soils	Critical concentration of extractable Cu		
	EDTA	DTPA	$Ca(NO_3)_2$
Wharminda			
Siliceous sands, solodized solonetz, solodic soils, solonized brown soils	0.4	0.1	0.005
Stokes			
Lateritic podzolic soils	0.6	0.15	0.010
Solodic soils	0.8	0.20	0.015

authors are unaware of any published critical concentrations for DTPA- or calcium nitrate-extractable Cu. Although calcium nitrate, DTPA and EDTA were equally useful as extractants for Cu, extraction and measurement of Cu with EDTA can be performed most simply and accurately, and this extractant is recommended for routine testing of soils.

McKenzie (1966) found that total soil Cu was as good as EDTA-extractable Cu as an index of availability in a group of soils that included krasnozems, calcareous sands, podzolic and solodic soils. However, in the present study, the grain yield response of wheat to applied Cu was poorly related to the total Cu contents of the soil.

CONCLUSIONS AND EXTRAPOLATION OF THE RESULTS

This investigation has shown that field experiments can be used in conjunction with soil maps to delineate areas actually or potentially deficient in trace elements. In the case of Cu, plant and soil analyses were able to distinguish most of the sites that were actually deficient. Analysis of the grain to detect Cu deficiency could be performed without reference to soil maps, but this would entail many more analyses than would be required where only plants from potentially deficient areas are sampled. Soil analysis can be used effectively only after differences in the soils have been recognized.

Because few experiments were performed on each soil, the conclusions drawn about the occurrence of deficiencies on particular soils or groups of soils should be regarded as tentative. Land units and soil associations similar to those at Stokes and Wharminda occur widely on Eyre Peninsula, and it is likely that deficiencies of Cu in particular, and also of Zn and Mn, are widespread in the region. The topographic relationships of the soils within the land units provide a basis for extending the investigation and for applying its results to areas of the Peninsula outside those surveyed by King & Alston (1975). The results might possibly be extended to other parts of southern Australia where similar patterns of soil and climate exist, e.g. in the Murray mallee (Potter *et al.* 1973), on Kangaroo Island (Northcote & Tucker 1948), and in the southern Mount Lofty Ranges (Rix & Hutton 1953).

ACKNOWLEDGEMENTS

The authors wish to thank Mr. J.E. Ellis, Mr. B.L. Glaetzer, Mr. G.R. MacPhie and Mr. D.J. Reuter of the Department of Agriculture, South Australia, and Dr. K.G. Tiller of the Division of Soils, CSIRO, for help with statistical and chemical analyses and with field work. The ready cooperation of the farmers on whose properties the experiments were conducted is greatly appreciated.

REFERENCES

ALLAN, J.E. (1961) - *Spectrochim. Acta 17* : 459

CALDWELL, T.C. (1971) - *In* "Trace Elements in Soils and Crops". Technical Bulletin, Ministry of Agriculture, Fisheries & Food, No. *21* : 73. HMSO : London

COX, F.R. & KAMPRATH, E.J. (1972) - *In* "Micronutrients in Agriculture" (J.J. Mortvedt, P.M. Giordano & W.L. Lindsay, eds.), p. 289. Soil Science Society of America : Wisconsin

FRENCH, R.J. (1958) - Department of Agriculture, South

Australia : Bulletin 457

HANNA, W.J. (1964) - *In* "Chemistry of the Soil" (F.E. Bear, ed.), 2nd Edn, p. 474. Reinhold : New York

KING, P.M. & ALSTON, A.M. (1974) - *Trans. 10th Int. Congr. Soil Sci., Moscow* (in press)

KING, P.M. & ALSTON, A.M. (1975) - Department of Agriculture, South Australia : Bulletin (in press)

LINDSAY, L. & NORVELL, W.A. (1969) - *Agron. Abstr. 1969* : 84

McKENZIE, R.M. (1966) - *Aust. J. exp. Agric. Anim. Husb. 6* : 170

NORTHCOTE, K.H. & TUCKER, B.M. (1948) - CSIRO Australia, Bulletin No. 233

NORTHCOTE, K.H. (1971) - "A Factual Key for the Recognition of Australian Soils". Rellim : Glenside, South Australia

POTTER, J.S., WETHERBY, K.G. & CHITTLEBOROUGH, D.J. (1973) - Department of Agriculture, South Australia, L.D.1

REITH, J.W.S. (1968) - *J. agric. Sci., Camb. 70* : 39

RICEMAN, D.S. (1943) - *J. Dep. Agric. S. Aust. 47* : 64

RIX, C.E. & HUTTON, J.T. (1953) - Land Tax Department, South Australia, Bulletin No. 1

STACE, H.C.T., HUBBLE, G.D., BREWER, R., NORTHCOTE, K.H., SLEEMAN, J.R., MULCAHY, M.J. & HALLSWORTH, E.G. (1968) - "A Handbook of Australian Soils". Rellim : Glenside, South Australia

TILLER, K.G., HONEYSETT, J.L. & DE VRIES, M.P.C. (1972) - *Aust. J. Soil Res. 10* : 165

TIVER, N.S. (1956) - *J. Dep. Agric. S. Aust. 59* : 100

VIRO, P.G. (1955) - *Soil Sci. 79* : 459

TRACE ELEMENT DISORDERS IN PINUS AND THEIR CORRECTION

M. Raupach

Division of Soils, CSIRO
Glen Osmond, South Australia 5064

INTRODUCTION

Australia's softwood resources are man-made forests, mostly conifers. Before about 1918, it was believed both here and elsewhere that the demands of these plantations could be satisfied by all soils down to the poorest. A consequence of this mistake was the poor performance or even failure of some plantations on infertile soils. Simple measures such as the application of superphosphate fertilizer sometimes allowed economic production, but gains took time to assess, and forest management, with a different outlook from agriculture, tended to be conservative in regard to the need for fertilizers.

Stoate (1950) has described how, in Western Australia, commencing in 1937, Zn was a corrective and preventative of a group of disorders which frequently resulted in the death of the leader in *Pinus radiata* plantations. He stated that Zn could be given to the tree in almost any manner; a Zn-coated nail driven into the trunk or the proximity of a wire-netting fence were among the means which, by design or accident, restored disordered trees to vigorous growth.

The use of 2½% zinc sulphate sprays on plantations of young trees (aged 2-3 years) is widespread in South and Western Australia, and in western Victoria in susceptible soils, e.g. sands and lateritic podzolics. Large improvements in tree performance frequently result. Only occasionally is it necessary to add further Zn later in the life of the rotation. This is presumably because the demand for Zn by young trees cannot be satisfied by their limited root systems on some soils.

BACKGROUND

In Australia, *P. radiata* is the main species of softwood planted in the southern states; in 1971 it occupied the following areas in hectares (per cent of total conifers in brackets): Victoria 90,000 (93%); South Australia 77,000 (92%); New South Wales 74,000 (74%); Tasmania 24,000 (99%); Western Australia 15,000 (46%); Australian Capital Territory 11,000 (91%); and Queensland 2,000 (2%). The main species in the subtropical areas (Bevege 1973) is *P. elliottii* (Slash Pine) and in the tropical areas, *P. caribaea* (Caribbean Pine). Of the latter there are three varieties, sometimes called "honduras", "cuban" and "bahamas", as well as hybrids between varieties and with Slash Pine. *Araucaria cunninghamii* (Hoop Pine) is also of significance. Additional tropical species may be of interest in the future. *P. pinaster* is grown in Western Australia, and to a lesser extent in South Australia, in situations where *P. radiata* would fail to give economic returns. This hardy species does not show nutritional stress so easily but has responded to Zn in Western Australia, where a suspected Mn deficiency is also under examination.

General reviews exist on micronutrient nutrition (Stone 1968), fertilizer treatment (Baule & Fricker 1970) and plant analysis (Leaf 1973) of forest trees. Raupach (1967) has summarized soil and fertilizer requires for forests of *P. radiata*.

Pinus taeda (Loblolly Pine) is restricted to a relatively small area of fertile, poorly drained sites (humic gley soils) of the coastal lowlands of south-east Queensland. Slash Pine is the usual species for poorly drained sites. However, because Loblolly Pine has a greater nutrient demand than Slash Pine (Bevege 1973), it has been used extensively as a test species. Suitable sites give spectacular responses to modest amounts of phosphate fertilizers. Copper and potassium responses have been observed in the glasshouse, but not yet in the field.

Slash Pine, the major plantation species of Queensland, is nutritionally less demanding. There is evidence of a response to a mixture of trace elements, added with P and N at one site on waterlogged gleyed podzolic and low humic gley soils (Bevege & Simpson 1973).

Pinus patula is confined to deep lateritic krasnozem soils. Trace element responses have not yet been reported.

PREDICTION AND ASSESSMENT OF POSSIBLE TRACE ELEMENT DISORDERS

It is sometimes necessary to judge whether nutrient elements, in the present instance particularly trace elements, are possible causes for poor forest performance. The poor performance may be associated with either a decline in growth or with sustained lack of vigour, but it is frequently accompanied by the appearance of symptoms of distress if trace elements are involved. Again, a new area on which forestry is contemplated may be situated on soils suspected to be trace element deficient, and advice may be required as to whether this is likely to restrict production.

Glasshouse experiments may be of some assistance for the new area, with the possible inclusion of a more demanding species. Glasshouse results are frequently of value for nursery sites where growth is in a depth of soil similar to that used in pot experiments. Glasshouse work with culture media and with soils has been of great value in describing the symptoms of trace element stress, as well as giving estimates of foliar levels of deficiency and toxicity. This has been documented by Stone (1968). The technique has to be used with care. For example, nutrient balances between major and trace elements may affect both levels and symptoms and, as Stone has said, 'nutrient cultures' have 'environments quite alien to those encountered in nature'. Even nursery work, economically desirable as it is, deals with juvenile plants under environments and treatments poorly related to field conditions.

Tamm (1964), Raupach (1967) and Bevege & Richards (1972) all agree that insufficient information is available on the nutrient levels of populations of trees. That is to say, the status of large numbers of trees, or forests, is of concern rather than individual trees.

The considerable variation of needle composition from year to year (see Humphreys & Kelly 1962 for *P. elliottii* and Humphreys *et al.* 1972 for *P. radiata*) has led to the suggestion that sampling should go on for a period of years to establish critical levels (Tamm 1964). On the other hand, by sampling from a wide area with large differences in climate and soils, rather than taking trees from the one stand, it might be possible to gather better information in one year than at one site sampled annually.

Soil analysis, although of use in special circumstances, does not yet compare favourably with plant analysis as an indicator of trace element status.

It has been pointed out that, although the establishment of critical levels is important, investigators should have a wider view than this when planning work or interpreting results. Leaf (1973) has stated this view:

For plant analysis to attain its full potentiality for

aiding in forest fertilization programs, it must be used for more than determining critical or deficient element levels as compared with non-deficient levels. Forest fertilization programs will become more of a managerial practice when the forest researcher provides information to improve a 'good productive' site to a 'very good productive' site.

This improvement will obviously give a larger economic return and may soon be possible for some of the major elements. It requires that a statistically significant modest relative response should be able to be evaluated in terms of foliar nutrients. This cannot be accomplished yet for the trace elements. Work should be directed towards it, especially in relation to nutrient element interactions (Raupach 1967), including those involving trace elements.

QUANTITIES OF TRACE ELEMENTS IN *PINUS*

The small quantities of trace elements in a whole tree plus roots have been estimated for *Pinus strobus* with a height of 15 m and a diameter (DBH) of 20 cm: B, 0.6 g; Zn, 3.6 g; Cu, 0.9 g and Mn, 12 g (Young *et al.* 1965). Fortescue & Marten (1970) have given more data for *Pinus*, as well as for a number of other forest trees.

Under Australian conditions in particular, a sub-optimal P status without the appearance of symptoms of distress is somewhat frequent (see Raupach *et al.* 1969, for the South-East of South Australia). Snowdon (1972) has pointed out that, for B, dry weight production is related to foliar concentration over a very small range. This may well hold for other trace elements as well, and would mean that sub-optimal ranges were generally narrow. More peculiar to B is Snowdon's observation that irreparable damage may take place if B is unavailable for as short a time as 48 hr.

TRACE ELEMENT LEVELS IN *PINUS* FOLIAGE

Foliar concentrations of some trace elements in mature, green, first-season foliage from large trees in the field (with needle position sometimes unknown) are given in Table 1, together with an indication of deficiency symptoms for the species of *Pinus* grown in Australia. The table has been compiled from the data of Stone (1968), with supporting information and many additions from recent work. Ranges for other *Pinus* spp. are given and general symptoms of deficiency and toxicity noted.

TABLE 1

Foliar symptoms of deficiency (Defic) and toxicity (Toxic) of Pinus *and levels of trace elements, where available, for the species most widespread in Australia and an overall range for other species. [Data from Stone (1968) and from references cited in the table and in the text.]*

Element & Symptoms	Intensity	*Pinus* spp. & Level (ppm)			
		radiata	*pinaster*	*elliottii*	Other (range)
ZINC	Defic	1-5	∿ 3[a]		
Defic: rosetting, short needles	Low	5-11	∿ 8[a]		
dieback	Inter[f]	19		15-75	12-84
stunting, yellow needles					
Toxic: chlorosis of Fe defic					
COPPER	Defic	1[b]-2.0[c]			
Defic: branch twist, bushy retarded tree	Low	2.8[c]			2.5-2.8
	Inter	3.8-6.0			3.5-33
necrotic needle tips					
Toxic: as Fe defic					
MANGANESE	Defic		5-16[a]		
Defic: needles yellow tipped,	Low	60[c]			7-10
bronzed	Inter	60[c]-900		330-346	35-1536
Toxic: chlorosis?					
IRON					
Defic: yellow or brown needles	Low	21-52			15-84
terminal bud necrosis	Inter				200[e]
BORON	Defic	3-9	5-6	4-10	8-18
Defic: shoot, tip or bud necrosis	Low	10[d]			
branch crooking	Inter	11-13	16	10-28	7-135
resin flow	High	50[d]			
Toxic: needles with brown, yellow, green zones towards base; needle cast	Toxic	< 2500			73-152[g]
MOLYBDENUM	Range	0.03-0.05			0.05-0.12

[a]Anonymous (1971)
[b]Ruiter (1969)
[c]Raupach *et al.* (1972)
[d]Snowdon (1972)
[e]Keay (1964)
[f]Intermediate
[g]1+ yr needles

PROBLEMS IN THE USE OF FOLIAR LEVELS

The real question to be asked is how useful are such data as those in Table 1 for advice on forestry plantations, and if they are helpful, can any steps be taken to improve them. Except perhaps for B, differences between species are not as great as might be expected from work with the major elements. As will be shown below, considerable differences in levels which give rise to foliar symptoms are to be found for Cu within a species. Snowdon (1972) found the same for B.

The difficulties of sample collection are greater for multinodal species.

Unless otherwise stated, the following discussion deals with the major species grown in southern Australia, *P. radiata*.

EFFECT OF AGE OF FOLIAGE AND POSITION IN THE CROWN

Briefly, groups of trees from four sites in the Central and South-Eastern Forest Reserves of South Australia were sampled at two-monthly intervals over one year. Fresh litter was also collected from beneath the trees. The needles were rinsed thoroughly with distilled water to remove dust; weights of 100 dry needles and foliar concentrations of Mn, Fe, Al, Cu, Zn and B were determined in addition to six major elements. The results did not suggest serious nutrient deficciencies at any of the sites. Samples were from 0-1 year old (A), 1-2 year old (B) or 2-3 year old (C) foliage at four positions numbered consecutively up the crown, as follows: (1) bottom whorl with only three ages of needles, (2) top whorl with three ages, (3) top whorl with two ages and (4) top or second whorl with one age of needles. The data have been examined to determine the most favourable time and position of sampling. The reciprocal of the coefficient of variation reflected the sensitivity of a point of collection. The best times are the most sensitive and the most constant in time. The sites × times mean square of the analyses of variance was not significant. The favoured positions have been tabulated elsewhere (Raupach, Clarke & de Vries 1972) without presentation of levels at the points sampled. The most generally favoured sampling positions were the three possible ones in the top two whorls, i.e. A4, A3 and B3. At least one of these was suitable for each of the elements listed except for Al, where no position was suitable.

Means over sites and times (Fig. 1) show that Cu, Zn and B behaved like P, which is concentrated in the younger needles towards the top of the tree. Iron, aluminium and to a large extent Mn behaved like Ca, which accumulated in the older needles; Mn and Ca both concentrated towards the base of the crown.

Mean results over sites and the top three positions (Fig. 2) show that in May, which was judged to be the best time of the year for sampling, Fe, Al and Mn were at their maximum concentrations, whereas the other trace elements were steady but not quite minimal. Dust contamination was shown to be negligible by analysis of some samples for Si and Ti. The shapes of curves showing the variation of concentration with time (Fig. 2) were similar for the other ages of needles sampled.

These results show that it is insufficient to collect the current year's foliage without regard to the position in the crown, as may have been done for some results in Table 1. Likewise, foliage from a branch in the upper part of the crown, where there could well be two or three ages of needles,

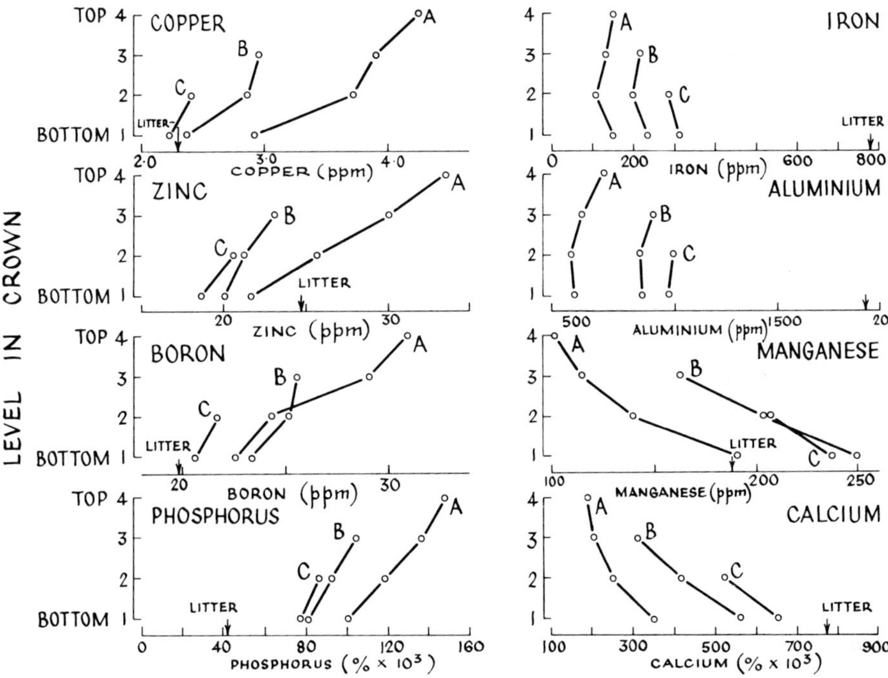

Fig. 1. Mean needle concentrations of Fe, Al, Cu, Zn, Mn, B, P and Ca for four sites and two-monthly samples taken over a year from four crown levels and the following needle ages: 0-1 year (A), 1-2 year (B) and 2-3 year (C). Means of the results for litter which had freshly fallen at the sampling times are indicated.

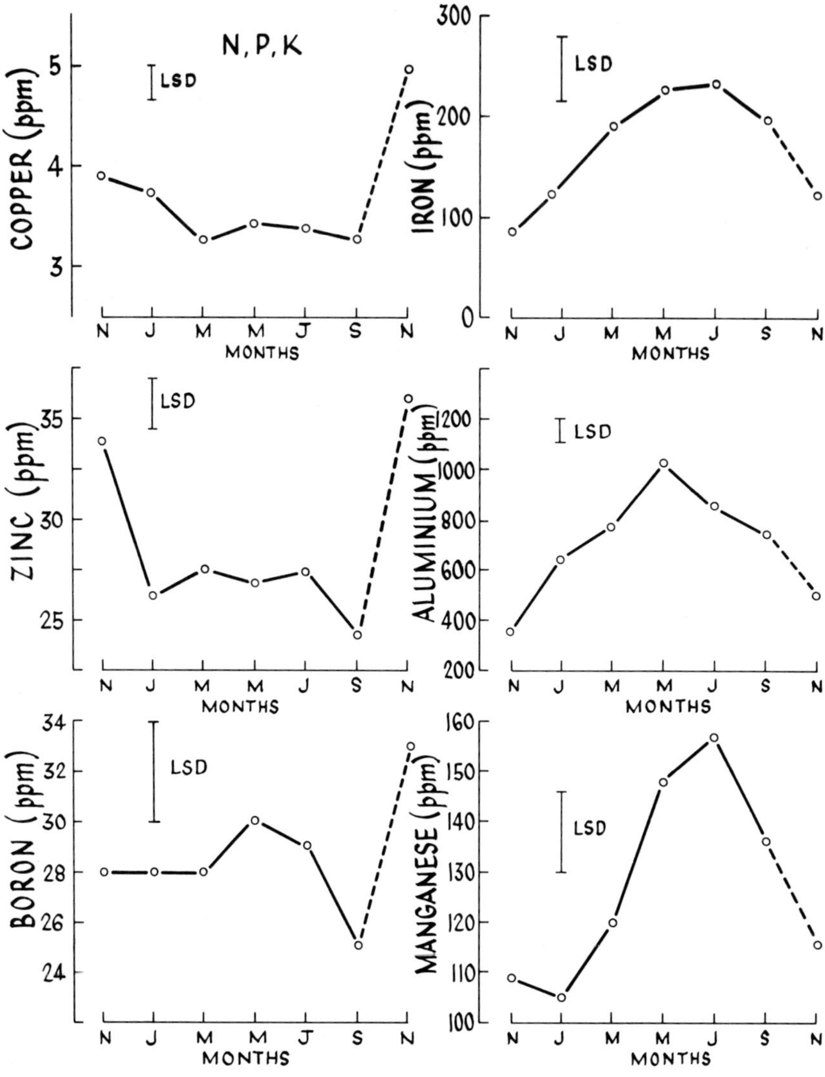

Fig. 2. Mean results of foliar concentrations of Cu, Zn, B, Fe, Al and Mn at the A4, A3 and B3 positions over four sites at two-monthly intervals. The solid line joins a sequence collected from the same lot of needles; the broken line joins the results for these needles to those collected higher up the tree in the equivalent position of the next season's growth. Curves showed the same trends at the other positions which are shown in Figure 1 and indicated in the text.

would be unsatisfactory. Sampling time, branch position and needle age should all be standardized.

Morrison (1972) has given results for Fe and Mn on *Pinus banksiana* Lamb. in a 30- to 33-year-old natural stand in Ontario. Morrison considered that such results could not be extrapolated from one species to another with any degree of certainty. Using N as an example, the following generally had concentrations higher in the upper part of the crown and diminishing down the crown: *P. sylvestris* L., *P. nigra* var. *calabrica* Schneid. and *P. pinaster* Ait. However, in *P. radiata* D. Don, *P. banksiana* and *P. resinosa* Ait. crown position had no effect on foliar N. Highest N concentrations were found in the lower part of the crown in *P. taeda* L. (see Morrison 1972). According to Raupach (1967), in *P. radiata*, N and P followed one another as was the case in the data used for Figure 1, thus placing *P. radiata* with the set containing *P. sylvestris*.

Morrison also found that P and Ca followed the distribution of Figure 1. Iron decreased and manganese increased with decreasing height in the crown; both trace elements increased in concentration with increasing needle age.

If foliar analyses are to be used for stand monitoring or element-deficiency prediction, the material sampled should be that giving the best relationship with growth. Leyton & Armson (1955) concluded that for *P. sylvestris* 'current needles from as near the terminal shoot as possible provide the best material for chemical analysis'.

Madgwick (1964), however, suggested for *P. resinosa* (Red Pine) that, because the greatest differences between well grown and poorly grown trees occurred at 'the base of the canopy', samples for foliar K, P and (to a lesser extent) N should be collected from that position rather than from near tree apices.

Tree-to-tree variability and the variability of foliage are the main causes of deviations among foliar results for the major elements. Methods of sample handling, drying and preparation for analysis, and the methods of analysis themselves are minor contributors for the major elements but could be larger for the trace elements. It is necessary to take precautions such as the processing of repeat samples in consecutive runs of analyses and the analysis of standard bulk samples from time to time.

RANGES OF TRACE ELEMENTS IN *P. RADIATA* IN AUSTRALIA

Distributions of trace elements have been determined from unpublished data at three positions on trees aged from 3 to

19 years; more than 200 sites from all mainland states except Queensland were sampled. Foliage was collected May-June from the following positions, using the previous coding: (A4) the tip 0-15 cm of the lowest whorl with only current year's needles; (A3) a similar length of tip from a lateral in the whorl below, i.e. the top whorl with two ages of needles; and (B3) a sample of the previous season's growth from this whorl. Six trees were sampled at each site in the one year.

The distributions of the Cu concentrations in the foliage at the three positions are shown in Figure 3. Figure 4 gives the distributions for other trace elements at position A4.

These figures set trace element problems in Australian forestry in perspective. Some levels of suspected deficiency of trace elements have arisen through empirical assessment of foliar data and, although these do not disagree with the 'Low' levels of Table 1 and are commonly used as a diagnostic criterion, they must be regarded with caution. If at fault, these levels tend towards high values for deficiency. The levels for the trace elements at position A4 in *P. radiata* are (ppm): Mn (30), Zn (10), B (12) and Cu (2). These values are indicated by arrows in Figures 3 and 4. It should not be supposed that the deficient levels, if in fact they limit growth, do so independently, as trace element deficiencies in *Pinus* frequently occur together.

RESULTS FROM A PLANTATION OF *P. RADIATA* UNDER TRACE ELEMENT STRESS

Results from one particular area (at Clover Hill near Mt. Gambier), which had been burnt in the first rotation and now

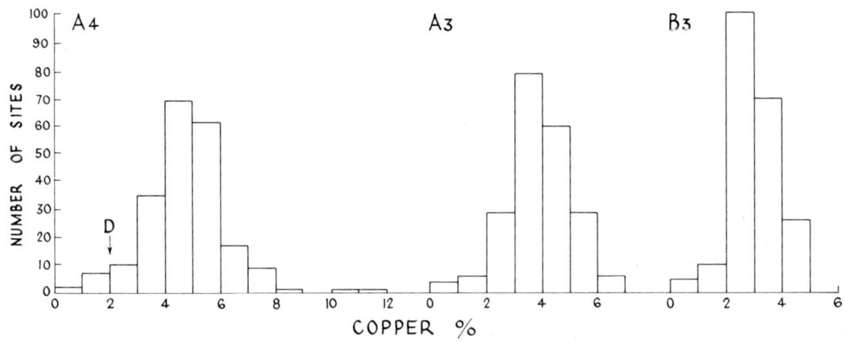

Fig. 3. Distribution of Cu in Australian P. radiata *at positions A4, A3 and B3 (see text).* D *is an approximate deficiency level.*

contained second-rotation trees, are worth further consideration. Many of these trees developed symptoms of trace element stress. A preliminary account has described the area (Raupach, de Vries & Ruiter 1972) and has explained the sampling of trees in pairs, one tree being apparently healthy and the other unhealthy. The soils varied from shallow to deep sands underlain by clayey subsoils over limestone; they included solonetzic soils and terra rossas. The pH ranged from 5 to 8. Some isolated portions were affected by excessive salt. The symptoms most related to trace element stress were distorted branches (classified into moderate and severe at the time of sampling, in winter) and pronounced yellow-tipping of the needles. Some sections had closed canopy, but much of the area was park-like, with tracts of moss, *Scirpus*, grass and bracken between the trees. Sampling was dictated somewhat by the foliage which could be gathered from the affected trees and from their relatively healthy partners less

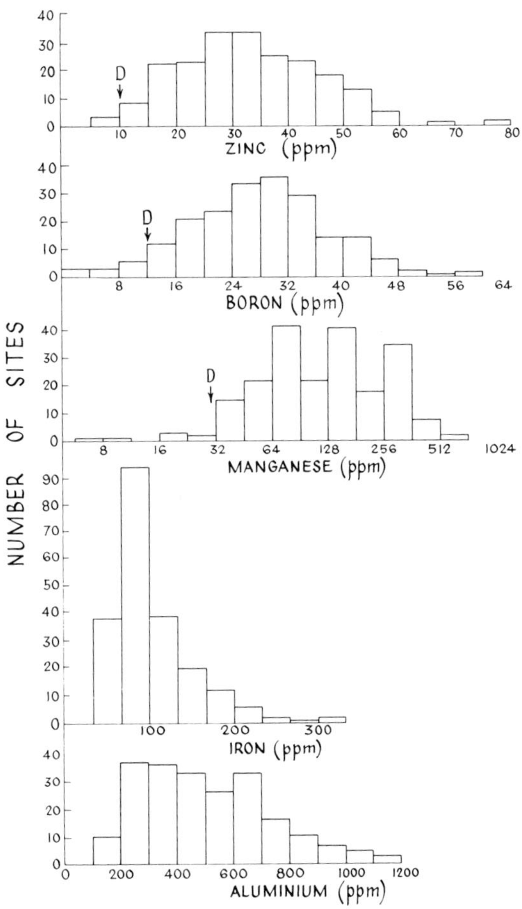

Fig. 4. Distribution of Zn, B, Mn, Fe and Al in Australian P. radiata *at position A4 (see text). D is an approximate deficiency level.*

than 20 m away. Collection was at three positions in each tree: (A4) foliage from the terminal 20 cm of four laterals from the topmost whorl; (2) a sample of the older foliage on these four laterals; and (B3) the 1- to 2-year-old needles from the first laterals down from the top of the tree, carrying two ages of needles. Pairs of trees which appeared healthy were also sampled to act as controls. Further atypical trees were observed in surrounding plantations.

The distortion of leaders and branches resembled that described by van Goor & Henkens (1966) in Douglas fir under Cu stress and by Ruiter (1969) in establishment trials of *P. radiata* in the Mt. Gambier area, where the trees had low Cu levels in the foliage.

The trees without distortion but showing the other symptoms had levels of foliar Cu, Zn and Mn similar to the control trees, as shown in Table 2. Severe distortion gave significant differences between the pairs for Cu and Mn, but not for Zn. Moderate distortion gave significant differences for Cu. The results, summarized in Table 3, were similar whether or not other symptoms were present with the distortion.

TABLE 2

A comparison of Cu, Zn and Mn concentrations (ppm dry weight of foliage) between means of 50 control trees and 36 pairs of healthy trees and distortion-free but unhealthy trees. No differences were significant.

Element	Sample position	Trees		
		Control	Healthy	Unhealthy
Copper	A4	3.07	3.35	3.68
	2	2.68	3.10	3.22
	B3	2.21	2.40	2.28
Zinc	A4	19.2	21.2	26.0
	2	13.6	15.0	16.4
	B3	8.9	9.3	8.4
Manganese	A4	66.8	62.9	61.9
	2	49.8	45.2	45.0
	B3	44.7	33.0	36.3

Histograms of the concentrations of the trace elements among the distorted and/or yellow-tipped trees are contrasted with the whole population of trees sampled in Figure 5. As expected from Table 3, Cu and Mn are distributed towards the low side of the histograms representing all the trees in the

TABLE 3

A comparison of Cu, Zn and Mn concentrations (ppm dry weight of foliage) between means of pairs of trees with one member healthy and the other either moderately (35 pairs) or severely (26 pairs) distorted

Element	Sample position	Distortion			
		Moderate		Severe	
		Healthy	Unhealthy	Healthy	Unhealthy
Copper	A4	2.78	1.85***	2.66	1.47***
	2	2.50	1.75**	2.25	1.29***
	B3	1.95	1.42***	2.18	1.20***
Zinc	A4	18.9	23.0*	19.0	18.9(ns)
	2	14.6	15.9(ns)	14.6	13.6(ns)
	B3	7.5	7.3(ns)	7.6	6.9(ns)
Manganese	A4	52.4	43.5(ns)	59.7	31.5***
	2	34.8	28.3(ns)	36.5	22.0*
	B3	24.9	21.6(ns)	28.6	20.1(ns)

Significance levels: $*P<0.05$, $**P<0.01$, $***P<0.001$, (ns) = not significant

presence of deficiency symptoms. The distributions are in marked contrast to those of Figures 3 and 4, which are for more normal situations. The results for the affected trees are in agreement with Table 1. However, the spread about the 'critical' levels appears somewhat broad. Many factors could

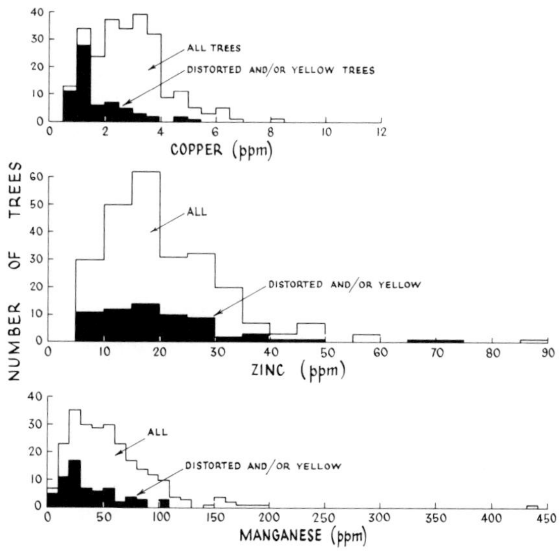

Fig. 5. Histograms of the populations of Cu, Zn and Mn concentrations in position A4 of all the trees sampled at clover Hill and of trees with the symptoms of distortion and/or yellowing of needles.

cause this, including other limiting nutritional effects as well as the selection (survival) of favourable genotypes.

MULTIVARIATE ANALYSIS AND FOLIAR ASSESSMENT

Of many examples in the literature on the application of multivariate techniques to foliar work, two will be described.

Discriminant analyses have been used by White & Mead (1971) to distinguish between trees of Slash Pine with "green" and "yellow" foliage. They analysed current near-terminal winter foliage for N, P, K, Ca, Mn, Fe, Al and Zn. A stepwise procedure was used so that, at each step, the variable producing the greatest improvement in the discriminating ability of the function was introduced. Nitrogen and zinc were the only useful discriminators, the resulting expression giving an overall misclassification of about 10%. Published "critical" levels, however, suggested that Cu and P were deficient in most of the trees. In acknowledging the well-known fact that the significance of useful discriminators does not necessarily imply a cause-and-effect relationship, the authors pointed out that the significance may be useful in formulating hypotheses about winter yellowing.

Mader & Howarth (1970) used multiple regression analysis in an investigation on Red Pine. Previous results from dominant and co-dominant 30- to 40-year-old trees had indicated a significant correlation of nutrients with growth characteristics of the stand across a range of plots (Ca with height growth and K with basal-area growth). Since the foliar nutrient levels and relationships were variable from one plot to another, the number of trees studied in each plot was enlarged and the plots analysed separately. Foliage was collected from one lateral branch at the topmost whorl of each of 35 sample trees in each plot in late autumn and early winter. Using three foliar nutrient measures (per cent, mg/100 needles and mg/branch), a stepwise multiple regression analysis was carried out of Ca, Mg, P, K, Mn and N on 11 dependent-growth variables. No variable was eliminated, since the results of stepwise regressions are dependent on the order of the inclusion of the independent variables and because of the well-known difficulties with highly interrelated independent variables. Of the examples given, regressions of nutrient concentrations on total height had values of the multiple correlation coefficient R^2 ranging from 0.149 to 0.579 over 10 plots. As dependent variables, both the five-year basal-area growth and the height were about equally well related to nutrient concentrations. However, nutrient contents, expressed as mg/branch, had a higher range of R^2, from 0.378 to 0.784. Influential nutrients varied from plot to plot, but Ca and Mg were important in more regressions than other nutrients. Manganese was influential in several regressions. The authors concluded that further study of this method as a diagnostic for the nutritional status of different sites was justified.

CORRECTION OF TRACE ELEMENT DISORDERS IN FORESTS

Where there is a dramatic response to a deficiency such as Zn, the missing element may be added in almost any form. Aerial spraying of zinc sulphate seems to be as economic and satisfactory as other methods which may be contemplated for one element. The quantity required is small since the recycling system of the tree and greater root exploration of underlying horizons (Tiller 1958) seem able to sustain the supply after an initial treatment.

For other trace elements such as B, deficiency may sometimes be corrected by changes in managerial practices (i.e. removal of grass) although responses have been obtained to borax sprays or to boronated superphosphate. On the other hand, it has been observed in our work that high-demand crops, together with the practice of burning, have induced deficiency of major nutrients. A similar situation could occur for trace elements, but it is perhaps less likely. Nevertheless trace elements should not be neglected in long-term studies on nutrient turnover.

Multiple trace element deficiencies, such as those at Clover Hill, may be corrected by a spray, either of a mixture of salts or a mixture of chelates, or by ground application, although soil conditions may reduce the effectiveness of the latter.

Although many forestry experiments of varying complexity of design have used trace element applications on *Pinus*, there has not been adequate reporting of much of this exploratory work. The results have frequently been confused by major and trace element interactions whose origins either in the plant or in the soil have been little understood. One of the ways by which indications may be gained as to these effects is by multiple regression analysis of foliage which has been sampled with care to minimize variations within and between trees and errors between sites. Whatever leads are obtained by this method, or by the extrapolation of results from field experiments, may give rise to work on the growth mechanisms in *Pinus*, with a special emphasis towards trace element-limiting situations. It scarcely seems justified to embark on these studies until a more complete appraisal of Australian softwood forests has been made using foliar techniques.

ACKNOWLEDGEMENTS

Thanks are expressed to Messrs. A.R.P. Clarke, M.P.C. de Vries, J.T. Hutton and J.H. Ruiter for assistance in analysis and sampling.

REFERENCES

ANONYMOUS (1971) - Biennial Report, Forests Department, Western Australia

BAULE, H. & FRICKER, C. (1970) - "The Fertilizer Treatment of Forest Trees". BLV : Munich

BEVEGE, D.I. (1973) - "Nutrition of Tropical Conifers". Queensland Department of Forestry : Brisbane

BEVEGE, D.I. & RICHARDS, B.N. (1972) - *Aust. Forest Tree Nutr. Conf., 1971* : 349 (Forestry & Timber Bureau : Canberra)

BEVEGE, D.I. & SIMPSON, J.A. (1973) - "Statement of Current Activities in Forest Nutrition". Queensland Department of Forestry : Brisbane

FORTESCUE, J.A.C. & MARTEN, G.C. (1970) - *In* "Analysis of Temperate Forest Ecosystems". Chapman and Hall : London

HUMPHREYS, F.R. & KELLY, J. (1962) - *Aust. For. 26* : 77

HUMPHREYS, F.R., TURNER, J. & WATT, A.J. (1972) - *Aust. Forest Tree Nutr. Conf., 1971* : 258 (Forestry & Timber Bureau : Canberra)

KEAY, J. (1964) - *J. Roy. Scottish For. Soc. 18* : 22

LEAF, A.L. (1973) - *In* "Soil Testing and Plant Analysis". Soil Science Society of America, Inc. : Madison, Wisconsin

LEYTON, L. & ARMSON, K.A. (1955) - *Forest Sci. 1* : 210

MADER, D.L. & HOWARTH, J.A. (1970) - *In* "Tree Growth and Forest Soils". Oregon State University Press : Corvallis

MADGWICK, H.A.I. (1964) - *Forestry 37* : 87

MORRISON, I.K. (1972) - *Can. J. Forest Res. 2* : 89

RAUPACH, M. (1967) - *Adv. Agron. 19* : 307

RAUPACH, M., BOARDMAN, R. & CLARKE, A.R.P. (1969) - CSIRO Aust. Soil Publ. No. 26

RAUPACH, M., CLARKE, A.R.P. & DE VRIES, M.P.C. (1972) - *Aust. Forest Tree Nutr. Conf., 1971* : 326 (Forestry & Timber Bureau : Canberra)

RAUPACH, M., DE VRIES, M.P.C. & RUITER, J.H. (1972) - *Aust.*

Forest Tree Nutr. Conf., 1971 : 344 (Forestry & Timber Bureau : Canberra)
RUITER, J.H. (1969) - *Pl. Soil 31* : 197
SNOWDON, P. (1972) - *Aust. Forest Tree Nutr. Conf., 1971* : 191 (Forestry & Timber Bureau : Canberra)
STOATE, T.N. (1950) - Forestry & Timber Bureau, Canberra, Bull. 30
STONE, E.L. (1968) - *In* "Forest Fertilization : Theory and Practice". Tennessee Valley Authority : Muscle Shoals, Ala.
TAMM, C.O. (1964) - *Intern. Rev. Forestry Res. 1* : 115
TILLER, K.G. (1958) - *J. Soil Sci. 9* : 225
VAN GOOR, C.P. & HENKENS, C.H. (1966) - *Ned. Boschb Tijdschr. 38* : 108
WHITE, E.H. & MEAD, D.J. (1971) - *Forest Sci. 17* : 425
YOUNG, H.E., CARPENTER, P.N. & ALTENBERGER, R.A. (1965) - Maine Agr. Exp. Sta. Tech. Bull. 20

THE DIAGNOSIS OF TRACE ELEMENT DEFICIENCIES IN THE GRAZING RUMINANT

A.R. Egan

Department of Agronomy
Waite Agricultural Research Institute
The University of Adelaide
Glen Osmond, South Australia 5064

INTRODUCTION

The patterns of appearance of signs and the clinically detectable metabolic changes which occur in ruminants where frank deficiencies of individual trace elements are imposed experimentally are well described in classical studies and have been detailed by Underwood (1971). Such physiological changes are discussed elsewhere in these proceedings by Underwood, by Gawthorne & Smith, and by Godwin. They may be signs which are clearly and uniquely indicative of a specific deficiency when observed in the field (e.g. the ataxia in lambs with Cu deficiency), or they may be non-specific changes which are virtually undetectable in the grazing animal or, if detectable, give no useful indication of the nature of the disorder (e.g. inappetence in Co deficiency). Where observed even at low frequency in the flock or herd, those more clearly indicative signs permit recommendation of supplementation with the indicated trace element by the practising veterinarian or animal health and nutrition advisor. However, the extent to which the occurrence of frank signs alone is effective in detecting economically important nutritional inadequacies and imbalances must be critically evaluated. In case histories of trace element-deficiency diseases in Australia and elsewhere, usually the existence of a specific trace element deficiency has come to attention because of frank signs in one or more flocks. When the nature of the disorder has been tentatively identified and responsiveness to supplements subsequently demonstrated, other flocks in nearby areas or areas of related soil type, where no unequivocal signs have been noted, frequently show a general improvement in production when given supplements of the same element. In delineating the extent of these 'halos' of marginal or subclinical inadequacies, in detecting the more isolated occurrences in areas not readily related to those where frank deficiencies exist, and in detecting trace element inadequacies which limit one specific phase of animal production at the level of supply where no clearly indicative signs are apparent, diagnostic aids are necessary.

THE NATURE OF TRACE ELEMENT DEFICIENCIES IN THE FIELD

It is becoming increasingly clear that the severity, the sequence, and the time course of development of signs or physiological impairments under field conditions are not necessarily like those seen in studies with semi-purified diets. In such studies a single trace element is almost devoid in the diet while all other nutritional requirements are supplied in supposedly adequate amounts and proportions. The characteristic metabolic disorder or the signs of deficiency are produced in periods of from 3 weeks (Zn deficiency in lambs) to 3 months (Mn deficiency in cattle), depending on the relative size of pre-existing stores and on the rate of turnover and endogenous loss of the element from the metabolically active site in cells.

In contrast, under grazing conditions, simple or induced deficiency states are often associated with seasonally variable patterns of trace element intake varying about the 'adequate' level. Here relative adequacy is usually determined by comparison of herbage analyses with dietary standards, in turn determined mainly in pen-feeding experiments. The field deficiency may be complicated by concurrent inadequacies of more than one element, which influences the ease of recognition of signs and possibly the nature and site of the metabolic lesions developed. The influence of other nutrients which bind, antagonize, or compete with, or which have a sparing effect on, the critical trace element may vary with time. Thus responsive conditions may be precipitated by changing proportionalities among the respective trace elements and other dietary factors. These may interact at a number of physiological loci (e.g. Suttle, this volume) so that different tissues or organs are differentially affected. The possibility exists that physiological function in a specific organ or tissue is affected adversely in chronic marginal deficiency or in acute deficiencies coincident with important stages of physiological development of, for example, the young *in utero*, in early life post-partum, or at puberty. These effects may be reversible or irreversible when subsequently a supply of the deficient element is given. If irreversible, detection of a responsive condition is made impossibly difficult without 'whole-of-life' studies. In all these respects, it may well be that reproductive function, with its many critical phases, may be the most vulnerable physiological aspect sensitive to transient inadequacies or imbalances of trace elements, and one which is most difficult to investigate adequately.

In this paper I wish to focus mainly on data from a single field experiment on trace element-responsive conditions in sheep. These data, and other information reviewed, are advanced to direct attention to four aspects of major importance where information is limiting in the identification and treatment of trace element disorders in grazing ruminants.

These aspects are: (1) the applicability of regional indices, soil tests, plant analysis and plant growth responses in planning surveys to delineate areas of likely trace element response in animals; (2) the seasonal patterns of trace element status and the identification of 'most likely' periods and length of periods of simple or induced inadequacy; (3) the extent of variability in status existing within a flock or herd; (4) the present reliability of tissue sampling and clinical tests in delineating situations where potential exists for response surveys.

THE EXISTENCE OF RESPONSIVE CONDITIONS IN ABSENCE OF CLINICAL SIGNS

There is abundant evidence of improved wool production and lamb growth rate in large areas of South Australia and Western Australia where increased amounts of Cu and Co have been supplied through the herbage or by animal supplementation, even though clear deficiency signs had not been recognized (e.g. Lee, Reuter, this volume). The studies of Thornton *et al.* (1969) in Britain and Ireland have demonstrated, in the case of Mo-induced Cu deficiency, that geochemical survey methods (such as stream sediment sampling) can be used to define areas in which unsuspected responsiveness to Cu supplementation occurs in animals without appearance of clear clinical signs. Supplementation studies in the field have also revealed areas where female reproductive performance may be improved by specific trace elements. Thus, in cattle, responses in terms of reduced returns to service have been reported following Mn supplementation (Wilson 1965). On a single property at Moorlands, South Australia, reproductive efficiency among Dorset Horn ewes, measured as numbers of mated ewes giving birth to lambs, was increased by 25-30% when Zn or Zn + Mn supplements were given (Egan 1972).

In pen studies, Underwood & Somers (1969) have shown the Zn requirement for testicular development in young male sheep to be greater than that for rapid body growth. Though not yet carried into field observations on reproductive efficiency, the implications of their studies are that large areas of Australian soils support plants with a Zn content which is less than required for normal testicular development. This possibility may or may not be shown to be of real economic significance, but supplementation studies in the field in this respect seem warranted.

DELINEATING AREAS WHERE TRACE ELEMENT-RESPONSIVE SITUATIONS MAY OCCUR

Regional reconnaissance on a number of bases has been shown to be effective in revealing trace element-responsive situations. Broad predictive bases, such as the association of Se deficiencies and toxicities with particular parent-rock/soil-type/climatic combinations (Allaway 1971), form one starting point. The sampling of stream bed sediments to give indications of the trace element status of rocks and soils in the upstream catchment areas has been used with notable effect. Thus, using this technique, areas of hitherto unsuspected Mo-induced Cu deficiency in sheep and cattle (Thornton *et al.* 1969; Thornton *et al.* 1972), Mn deficiency in cattle (Thornton & Webb 1970) and Co deficiency in sheep (Thornton 1972) have been revealed. These conditions have been verified by tissue analysis and supplementation studies with animals in these areas.

Information on soil type, together with any associated indication of plant growth response or increased mineral content of herbage following trace element application, is a useful basis for identification of possible areas of responsiveness of animals to some trace elements. On soil types where plant growth responses to Cu, Zn and Mn have been observed, but application of these elements in fertilizer has not been practised, animal responses cannot be predicted with certainty. For the grazing ruminant, the likelihood of responses in growth rate, or the appearance of visible signs of frank deficiency for Zn and Mn, is correctly assessed as being very low (Underwood 1971). Even where plant growth responds to Zn or Mn application, the content of these elements in plants is not often below the concentration accepted as adequate for normal animal growth (Hodgson *et al.* 1962). In contrast, the grazing animals' requirement for Cu is not always met even when plant concentrations are sufficiently high to indicate little likelihood of a plant growth response to Cu application. With I, Co and Se, plant concentration can fall far below the range accepted as necessary to satisfy animal needs without any adverse effect on plant growth (Whitehead 1966). Herbage concentrations at which requirements for growth of young sheep and cattle are expected to be met are: Cu, 6 ppm; Mo, 1 ppm; Co, 0.08 ppm; I, 0.6 ppm; Mn, 30 ppm; Zn, 20 ppm; and Se, 0.05 ppm.

However, because of great variation in the availability of the respective trace elements in feed materials, values for 'requirements' for animals expressed simply as a concentration of the element in the diet are particularly open to challenge. Thus herbage analysis can be misleading if interpreted along lines that, because a specific element is present at greater concentrations than the specified 'requirement', no possibility of response to supplementation exists.

Interactions with other nutrient fractions such as protein, lipid, or other elements can affect greatly the availability for digestion, absorption and retention of the respective elements. When pasture content falls below the above-stated concentrations, supplementation studies with animals grazing those pastures are increasingly justifiable where poor productivity and non-specific disorders are consistent with physiological effects of particular trace element inadequacies.

The suggestion is often made that selective grazing or soil ingestion by the animal can provide a different pattern of trace element intake to that indicated by analyses of herbage sampled by hand. Often this argument is presented with an inference that there is consequently an even lower likelihood of inadequate intakes of trace elements in cases of marginal herbage content. However, selective grazing does not necessarily result in ingestion by the animal of a selected herbage mixture providing higher content of an otherwise inadequately supplied element. The bases of dietary selection are not clearly understood. There is recorded evidence that grazing animals tend to select higher protein, more highly digestible fractions from the pasture available (Arnold 1964). In conditions of reduced availability of trace elements from the soil, the younger leaves of growing plants, which constitute such a highly digestible fraction, are often the leaves of lowest content of most trace elements, which are poorly or only slowly mobilized within the plant (Loneragan, this volume).

Again, the possibility of ingestion of additional mineral in soil has been referred to as an unknown factor likely to supplement any deficit in herbage trace element content. Yet, with soil ingestion, not only is more of the potentially limiting trace element taken in but, along with this, possibly larger quantities of interacting elements or substances. The net effect upon overall availability *in vivo* of trace elements has not been assessed for any set of conditions of soil ingestion, though estimates of soil intake range up to 400 g soil/day in sheep grazing heavily stocked pasture at the Waite Institute (E.D. Carter, personal communication). It cannot yet be assumed that soil ingestion will decrease the likelihood of trace element inadequacies where herbage concentration is marginal.

Because of these uncertainties, it is the author's view that, in areas where plant growth and herbage trace element concentration respond to Cu, Zn and Mn application, situations will exist which warrant additional animal studies. Animal supplementation studies with reproductive efficiency, longevity and performance of offspring born to supplemented animals as criteria of response are necessary. In the Moorlands study (Egan 1972), herbage analysis indicated few periods of the year when herbage samples (harvested by hand from quadrats) contained concentrations of Cu, Zn and Mn low enough to be suspect. Samples harvested at a particular time rarely contained less than 6 ppm Cu, 30 ppm Mn or 20 ppm Zn. Subsequent plant growth studies with *Medicago truncatula* cv. Gemalong (A.R. Egan & A.M. Alston, unpublished) showed a growth response to Cu and Zn ($P < 0.001$) with a small additional influence of Mn ($P < 0.05$). The concentration of these elements was also increased by the respective fertilizer applications. Reproductive responses of ewes grazing mixed pastures were recorded with Zn (two consecutive years) and Mn (one year only) but not with Cu supplementation.

ANIMAL TISSUE SAMPLING AND CLINICAL TESTS

Though it is frequently stated [e.g. Andrews (1956), speaking about Co insufficiency] that response to supplementation is the only adequate test of the occurrence of a trace element deficiency, supplementation studies are costly in time and resources if conducted with adequate control and assessment. Since each investigation of a suspected subclinical or marginal deficiency is experimental in nature, control groups of animals must be run in common with treated groups but denied access to the supplement. Injectable Cu and I preparations and Co and Se bullets provide adequate insurance of effective supply of these elements to specific animals, with total exclusion of supply to control animals. As yet no satisfactory method of supply of Zn and Mn is commercially available, and experimental treatments have been imposed by the more time-consuming but less reliable method of drenching. To reduce initial work-load, therefore, supporting information such as animal tissue analysis is needed in order to (1) reduce the range of possibilities and (2) identify the best focal site for experimental work in those areas where, on regional assessment of soil and plant criteria, an animal response may be considered possible. Criteria on which trace element status of animals may be judged are well developed for Cu (Hartmans 1969). For other trace elements, either tissue concentrations are so low or show so little variability that assay as a routine is not yet practical, or standard values or comparative ranges for adequacy and inadequacy are virtually undeveloped. Thus Co and Se status are not yet generally assessed on analytical bases. Assessment of I status requires more specialized techniques than are applicable in rapid survey procedures. Tissue Zn and Mn concentrations which are reliably indicative of inadequacy have not been established because few responsive situations have been found. For these two elements, it is essential that blood and tissue analyses be undertaken for supplemented and unsupplemented animals in field experiments, whether the outcome results in identification of a responsive situation or not. Only in this way can adequate delineation be made. Such assays, previously prohibited by slow and difficult techniques, are now possible on survey scale using atomic absorption spectrophotometry and other new assay techniques.

In 1969-70, in the Moorlands study, sets of 4 Dorset Horn ewes (3-5 years old) from each of the 5 respective treatment groups (control; +Cu; +Zn; +Mn; and Zn + Mn) were slaughtered at approximately 2-monthly intervals at a commercial abbatoirs. The aim was to determine the seasonal pattern of concentrations of Cu, Zn and Mn in blood, liver, kidney, ovaries, uterine wall, spleen, bone and skeletal muscle. For each of these elements, information was sought on the order of variability in concentration in the various tissues and on the order of variability between animals within treatments.

Seasonal patterns of concentration of the respective elements in liver are shown for Cu, Zn and Mn in Figures 1, 2 and 3 respectively, together with concentrations in liver of control sheep. Results indicate minimum values in control animals in November (Cu) and February (Zn, Mn). Data for specific tissues at those times are shown in Tables 1, 2 and 3 for Cu, Zn and Mn respectively.

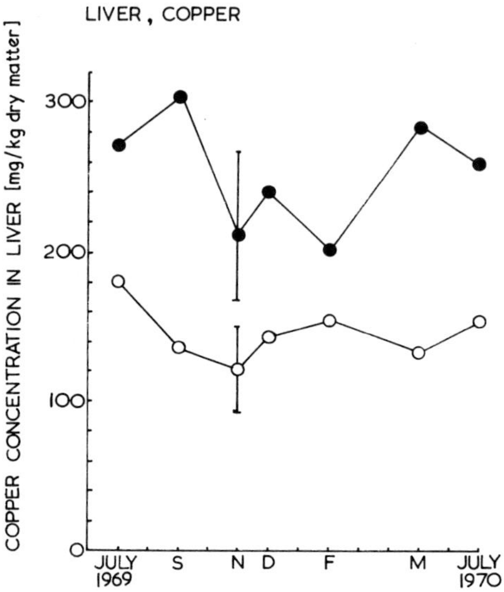

Fig. 1. Seasonal pattern of concentration of Cu in liver from control (o) and Cu-supplemented (●) ewes (mean of 4/treatment).

COPPER

The data in Table 1 indicate significant differences in Cu concentration between control and Cu-treated animals for liver and for whole blood but not for other tissues. Values indicate adequacy in all animals at all times, and no Cu response was observed in the treated animals. Copper status is widely assessed on the basis of blood samples or liver biopsy samples [e.g. Committee on Mineral Nutrition (Netherlands) 1973]. Wide variation in concentration occurs within flocks or herds, though usually the range indicates the probable adequacy or

TABLE 1

Copper concentration in blood and in tissues (dry basis) of sheep (mean of 4/treatment) at time of lowest values for control group (November)

Tissue	Control	+Cu	+Zn
Liver (mg/kg)	120	210*	150
Kidney (mg/kg)	19.5	22.4	15.8
Ovary (mg/kg)	6.8	8.5	7.2
Whole blood (μg/100 ml)	69	90*	75

*Significantly different from control ($P < 0.05$)

inadequacy of Cu samples. An empirical judgment is that where liver Cu concentration for an estimated 25% of the flock or herd falls below 25 mg/kg at any time of the year, but particularly in late spring, Cu supplementation studies are warranted. Blood samples are less reliable and show marked fluctuations between sheep and between consecutive samplings in the same sheep only weeks apart (A.R. Egan, unpublished data).

ZINC

The uniformity and relative constancy of Zn concentration in various tissues of animals subjected to diets varying in Zn content have been emphasized by Underwood (1971) and by Smith & Gawthorne (this volume). However, Table 2 shows significant differences in Zn content of kidney, ovary and whole blood between animals receiving and not receiving supplemental Zn. Plasma concentration did not differ significantly and packed cell volume was not significantly different; difference in the concentration in the erythrocytes was indicated. High variability in liver Zn concentration occurred within the supplemented group. Responses to Zn included a small response in wool growth, improved crimp definition in wool, a significant increase in percentage of mated ewes lambing, and an increased lambing percentage. This experience suggests that whole blood values may prove useful and that there is a need for supplementation studies where values lie around 160 μg Zn /100 ml whole blood. Attention is drawn, however, to the

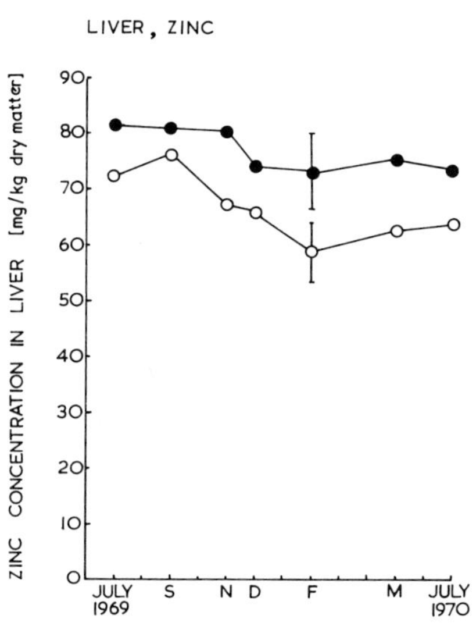

Fig. 2. Seasonal pattern of concentration of Zn in liver from control (○) and Zn-supplemented (●) ewes (mean of 4/treatment).

TABLE 2

Zinc concentration in blood and in tissues (dry basis) of sheep (mean of 4/treatment) at the time of lowest values for control group (February)

Tissue	Control	+Zn	+Mn
Liver (mg/kg)	56	73	58
Kidney (mg/kg)	90	120*	82
Ovary (mg/kg)	70	85*	74
Plasma (μg/100 ml)	64	78	75
Whole blood (μg/100 ml)	158	220*	178
Wool (mg/kg)	198	264	182

*Significantly different from control ($P < 0.05$)

fact that low blood Zn levels are observed after injury and in many disease states.

Plasma values are less valuable. Grace (1972) has reported plasma concentrations for grazing sheep in New Zealand in the order of 68 μg Zn/100 ml plasma. This lies in the low end of the range of plasma values found in the Moorlands experiment and is markedly lower than the value of about 100 μg Zn/100 ml plasma commonly accepted as 'normal'. However, in Grace's studies, none of the farms from which samples were collected were experiencing problems in terms of reproduction or wool production and were observing good rates of weight gain in young sheep (N. Grace, personal communication). Testicular development and concentration of Zn in semen may prove another useful index when adequate survey work has been done.

MANGANESE

There is no evidence indicating a specific metabolic step in any tissue which becomes limiting in Mn insufficiency (Smith & Gawthorne, this volume). Furthermore, Mn is not specifically concentrated in any one tissue or organ. However, cattle apparently hold some reserve of Mn since, in pen trials, signs of Mn deficiency develop only after 2-4 months. As yet no practical criterion exists for a tissue assay to detect Mn inadequacy, and indeed Mn deficiency is rarely suspected in grazing animals (Underwood 1971). However, in the Moorlands experiment, liver analysis indicates a significant difference between control and supplemented animals, and seasonal changes are apparent. Concentrations in the liver were generally lower than those reported in the literature (Fore &

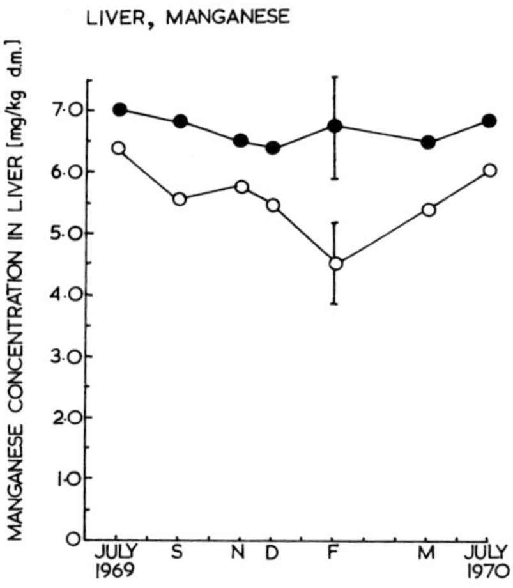

Fig. 3. Seasonal pattern of concentration of Mn in liver from control (o) and Mn-supplemented (●) ewes (mean of 4/treatment).

Morton 1952; Underwood 1971) and showed a significant elevation in Mn-treated sheep. Concentration in the ovary was also affected by supplementation.

More data are clearly essential, but on these experiences, where a significant reduction of returns to service occurred in one year, it is suggested that where liver Mn levels of less than 6 mg/kg dry basis are observed, Mn supplementation studies are warranted. Conception rate of ewes and bone abnormalities in newborn lambs are the most likely bases for assessment of response. Wool samples do not provide a useful basis for assessment of Mn status in sheep, any more than do hair samples in cattle (Hartmans 1968).

TABLE 3

Manganese concentration in blood and in tissues (dry basis) of sheep (mean of 4/treatment) at time of lowest values for control group (February)

Tissue	Control	+Mn	+Zn
Liver (mg/kg)	5.80	6.70*	6.04
Kidney (mg/kg)	3.12	4.23	3.28
Ovary (mg/kg)	1.38	2.04**	1.62
Whole blood (µg/100 ml)	3.2	4.2	2.9
Wool (mg/kg)	0.58	0.60	0.52

Significantly different from control (*$P < 0.05$; **$P < 0.01$)

COBALT

Tissue Co levels are so low and show so little variability in the deficient state that there is little prospect of a practical assay for definition of Co status by blood or tissue Co analysis. Liver vitamin B_{12} content may prove a useful assay for limited numbers of samples in experimental situations, but it is unlikely to become a practical standard test. Rate of clearance of injected propionate (Smith & Marston 1971) and rate of excretion of methylmalonic acid in urine (Smith *et al.* 1969) may provide diagnostic criteria useful in some circumstances. At present, response to supplementation is the only effective test of Co inadequacy.

SELENIUM

Situations in which Se inadequacy may be expected and the physiological abnormalities associated with such inadequacy are discussed fully in the papers by Johnson, Godwin, Reuter and Judson & Obst in this volume. No individual test or tissue analysis has yet been accepted in practice to identify situations of likely Se responsiveness. Whole blood Se values and liver and kidney Se contents show variations which, with increasing availability of data, may form future bases for reliable diagnosis of Se inadequacy. As with Co, supplementation studies remain the only effective test of Se-responsive conditions where reproduction and growth rate of lambs are poor.

IODINE

Since either low I intake or a high intake of goitrogenic substances can result in poor I status in the grazing ruminant, supplementation with I may reveal responses where I content of pastures is apparently adequate. Thyroid examination and blood thyroid hormone titres are specialist criteria on which I insufficiency may be detected.

CONCLUSION

As more tissue analysis results become available from field supplementation studies, it will be possible to determine from the range of values observed the number of animals which should be sampled and the period of the year most suitable for sampling. Generally spring and early summer in the Mediterranean environment and mid-summer in the tropical environment are the periods in which trace element status in grazing animals is most likely to commence falling. Supplementation trials will be the final arbiter of existence of trace element-responsive conditions. Wherever supplementation trials are undertaken, it is important that these be coordinated in a manner analogous to the programme developed in the Netherlands [Committee on Mineral Nutrition (Netherlands) 1973]. A strategic programme of blood and tissue sampling should be undertaken. Such a programme will provide needed background information on the variation which occurs between animals in the same flock and in different flocks where responses have or have not been observed. Only then can recommendation for supplementation be reliably based on tissue analysis as a means of indicating trace element status in grazing ruminants for elements other than Cu.

ACKNOWLEDGEMENTS

The author acknowledges with thanks the cooperation of Mr. M. Piggott, Illoura Stud, Moorlands, South Australia; the Manager and staff of the Murray Bridge Abbatoirs; and Mr. J.P. Hantrais, Mr. J. Scarvellis and Miss J. Barker in studies referred to in this paper.

REFERENCES

ALLAWAY, W.H. (1971) - *In* "Selenium in Nutrition". Report of the Sub-Committee on Selenium, Committee on Animal Nutrition, Agriculture Board, National Research Council. National Academy of Sciences : New York

ANDREWS, E.D. (1956) - *N.Z. Jl Agric. 92* : 239

ARNOLD, G. (1964) - *In* "Grazing in Terrestrial and Marine Environments" (D. Crisp, ed.). Blackwell : Oxford

COMMITTEE ON MINERAL NUTRITION (NETHERLANDS) (1973) - "Tracing and Treating Mineral Disorders in Dairy Cattle". Centre for Agricultural Publishing and Documentation : Wageningen

EGAN, A.R. (1972) - *Aust. J. exp. Agric. Anim. Husb. 12* : 131

FORE, H. & MORTON, R.A. (1952) - *Biochem. J. 51* : 594

GRACE, N. (1972) - *N.Z. Jl agric. Res. 15* : 284

HARTMANS, J. (1968) - *Meded. Inst. biol. scheik. Onderz. LandbGewass. 352* : 113

HARTMANS, J. (1969) - *Agric. Digest 18* : 42

HODGSON, J.F., LEACH, R.M. Jnr. & ALLAWAY, W.H. (1962) - *J. agric. Fd Chem. 10* : 171

SMITH, R.M. & MARSTON, H.R. (1971) - *Br. J. Nutr. 26* : 41

SMITH, R.M., OSBORNE-WHITE, W.S. & RUSSELL, G.R. (1969) - *Biochem. J. 112* : 703

THORNTON, I. (1972) - *Billiton-Venture, Christmas 1972,* p. 22

THORNTON, I., KERSHAW, G.F. & DAVIES, M.K. (1972) - *J. agric. Sci., Camb. 78* : 151

THORNTON, I., MOON, R.N.B. & WEBB, J.S. (1969) - *Nature, Lond. 221* : 457

UNDERWOOD, E.J. (1971) - "Trace Elements in Human and Animal Nutrition", 3rd Edn. Academic Press : New York

UNDERWOOD, E.J. & SOMERS, M. (1969) - *Aust. J. agric. Res. 20* : 889

WHITEHEAD, D.C. (1966) - "Nutrient Minerals in Grassland Herbage". Commonwealth Agriculture Bureau Mimeographed Publication No. 1/1966 CAB, Farnham Royal, Bucks.

WILSON, J.G. (1965) - *Vet. Rec. 77* : 489

DIAGNOSIS AND TREATMENT OF SELENIUM INADEQUACIES IN THE GRAZING RUMINANT

G.J. Judson* and J.M. Obst†

*Institute of Medical and Veterinary Science
†South Australian Department of Agriculture
Adelaide, South Australia 5000

INTRODUCTION

Selenium is a nutritionally essential element, and disorders resulting from Se inadequacy have been recognized in practically all the major livestock-producing countries of the world. The incidence of these disorders varies widely in different countries and regions, but is increasing. Areas of Se deficiency are now much more extensive than those of Se toxicity (Wolf *et al.* 1963; Oldfield *et al.* 1971; Jenkins & Hidiroglou 1972).

The most widely recognized and clearly defined Se-responsive disorder is nutritional muscular dystrophy (NMD), also referred to as white muscle or stiff lamb disease. It is a degenerative disease of the skeletal and cardiac muscle, being most common in suckled lambs and calves at about 1-3 months of age (see reviews by Hartley & Grant 1961; Gardiner 1962; Oksanen 1967). In New Zealand, a congenital form of the disease is also found in sheep (Hartley & Grant 1961). Nutritional muscular dystrophy is relatively rare in mature animals but has been observed in hoggets in New Zealand (Dodd 1954; Hartley & Dodd 1957) and Scotland (Graham *et al.* 1956) and in cattle in Canada (Doig 1970) and the Scandinavian countries (Andersson 1960; Oksanen 1965).

New Zealand workers have described a variety of other Se-responsive disorders in livestock, including infertility and periodontal disease in ewes and unthriftiness and scouring in sheep and cattle (Andrews *et al.* 1968). Some of these disorders have been observed in other countries. Growth responses to Se supplementation have been reported for sheep in

Australia (Walker *et al.* 1961; Gardiner *et al.* 1962; Setchell *et al.* 1962; Skerman 1962; Pulsford *et al.* 1966), Scotland (Blaxter 1963) and the U.S.A. (Oldfield *et al.* 1960), and for cattle in the U.S.A. (Burroughs *et al.* 1963). Increased wool production with added Se has also been noted by Slen *et al.* (1961) in Canada, by Quarterman *et al.* (1966) in Scotland and by Gabbedy (1971) in Western Australia. The beneficial effects of Se for correcting ewe infertility or increasing fecundity have been supported, in part, by studies in Scotland (Mudd & Mackie 1973) and in South Australia on Kangaroo Island (Godwin *et al.* 1970). Recently, Se therapy has also been implicated as a factor which reduces the incidence of retained placentae in dairy cows (Trinder *et al.* 1969; Trinder *et al.* 1973).

The concern of livestock producers is to ensure that each animal obtains sufficient Se to prevent lowered production resulting from clinical or sub-clinical Se deficiency. Our intention in this paper is therefore to consider the Se requirements of ruminants and the methods for diagnosing and treating Se inadequacy.

SELENIUM INADEQUACY ON A REGIONAL BASIS

A comprehensive study of the concentration of Se in crops in the U.S.A. has been reported in the form of a map by Kubota *et al.* (1967). Broad areas where Se concentration in plants is generally below 0.6 μmol/kg have been delineated, and these correspond approximately with areas in which Se-responsive disorders in livestock are common (Muth & Allaway 1963; Allaway & Hodgson 1964). In New Zealand it has been shown that Se concentrations in soils, pastures and blood of grazing livestock correlate closely with each other and with areas within which Se-responsive disorders would or would not be expected (Andrews *et al.* 1968). These studies suggest that the disorders may be encountered when the Se concentrations in the soil, pasture, liver, kidney cortex and blood of sheep were less than about 5.7, 0.4, 4.0, 40 μmol/kg of dry material and 0.13 μmol/ℓ respectively. Similarly, in Sweden, Lindberg & Lannek (1970) reported that NMD in livestock was associated with Se concentrations of less than 0.32 μmol/kg of dry feed and 0.25 μmol/ℓ of blood. Where Se-responsive disorders have been observed in livestock in the eastern provinces of Canada, the concentrations of Se in skeletal muscle and kidney cortex of calves and lambs were considerably less than those found in tissues of livestock from the western provinces, being about 3.5 and 30 μmol/kg of dry tissue respectively (Hoffman *et al.* 1973).

In a survey of Se in pastures in the south-western areas of Western Australia, Gardiner & Gorman (1963) found that areas in which Se concentrations were usually below 0.6 μmol/kg corresponded well with areas where Se-responsive disorders in sheep have been observed (Gardiner 1962; Gardiner *et al.* 1962). These areas were, in general, in the higher rainfall districts (greater than 500 mm/year) and associated with soils of sands or sandy gravels overlying lateritic podsols (Gardiner 1969; Gabbedy 1971).

Apart from these studies by Gardiner and associates, little attempt has been made to examine Se inadequacy for livestock in other agriculturally productive areas of Australia, despite the sporadic occurrence of Se-responsive disorders in most states (Walker *et al.* 1961; Christie 1962; Setchell 1962; Skerman 1962; Smith 1962). In South Australia, Se-responsive disorders have been observed in sheep grazing near Keith and Yahl in the South-East, in the Adelaide Hills and on Kangaroo Island (Smith 1962; Pulsford *et al.* 1966; Godwin 1968; Godwin *et al.* 1970; Sheriff *et al.* 1973). The incidence of Se-responsive disorders on Kangaroo Island has been restricted to the central and western portions of the Island (R.J. Clare, personal communication). These areas are also indicated from a survey of blood Se concentrations in grazing stock (Fig. 1). The low concentrations of Se in blood are associated with the laterite soils, suggesting that the soil is either inherently low in Se or the Se is of low availability to plants.

The surveys mentioned above are useful primarily in the detection of regional patterns or pronounced differences in Se adequacy for grazing livestock. Surveys based solely on Se content of soils or pastures may be misleading. Selenium is present in soils in different chemical forms that have widely varied availability to plants (Lakin 1961) and, of the available Se, uptake by the plant may be influenced by other elements such as S. With the sampling of pasture for Se content, there is the problem of relating the material analysed to that actually eaten by animals with selective grazing habits. Despite these limitations, these surveys are frequently a practical first approach to delineating areas likely to be Se inadequate for grazing livestock.

FACTORS AFFECTING THE DEVELOPMENT OF SELENIUM-RESPONSIVE DISORDERS

Andrews *et al.* (1968), in reviewing the New Zealand evidence for Se requirements for grazing sheep, concluded that the critical level for dietary Se, below which deficiency

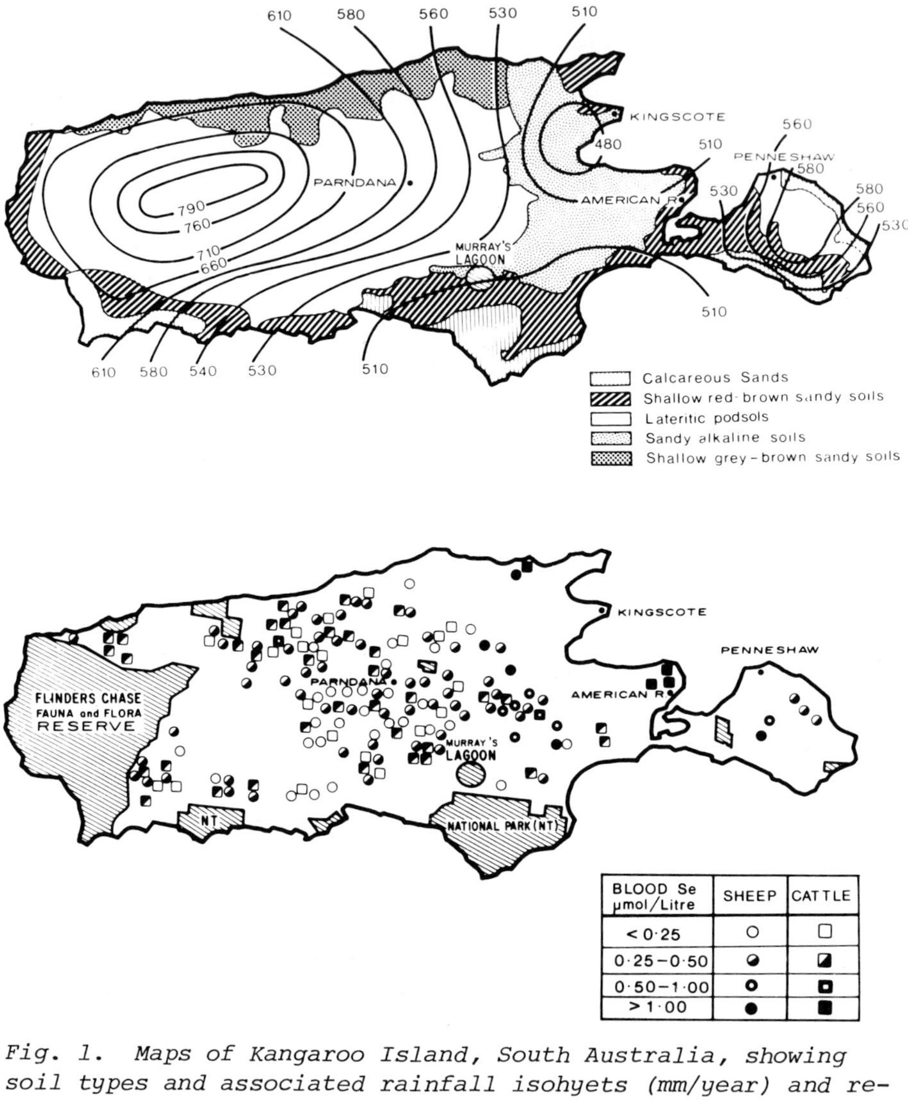

Fig. 1. Maps of Kangaroo Island, South Australia, showing soil types and associated rainfall isohyets (mm/year) and regional pattern of Se concentrations in blood of grazing livestock. At least 5 animals were bled from each property and the position of the blood Se symbol approximates the location of the property. Bloods taken at frequent intervals over a period of about 2 years from selected properties on the laterite soils showed little variation in Se content.

symptoms are observed, was about 0.35 μmol/kg of dry matter and that symptoms did not occur on pastures containing more than about 0.4 μmol Se/kg. These criteria generally agree with values for feeds producing Se-responsive disorders in sheep and cattle in other countries (see Tables 1 and 2), although in a few studies higher concentrations of Se were observed. A minimum requirement of about 0.8-1.1 μmol Se/kg of dry feed for the prevention of NMD in livestock is indicated from the American studies (Burton *et al.* 1962; Oldfield *et al.* 1963). Allaway & Hodgson (1964) have suggested that a dietary intake of about 1.3 μmol Se/kg of dry feed provides a satisfactory margin of safety against any dietary variables or environmental stresses likely to be encountered by the grazing ruminant.

The minimum requirement for Se may vary with the form of the Se ingested and the nature of the rest of the diet, particularly its content of α-tocopherol. It is well documented that tocopherol administration reduces the incidence of NMD in young ruminants (Blaxter 1962; Oksanen 1967,1973; Jenkins & Hidiroglou 1972), although it apparently will not prevent it if dietary Se concentrations are very low (Drake *et al.* 1960). The variable responses obtained to tocopherol may be also due, in part, to the small doses used and the limited placental transfer of tocopherol when given to the pregnant animal (Muth *et al.* 1958; Paulson *et al.* 1968). Selenium is not only more efficiently transferred across the placenta and secreted in the milk than tocopherol but it is also more actively retained by the tissues of the young (see Oldfield *et al.* 1971). The precipitating factor in NMD in hoggets in New Zealand is thought to be tocopherol deficiency, although concurrent Se deficiency is also involved (Cousins & Cairney 1961). Tocopherol, or the synthetic antioxidant ethoxyquin, does not appear to be effective against the unthriftiness or infertility syndromes in sheep (Hartley 1967).

A high intake of unsaturated fats (Hidiroglou *et al.* 1973), a low intake of sulphur-amino acids or the presence of some unidentified factor in the diet (Gardner & Hogue 1967) may increase the need for Se or tocopherol, particularly in the suckled animal. The effects of these substrates are minimized in the animal with a functioning rumen on normal diets, since the microflora will hydrogenate ingested unsaturated fats and moderate the availability of sulphur-amino acids. However, an active microbial population in the rumen can destroy ingested tocopherols (Alderson *et al.* 1971) and convert pasture Se to less readily available forms (Cousins & Cairney 1961; Peterson & Spedding 1963). Besides the above factors, some evidence has been reported that high sulphate intakes may increase the Se requirement of sheep (Schubert *et al.* 1961; Hintz & Hogue 1964). When pregnant ewes were fed Se-deficient diets supplemented with sulphate, the number of offspring with cardiac lesions increased (Whanger *et al.* 1969) but this effect of sulphate was prevented by additions of Se to the ewe ration (Whanger *et al.* 1970). In beef herds, variations in the order of 22 to 47 mmol S/kg of dystrophogenic hay did not have any effect on the incidence of NMD in the calves (Hidiroglou *et al.* 1965).

Gardiner (1962,1969) considers that certain stress factors operating in association with pasture Se concentrations of less than 0.6 μmol/kg of dry matter are fundamentally responsible for the appearance of clinical manifestations of Se deficiency in grazing weaners in Western Australia. Apart from the physical stress of mustering and driving, Gardiner suspected chemical stresses such as stock feed containing cod liver oil, dosing with phenothiazine or organic phosphate or the administration of Co pellets as possible factors precipitating NMD. Diarrhoeal states causing electrolyte disturbances, particularly those associated with infestations of *Chabertia ovina* and intestinal *Trichostrongylus* spp., were also a frequent precipitating cause of NMD. Excessive exercise and other forms of stress have also been important in weaner and hogget NMD in New South Wales (Walker *et al.* 1961) and New Zealand (Hartley & Dodd 1957) and in calf NMD in Canada (Hidiroglou & Jenkins 1968).

Interactions between Se and other microelements may also affect the requirements for Se. In New Zealand, Andrews *et al.* (1964) reported that Co-deficient sheep accumulated more Se in the kidneys than Co-adequate sheep and that weight responses to Se were only expressed when Co was given. Interactions between Se and Co in terms of liveweight changes in sheep have also been observed by Hartley *et al.* (1959). Wise *et al.* (1968) reported that in one trial the kidney Se concentrations of lambs with NMD and of lambs in the same flock that did not develop NMD were the same, whereas the kidney Co concentration was significantly lower in the animals showing NMD lesions. Gardiner & Nairn (1969) in Western Australia obtained some indication of a possible relationship between Se and Co in ewes grazing Yarloop clover of high oestrogenic content, in which a depression in lambing percentage observed with Co treatment was partly relieved by treatment with Co plus Se. In addition, sheep on a marginal Co intake appeared more susceptible to selenosis from high intakes of Se than sheep on adequate Co (Gardiner 1966). Similarly, a Se and Cu interaction in terms of liveweight changes, wool weight, fecundity responses (Hill *et al.* 1969) and tissue Cu concentration (Thomson & Lawson 1970) has been reported.

Strain differences in the response of Blackface sheep, grazing on pastures of apparently low Se content, to injections of Se were reported by Quarterman *et al.* (1966). Whether this effect is due to differences in Se requirements is not clear as the grazing behaviour of the strains differed, suggesting that Se intake may also differ between strains. Obst *et al.* (1974) suggested that the response of ewes to Se administration may be governed, in part, by their blood haemoglobin type. Ewes with type BB responded to Se with an increase in fecundity, whereas the fecundity of types AA and AB was reduced by Se administration. Physiological factors involved in this interaction are currently being investigated on Kangaroo Island.

CLINICAL ASPECTS AND PATHOLOGY OF NUTRITIONAL MUSCULAR DYSTROPHY

Clinical signs of NMD in ruminants are variable and depend on the tissues affected and the extent of the injury: the age of the animal may also influence the symptoms (Muth 1955; Hartley & Grant 1961; Oksanen 1967). In the acute form there is dullness and respiratory distress with a frothy nasal discharge and sudden death. In the subacute form, lambs and calves affected show a stiff, stilted gait and an arched back, and are disinclined to move; those affected with severely degenerated skeletal muscle may not be able to stand. When driven, some with severe heart involvement die suddenly. Arrhythmia and tachycardia are more common in younger than older animals. Characteristic abnormalities in the electrocardiograms of lambs and calves with NMD have been demonstrated by Godwin (1968) and Hidiroglou & Jenkins (1968) respectively. The changes in the electrocardiogram pattern develop early and become very marked as death approaches. Myoglobinuria is rare in the young animal suffering from NMD although it is a common symptom in cattle (Oksanen 1965) and hoggets (Hartley & Grant 1961). The myoglobin content of the muscles of the very young is lower than in older animals.

A characteristic of NMD is its bilateral symmetrical distribution. Young & Keeler (1962*b*) observed that the symmetry frequently extends to the distribution of lesions within the muscles. Lesions are indicated by a lightening in the colour of affected muscle which varies considerably from a slight bleaching to a distinct calcification, often appearing as white striations in skeletal muscle. The striations are more often observed in lambs than calves (Oksanen 1967). Almost all muscles in the body may be involved but those of the neck, legs and trunk (i.e. the more active muscles) are more frequently affected. The variable distribution of the lesion accounts for the variety of clinical signs. Cardiac lesions in lambs, when present, are similar to those seen in the congenetal form of NMD. In the congenital form, clear fluid and fibrin strands are usually present in the body cavities and the liver is congested. Lesions in the heart appear as subendocardial grayish-white plaques or as more diffuse lesions of a similar colour extending up to about 1 mm into the myocardium. The lesions are most commonly found in the right ventricle of lambs (Gardiner 1962) and the left ventricle of calves (Oksanen 1967). In subclinical cases of NMD in which muscular degeneration is very mild, the disease may be diagnosed histopathologically or by chemical changes in the blood.

Histopathological examination of muscle samples from lambs and calves with NMD can reveal all stages of muscle degeneration. According to Gardiner (1962) and Jenkins *et al.* (1970), the disease is characterized by an initial swelling of the muscle fibres with the loss of cytoplasmic striations. Hyaline degeneration is followed by condensation of the sarcoplasm and splintering into fragments. Proliferation of sarcolemmal nuclei and infiltration into the empty sarcolemmal tube by adventitial cells, histiocytes and polymorphonuclear leucocytes followed. In some cases, calcification occurs in the degenerated sarcoplasma. Gardiner (1962) regarded the intense proliferation of the sarcolemmal nuclei as the most characteristic histological change in the disease. Alteration in the electrophoretic pattern of affected muscles is also a consistent feature of the disease (Young & Keeler 1962*a*).

Further characterization of NMD is often indicated by subnormal concentrations of Se (see later) and tocopherol (Sharman 1954) in blood and tissues and by alteration in serum enzymes. Increases in serum enzyme activities are, with some exceptions (Gerber 1969), attributed to leakage of enzymes from damaged organs, and these increases can usually be detected before any pathological changes or clinical signs appear. Marked increases in the activities of serum aspartate amino transferase (AST) and lactic dehydrogenase (LD) (Blincoe & Dye 1958; Blincoe & Marble 1960) and creatine phosphokinase (CK) (Whanger *et al.* 1970) have been observed in lambs with NMD. Creatine phosphokinase is a more sensitive indicator of NMD because it is markedly predominant in striated muscle, whereas AST and LD are ubiquitous enzymes (Freedland & Kramer 1970). However, in view of the instability of CK in serum when stored at room temperature for short periods (Keller 1973), its usefulness as a diagnostic aid for field investigations of NMD in animals at pasture may be restricted. The activities of LD and AST are relatively stable during storage in serum (Tollersrud 1969) or blood (G.J. Judson, unpublished observations) for up to 5 days at room temperature, although haemolysis will give false positive values, especially for LD. Oksanen (1967) indicated that serum AST values, properly applied, can be of considerable diagnostic significance, particularly in subclinical cases of NMD. However, enzyme values have their limitations, as pointed out by Oksanen for serum AST. This enzyme may be only moderately increased in animals with marked clinical symptoms caused by extensive subacute or chronic degenerative processes, and a myocardial degeneration, even an acute one, may also only produce a moderate increase in the enzyme, although sudden death is often the result. Hidiroglou & Jenkins (1968) observed no appreciable rise in serum AST activity in calves with degeneration of the skeletal muscle and they suggested that this may be due to the extreme rapidity with which the myopathy developed and subsequent death ensued. In the case of serum LD, the determination of its isoenzymes may give more detail about the source of leakage and possibly make a differentiation between heart and skeletal muscle damage (Paulson *et al.* 1966).

CLINICAL ASPECTS AND PATHOLOGY OF OTHER SELENIUM-RESPONSIVE CONDITIONS

Selenium-responsive infertility in ewes tends to occur in those areas of New Zealand in which congenital NMD is found in lambs (Andrews *et al.* 1968). Similarly, in Australia, Godwin *et al.* (1970) also observed Se-responsive infertility in ewes on properties where NMD had been observed in lambs. However, the level of infertility was not as low as levels recorded in New Zealand, where on some properties lambing percentages at marking may be as low as 25. Apart from the infertility problem in ewes, there are no clear-cut symptoms with this disease. According to Hartley & Grant (1961), the affected ewes may be of any age and are usually in fair to good condition at mating.

Selenium-responsive unthriftiness is a disease of cattle, and especially sheep. In New Zealand it is regarded as probably the most widespread and economically important of all the Se-responsive diseases (Andrews *et al.* 1968). The symptoms in livestock vary from subclinical inability to maintain optimum growth rate to clinical unthriftiness which may, in bad seasons, lead to a heavy mortality (Hartley & Grant 1961; Robertson & During 1961; Grace *et al.* 1974). Although it is usually confined to young livestock, in some areas and in some seasons, adult livestock are involved. Diarrhoea is not a constant feature of the disease in sheep, but in cattle the severe form of the disease is usually associated with profuse diarrhoea. On post-mortem examination of affected sheep, the only significant findings are advanced emaciation and osteoporosis. In some cases, the remaining fat may be yellow and when diarrhoea is present, a high gastrointestinal worm burden is usually observed (Andrews *et al.* 1968).

Histological examinations of tissues from New Zealand sheep affected with Se-responsive infertility or Se-responsive unthriftiness have not provided any evidence of muscular degeneration characteristic of NMD, and the serum enzyme, AST, is not raised during these diseases (Hartley 1967; Andrews *et al.* 1968).

Selenium content of blood and most tissues of ruminants are responsive to changes in Se intake (Cousins & Cairney 1961; Kuchel & Buckley 1969; Lindberg & Jacobsson 1970; Oldfield *et al.* 1971); the Se content appears to approach equilibrium with any given regime of dietary Se in about 10-30 days. Cousins & Cairney (1961) suggested that in sheep the Se concentration in the kidney provides a sensitive index of the animal's Se status, although others (Andrews *et al.* 1964; Allaway *et al.* 1966; Allaway *et al.* 1967) contend that the liver Se concentration is a more reliable indicator of the Se stores. Studies by Hidiroglou, Jenkins, Carson & Mackay (1968) support the latter view, as their results with sheep suggest that liver Se stores may be more readily available to the animal than kidney Se stores. The kidney cortex is often analysed instead of the whole kidney because of its higher Se concentration, being 2- to 6-fold greater than the kidney medulla (Hidiroglou, Carson & Brossard 1968; Hidiroglou, Jenkins, Carson & Mackay 1968; Hoffman *et al.* 1973).

Tables 1 and 2 present some examples of Se concentrations in liver, kidney and blood of ruminants reported to be Se deficient or Se sufficient. It is evident from these results that diagnosis based only on Se concentrations may prove suspect, particularly in marginal deficiencies of Se.

Erythrocytes from sheep deficient in Se have been shown to take up a larger amount of radioactive Se during *in vitro* incubation than those from animals with adequate Se (Wright & Bell 1963; Weswig *et al.* 1966; Lopez *et al.* 1968). It was suggested that this test may be a useful diagnostic procedure in anticipating Se-responsive diseases. This uptake of Se by erythrocytes, however, probably simply reflects changes in the Se content of the cell. We have found for sheep and cattle on different Se intakes that about 65-80% of the blood Se

TABLE 1

Selenium Concentration in Feed and Tissue Dry Matter and Blood of Sheep Raised in Different Countries

Se Status	Se Intake* (μmol/kg)		Kidney (μmol/kg)		Liver (μmol/kg)		Blood (μmol/ℓ)		Country
	Deficient	Normal	Deficient	Normal	Deficient	Normal	Deficient	Normal	(Reference)
Ewes	0.2	Se	-	-	-	-	0.29-0.38	-	Australia (1)
Ewes**	0.3-0.9	Se	-	-	-	-	0.28-0.67	0.41-4.52	Australia (2)
Ewes	0.2-0.6	Se/VE	26-52	31-38	1.9-6.5	3.4-7.2	0.32-0.99	0.36-1.89	Canada (3)
Lambs	(0.18-0.13)	(0.10-0.23)	25-35	24-45	1.5-4.8	1.9-5.0	0.37-0.98	0.42-1.91	
Ewes	0.0-0.6	-	-	-	-	-	0.15-0.71	-	Canada (4)
Lambs	(0.06-0.14)	-	-	-	-	-	0.14-0.62	-	
Lambs**	0.1-0.4	0.7	0-19#	16-95#	0.0-3.4	0.0-44.0	-	-	Canada (5)
Wethers	-	Se/VE	-	-	-	-	0.53-0.89	3.39-4.04	
Lambs	-	-	5-83#	66-89#	-	-	-	-	Canada (6)
Young sheep$^{\phi}$	< 0.3	> 0.4	< 19#	> 38#	< 1.9	> 3.8	< 0.10	> 0.13	New Zealand (7)
Lambs	-	-	-	-	-	-	0.05-0.57	0.89-1.90	Norway (8)
Ewes	0.1-0.6	-	-	-	0.7-2.9	-	0.06-1.30	-	Sweden (9)
Lambs	-	-	-	-	1.0-5.1	-	-	-	
Lambs**	0.1-0.5	5.4-7.2	3-14	15-68	0.0-2.0	28.1-73.7	-	-	U.S.A. (10)
Lambs**	0.4 (0.11-0.13)	Se (0.22-0.43)	19-42	38-76	0.3-4.6	2.5-21.1	0.10-0.66	0.84-3.72	U.S.A. (11)

* Se or Se/VE indicates normal animals given deficient diet plus Se or Se plus tocopherol (Se/VE) supplements. For lambs, Se concentration is given for maternal diet with maternal milk Se (μmol/ℓ) in parentheses. Lambs usually had access to maternal diet.

**Range for tissues calculated from mean value ± 2 standard deviations.

ϕ Values converted from fresh weight to dry weight basis, assuming 67% moisture in tissues.

\# Kidney cortex.

References - (1) Handreck & Godwin 1970; (2) Godwin *et al.* 1970; (3) Jenkins *et al.* 1974; (4) Hidiroglou *et al.* 1971; (5) Hidiroglou, Jenkins, Carson & Mackay 1968; (6) Hoffman *et al.* 1973; (7) Andrews *et al.* 1968; (8) Lunde & Odegaard 1972; (9) Lindberg & Jacobsson 1970; (10) Allaway *et al.* 1966; (11) Paulson *et al.* 1968.

TABLE 2

Selenium Concentration in Feed and Tissue Dry Matter and Blood of Cattle Raised in Different Countries

	Se Intake* (μmol/kg)		Kidney (μmol/kg)		Liver (μmol/kg)		Blood (μmol/ℓ)		Country
Se Status	Deficient	Normal	Deficient	Normal	Deficient	Normal	Deficient	Normal	(Reference)
Beef cows	0.2	Se/VE	-	-	-	-	0.12-0.23	0.22-0.82	Canada (1)
Beef calves**	(0.04-0.15)	(0.05-0.27)	9-45	2-50	0.3-2.8	0.5-3.4	0.15-0.33	0.39-0.87	
Beef calves**	-	-	14-23#	33-34#	1.0-2.2	9.4-17.0	-	-	Canada (2)
Beef calves	-	-	9-58#	52-88#	-	-	-	-	Canada (3)
Beef calves	0.6-1.1	1.6-3.7	-	-	-	-	-	-	Canada (4)
Beef calves**	0.3-0.6 (0.03-0.08)	-	9-29#	-	0.9-2.6	-	< 0.35	-	Canada (5)
Dairy cows	- (0.04-0.13)	2.5		-	-	-	-	-	France (6)
Dairy calves**	- (0.40)	1.1-5.1	10-37	24-117	1.9-9.5	9.9-23.6	-	-	France (6)
Calvesϕ	- (0.40)	- (0.08)	10-18#	-	0.9-1.6	-	-	-	New Zealand (7)
Beef cows	< 0.2	Se/VE	-	-	-	-	0.02-0.06	0.22-1.67	Sweden (8)
Beef calves	- (0.05-0.06)	- (0.13-0.25)	-	-	-	-	0.06-0.10	0.38-0.77	
Dairy cows	0.5-0.7	1.0	-	-	-	-	0.58-1.27	0.65-1.43	Scotland (9)

* Se/VE indicates normal animals given deficient diet plus Se or Se plus tocopherol (Se/VE) supplements. For calves, Se concentration is given for maternal diet with maternal milk Se (μmol/ℓ) in parenthesis. Calves had access to maternal diet except in references (5) and (7).

**Range for tissues calculated from mean value ± 2 standard deviations.

ϕ Values converted from fresh weight to dry weight basis, assuming 67% moisture in tissues.

\# Kidney cortex.

References - (1) Jenkins *et al*. 1974; (2) Hidiroglou, Jenkins, Carson & Mackay 1968; (3) Hoffman *et al*. 1973; (4) Hidiroglou *et al*. 1965; (5) Hidiroglou, Carson & Brossard 1968; (6) Lamand 1972; (7) Hartley 1967, Andrews *et al*. 1968; (8) Jacobsson *et al*. 1970; (9) Trinder *et al*. 1973.

is in the blood cells. Thus, apart from its greater ease of determination, this radioisotope technique has little to offer over blood Se estimations.

Factors that affect the deposition of Se in animal tissues are not well understood, nor is the chemical form of the Se in tissues and their relative potencies in protecting against the various Se-responsive disorders. Recent studies by Rotruck and his associates (Rotruck *et al.* 1971; Rotruck *et al.* 1972; Rotruck *et al.* 1973) have shown that a specific function of Se in animals is as a component of glutathione peroxidase. This work was supported by findings that the specific activity of glutathione peroxidase in chicken (Noguchi *et al.* 1973) and rat (Chow & Tappel 1974) tissue is a function of dietary Se and that the enzyme from ovine blood (Oh *et al.* 1974) and bovine blood (Flohe *et al.* 1973) contains 4 moles Se/mole of enzyme. The assay of the activity of this enzyme in blood or tissue of ruminants may well prove a sensitive diagnostic aid for detecting Se inadequacy (K.O. Godwin, personal communication; Hoekstra 1974). Preliminary investigations indicate that this enzyme is sufficiently stable in sheep blood cells, when stored at room temperature, to permit its use in assessing the Se status of grazing animals (P.N. Wilson & G.J. Judson, unpublished observations).

PREVENTION AND TREATMENT OF SELENIUM INADEQUACY

Alternative methods for providing sufficient but non-toxic amounts of Se to animals grazing Se-deficient areas have received considerable attention. The minimum toxic levels of Se in pasture are not known, although Underwood (1971) suggested a level of about 60 µmol/kg. Signs of chronic Se toxicity in livestock have been observed when animals consume seleniferous plants containing about 40-250 µmol of Se/kg over a prolonged period (Oldfield *et al.* 1971). With a maximum value of 50 and a minimum value of 1.3 (see above) µmol of Se/kg of feed, this would indicate that a mature ewe and cow could ingest respectively 1.3-50 and 13-500 µmol of Se/day (assuming respective daily intakes of 1-10 kg of dry feed) without ill-health.

Oral or intramuscular administration of Se, as sodium selenite or sodium selenate, is widely used for controlling Se-responsive diseases in sheep and cattle. In New Zealand, one common practice with cattle is to inject Se in doses of about 130 µmol of Se for calves and 390 µmol of Se for adults at 3-monthly intervals (Andrews *et al.* 1968). In sheep, the Se compounds are given as a drench. Ewes are given 65 µmol of

Se about 1 month before mating and a month before lambing to prevent infertility and congenital NMD, and at 1 month after lambing. The lambs are given 13 μmol Se at about 1 month of age and at 2- to 4-monthly intervals thereafter to prevent delayed NMD and Se-responsive unthriftiness (Andrews 1968).

Dose rates of Se recommended in Australia for sheep are of the order suggested by the New Zealand workers (see Gabbedy 1970). In Western Australia, it is generally advised (Gardiner 1963) that Se treatment of pregnant ewes, grazing dominant clover pastures and due to begin lambing in winter or spring, should be carried out from 1 to 2 months before lambing. This is followed by either oral dosing or intramuscular injection of 13-26 μmol at marking time, or an oral dose of 65 μmol at weaning and, in districts with a high incidence of NMD, a midsummer oral dose of 65 μmol to the hoggets.

The above dosing or injecting programmes have proved effective in preventing Se-responsive disorders, but under practical farm conditions the amount of labour required is a factor to be considered and there is a tendency to use heavier and more frequent doses of Se (see Gardiner 1966). These disadvantages prompted the development of an alternative means of supplying Se. Kuchel & Buckley (1969) and Handreck & Godwin (1970) in Australia investigated the use of heavy pellets for supplying Se to grazing sheep. The pellets, consisting of Fe and elemental Se compressed together, were placed in the reticulo-rumen and found to release physiological quantities of Se to the tissues. Similar pellets have been shown by Andrews *et al.*(1974) in New Zealand to increase the Se status and weight gains of young Romney wethers grazing pastures in Se-deficient areas. However, these workers reported that about 15% of the treated animals rejected the pellets within a 12-month period, and, in varying degrees, the pellets acquired deposits of calcium phosphate. This high rejection has not been observed in Merino sheep and placing grinders in the reticulo-rumen with the pellet has been shown to largely overcome any problem with salt deposition on the pellet (R. Kuchel, personal communication).

In Canada, attempts also have been made to develop a means for providing a continuous, low level of Se to ruminants. Selenium pellets, containing 560 μmol of Se as sodium selenite in an inert vehicle, implanted subcutaneously in Se-deficient ewes at mid-pregnancy, prevented the development of NMD in the lambs (Hidiroglou *et al.* 1971), presumably by supplying adequate amounts of Se to the developing foetus and to the lamb *via* the milk. Implantation of the Se pellets, containing 190 μmol of Se, directly into Se-deficient beef calves soon after birth also was effective in preventing NMD (Hidiroglou, Jenkins, Wauthy & Proulx 1972). One of the main disadvantages of this method, however, is the tendency for the relatively high concentrations of Se to remain in the area of implantation. Implantation of Se-deficient lambs with pellets containing 990 IU of tocopherol was shown to be almost completely effective against NMD (Hidiroglou, Jenkins & Corner 1972).

The addition of inorganic Se as sodium selenate to trace-mineralized salts offered free-choice to ewes was shown to be effective in preventing NMD in their lambs (Paulson *et al.* 1968; Rotruck *et al.* 1969). Levels of about 330 and 1700 μmol Se/kg in the salt supplied approximately 2 and 11 μmol/kg respectively, as added Se to the total diet of the ewes. Supplementation of Se at these levels on a continuous basis for 4 years did not result in excessive accumulation of Se in the tissues of the ewes or their lambs. In Finland, Westermarck (1964) utilized selenized salt successfully for the prevention of NMD in cattle for many years, and in Canada the inclusion of Se (190 μmol/kg) and tocopherol (2700 IU/kg) in a mineral supplement for Se-deficient beef cows was also shown to prevent NMD in their offspring (Jenkins *et al.* 1974). This method affords a simple means of providing Se supplementation as desired, and with suitable protection from the weather, utilizable under grazing conditions. As livestock, when given free access to the selenized salt, apparently regulate their intake of the salt, the only major disadvantage with this method is that some stock may not receive the supplement.

Alternative methods of providing Se directly to livestock include the addition of inorganic Se to the feed and the supplementation or partial replacement of the diet with feeds of higher Se content (see Oldfield *et al.* 1971). Neither method appears suitable for providing Se to the grazing ruminant. Even in intensive production of livestock, major problems likely to be encountered would include the practical problem of mixing minute amounts of Se with large quantities of feed and of obtaining feed of sufficiently high Se content. This would also necessitate the analysis of each new batch of feed to ensure sufficient but not toxic amounts of Se were fed to animals.

Addition of Se, as selenite, to the soil has been used as a means for raising the Se content of the herbage. Allaway *et al.* (1966) in the U.S.A. reported that an application of about 5.7 mol Se/hectare, injected into deficient soil in the form of a solution of sodium selenite, resulted in lucerne that contained protective levels of Se for sheep for at least 3 years following application. Applications of approximately

90-180 mmol Se/hectare, either sprayed on or mixed with superphosphate and top-dressed on pasture, has been effective in preventing Se-responsive disorders in lambs in New Zealand (Hartley 1967). Superphosphate as a carrier for Se would probably be the most economical method of correcting Se content of soils in many of the areas grazed by ruminants in Australia. However, in terms of the amount of Se required, the addition of Se to the soil is an inefficient method of meeting the Se requirements of animals (Allaway *et al.* 1966) and its general adoption may be discouraged if large reserves of Se are not discovered. Rates of application of Se will also need to vary, being higher for acid soils than for alkaline soils or for very sandy soils. With the acidic laterite soils on Kangaroo Island, substantial quantities of Se may be required, since a ferric iron-selenite complex could be formed and the Se in this form is only slightly available to plants (see Allaway *et al.* 1967). On such soils this method is probably needlessly hazardous as there is a danger of some plants absorbing amounts of Se which are toxic to animals (see Grant 1965).

CONCLUDING REMARKS

In areas of Se deficiency one reason for the prevalence of Se-responsive disorders in grazing livestock may be that these, more than any other domestic species, are likely to consume the native herbage as virtually their entire diet. The incidence of these disorders has increased over the past few years, particularly in North America (Jenkins & Hidiroglou 1972). Several factors may have contributed to this, including changes in agricultural practices leading to increased production and an increased awareness of the Se-responsive disorders in livestock.

In New Zealand, the occurrence of Se-responsive diseases is commonly associated with lush spring growth of improved pastures resulting from the liberal use of superphosphate (Andrews *et al.* 1968). This association has also been noted by Walker *et al.* (1961) in New South Wales and by Gardiner & Gorman (1963) in Western Australia. Under such management systems, the competition between S and Se in absorption by plants and the replacement of low fertility grasses by improved grasses or legumes which may not be able to absorb as much soil Se have been suggested as possible factors reducing the availability of Se to livestock (see Gardiner 1969). Further, Peterson & Spedding (1963) have postulated that the conversion of organo-selenium to insoluble inorganic forms in the alimentary tract of the grazing ruminant could, over a number of years, reduce the available soil Se and that this process may be accelerated by the increased stocking rate which often accompanies pasture improvement. Irrigation of pastures, with possible leaching of Se from the soil (see Lakin 1961), and changes in sources of supplementary feeds for livestock have also been implicated as causes of the increasing incidence of Se-responsive disorders (Hoffman *et al.* 1973).

Diagnosis of Se inadequacy is aided by a knowledge of the factors which affect the variation between and within animal species in the manifestation of signs of Se inadequacy. There is little chance at present of diagnosing disorders of subclinical Se inadequacy on an individual animal basis or even on a group basis except by measuring the response to administered Se. The variable responses in growth rate (see Gardiner 1969) and the absence of responses in reproductive performances (Davies 1966; Maxwell 1972; Obst *et al.* 1974; Oldfield 1974) to Se supplementation in areas where NMD has been observed, however, require further investigation.

ACKNOWLEDGEMENTS

We wish to thank Mr. R.J. Clare of the South Australian Department of Agriculture for the collection of bloods from livestock on Kangaroo Island and Mrs. R.J. Thompson and Miss L.A. Phillipps of the Institute of Medical and Veterinary Science for measuring the Se content of these bloods.

REFERENCES

ALDERSON, N.E., MITCHELL, G.E., Jr., LITTLE, C.O., WARNER, R.E. & TUCKER, R.E. (1971) - *J. Nutr. 101* : 655

ALLAWAY, W.H. & HODGSON, J.F. (1964) - *J. Anim. Sci. 23* : 271

ALLAWAY, W.H., CARY, E.E. & EHLIG, C.F. (1967) - *In* "Symposium: Selenium in Biomedicine" (O.H. Muth, ed.), p. 273. AVI Publishing Co. Inc. : Westport

ALLAWAY, W.H., MOORE, D.P., OLDFIELD, J.E. & MUTH, O.H. (1966) - *J. Nutr. 88* : 411

ANDERSSON, P. (1960) - *Acta path. microbiol. scand. 48* : Suppl. 134

ANDREWS, E.D., GRANT, A.B. & BRUNSWICK, L.F.C. (1974) - *N.Z. vet. J. 22* : 46

ANDREWS, E.D., GRANT, A.B. & STEPHENSON, B.J. (1964) - *N.Z. Jl agric. Res. 7* : 17

ANDREWS, E.D., HARTLEY, W.J. & GRANT, A.B. (1968) - *N.Z. vet. J. 16* : 3
BLAXTER, K. (1962) - *Vitam. Horm. 20* : 633
BLAXTER, K.L. (1963) - *Br. J. Nutr. 17* : 105
BLINCOE, C. & DYE, W.B. (1958) - *J. Anim. Sci. 17* : 224
BLINCOE, C. & MARBLE, D.W. (1960) - *Am. J. vet. Res. 21* : 866
BURROUGHS, W., KOHLMEIER, R., BARRINGER, R., KAWASHIMA, R. & TRENKLE, A. (1963) - *J. Anim. Sci. 22* : 929
BURTON, V., KEELER, R.F., SWINGLE, K.F. & YOUNG, S. (1962) - *Am. J. vet. Res. 23* : 962
CHOW, C.K. & TAPPEL, A.L. (1974) - *J. Nutr. 104* : 444
CHRISTIE, B. (1962) - *Aust. vet. J. 38* : 392
COUSINS, F.R. & CAIRNEY, I.M. (1961) - *Aust. J. agric. Res. 12* : 927
DAVIES, H.L. (1966) - *J. Aust. Inst. agric. Sci. 32* : 216
DODD, D.C. (1954) - *N.Z. Jl Agric. 89* : 369
DOIG, P.A. (1970) - *Can. vet. J. 11* : 24
DRAKE, C., GRANT, A.B. & HARTLEY, W.J. (1960) - *N.Z. vet. J. 8* : 4
FLOHE, L., GUENZLER, W.A. & SCHOCK, H.H. (1973) - *FEBS Lett. 32* : 132
FREEDLAND, R.A. & KRAMER, J.W. (1970) - *Adv. vet. Sci. comp. Med. 14* : 61
GABBEDY, B.J. (1970) - *Aust. vet. J. 46* : 223
GABBEDY, B.J. (1971) - *Aust. vet. J. 47* : 318
GARDINER, M.R. (1962) - *Aust. vet. J. 38* : 387
GARDINER, M.R. (1963) - *J. Agric. West. Aust. 4* : 632
GARDINER, M.R. (1966) - *Aust. vet. J. 42* : 442
GARDINER, M.R. (1969) - *Outl. Agric. 6* : 19
GARDINER, M.R., ARMSTRONG, J., FELS, H. & GLENCROSS, R.N. (1962) - *Aust. J. exp. Agric. Anim. Husb. 2* : 261
GARDINER, M.R. & GORMAN, R.C. (1963) - *Aust. J. exp. Agric. Anim. Husb. 3* : 284
GARDINER, M.R. & NAIRN, M.E. (1969) - *Aust. vet. J. 45* : 215
GARDNER, R.W. & HOGUE, D.E. (1967) - *J. Nutr. 93* : 418
GERBER, H. (1969) - *Equine vet. J. 1* : 129
GODWIN, K.O. (1968) - *Nature, Lond. 217* : 1275
GODWIN, K.O., KUCHEL, R.E. & BUCKLEY, R.A. (1970) - *Aust. J. exp. Agric. Anim. Husb. 10* : 672
GRACE, N.D., MAUNSELL, L.A. & SCOTT, D. (1974) - *N.Z. Jl exp. Agric. 2* : 99
GRAHAM, M.T., SHARMAN, G.A.M. & BLAXTER, K.L. (1956) - *Vet. Rec. 68* : 408
GRANT, A.B. (1965) - *N.Z. Jl agric. Res. 8* : 681
HANDRECK, K.A. & GODWIN, K.O. (1970) - *Aust. J. agric. Res. 21* : 71
HARTLEY, W.J. (1967) - *In* "Symposium: Selenium in Biomedicine"

(O.H. Muth, ed.), p. 79. AVI Publishing Co. Inc. : Westport
HARTLEY, W.J. & DODD, D.C. (1957) - *N.Z. vet. J. 5* : 61
HARTLEY, W.J., DRAKE, C. & GRANT, A.B. (1959) - *N.Z. Jl Agric. 99* : 259
HARTLEY, W.J. & GRANT, A.B. (1961) - *Fedn Proc. 20* : 679
HIDIROGLOU, M., CARSON, R.B. & BROSSARD, G.A. (1965) - *Can. J. Anim. Sci. 45* : 197
HIDIROGLOU, M., CARSON, R.B. & BROSSARD, G.A. (1968) - *Can. J. Physiol. Pharmac. 46* : 853
HIDIROGLOU, M., HOFFMAN, I., JENKINS, K.J. & MACKAY, R.R. (1971) - *Anim. Prod. 13* : 315
HIDIROGLOU, M. & JENKINS, K.J. (1968) - *Can. J. Anim. Sci. 48* : 7
HIDIROGLOU, M., JENKINS, K.J., CARSON, R.B. & MACKAY, R.R. (1968) - *Can. J. Anim. Sci. 48* : 335
HIDIROGLOU, M., JENKINS, K.J. & CORNER, A.H. (1972) - *Can. J. Anim. Sci. 52* : 511
HIDIROGLOU, M., JENKINS, K.J. & MACKAY, R.R. (1973) - *Can. J. Anim. Sci. 53* : 297
HIDIROGLOU, M., JENKINS, K.J., WAUTHY, J.M. & PROULX, J.E. (1972) - *Anim. Prod. 14* : 115
HILL, M.K., WALKER, S.D. & TAYLOR, A.G. (1969) - *N.Z. Jl agric. Res. 12* : 261
HINTZ, H.F. & HOGUE, D.E. (1964) - *J. Nutr. 82* : 495
HOEKSTRA, W.G. (1974) - *In* "Trace Element Metabolism in Animals, 2" (W.G. Hoekstra, J.W. Suttie, H.E. Ganther & W. Metz, eds.), p. 861. University Park Press : Baltimore
HOFFMAN, I., JENKINS, K.J., MERANGER, J.C. & PIGDEN, W.J. (1973) - *Can. J. Anim. Sci. 53* : 61
JACOBSSON, S.O., LIDMAN, S. & LINDBERG, P. (1970) - *Acta vet. scand. 11* : 324
JENKINS, K.J. & HIDIROGLOU, M. (1972) - *Can. J. Anim. Sci. 52* : 591
JENKINS, K.J., HIDIROGLOU, M., MACKAY, R.R. & PROULX, J.E. (1970) - *Can. J. Anim. Sci. 50* : 137
JENKINS, K.J., HIDIROGLOU, M., WAUTHY, J.M. & PROULX, J.E. (1974) - *Can. J. Anim. Sci. 54* : 49
KELLER, P. (1973) - *N.Z. vet. J. 21* : 221
KUBOTA, J., ALLAWAY, W.H., CARTER, D.L., CARY, E.E. & LAZAR, V.A. (1967) - *J. agric. Fd Chem. 15* : 448
KUCHEL, R.E. & BUCKLEY, R.A. (1969) - *Aust. J. agric. Res. 20* : 1099
LAKIN, H.W. (1961) - *In* "Selenium in Agriculture", Agriculture Handbook 200, p. 3. U.S. Department of Agriculture : Washington, D.C.
LAMAND, M. (1972) - *Annls Zootech. 21* : 29
LINDBERG, P. & JACOBSSON, S.O. (1970) - *Acta vet. scand. 11*:49

LINDBERG, P. & LANNEK, N. (1970) - *In* "Trace Element Metabolism in Animals" (C.F. Mills, ed.), p. 421. E. & S. Livingston : Edinburgh

LOPEZ, P.L., PRESTON, R.L. & PFANDER, W.H. (1968) - *J. Nutr. 94* : 219

LUNDE, G. & ODEGAARD, S.A. (1972) - *Nord. VetMed. 24* : 484

MAXWELL, J.A.L. (1972) - *Aust. vet. J. 48* : 598

MUDD, A.J. & MACKIE, I.L. (1973) - *Vet. Rec. 93* : 197

MUTH, O.H. (1955) - *J. Am. vet. med. Ass. 126* : 355

MUTH, O.H. & ALLAWAY, W.H. (1963) - *J. Am. vet. med. Ass. 142* : 1379

MUTH, O.H., OLDFIELD, J.E., REMMERT, L.F. & SCHUBERT, J.R. (1958) - *Science 128* : 1090

NOGUCHI, T., CANTOR, A.H. & SCOTT, M.L. (1973) - *J. Nutr. 103* : 1502

OBST, J.M., McGOWAN, C.J. & WALKER, S.K. (1974) - *Proc. Aust. Soc. Anim. Prod. 10* : 135

OH, S.-H., GANTHER, H.E. & HOEKSTRA, W.G. (1974) - *Biochemistry 13* : 1825

OKSANEN, H.E. (1965) - *Acta vet. scand. 6* : Suppl. 2

OKSANEN, H.E. (1967) - *In* "Symposium: Selenium in Biomedicine" (O.H. Muth, ed.), p. 215. AVI Publishing Co. Inc. : Westport

OKSANEN, H.E. (1973) - *Acta Agric. scand., Suppl. 19* : 22

OLDFIELD, J.E. (1974) - *N.Z. vet. J. 22* : 85

OLDFIELD, J.E., ALLAWAY, W.H., DRAPER, H.H., FROST, D.V., JENSEN, L.S., SCOTT, M.L. & WRIGHT, P.H. (1971) - "Selenium in Nutrition". Natn. Acad. Sci. : Washington, D.C.

OLDFIELD, J.E., MUTH, O.H. & SCHUBERT, J.R. (1960) - *Proc. Soc. exp. Biol. Med. 103* : 799

OLDFIELD, J.E., SCHUBERT, J.R. & MUTH, O.H. (1963) - *J. agric. Fd Chem. 11* : 388

PAULSON, G.D., BRODERICK, G.A., BAUMANN, C.A. & POPE, A.L. (1968) - *J. Anim. Sci. 27* : 195

PAULSON, G.D., POPE, A.L. & BAUMANN, C.A. (1966) - *Proc. Soc. exp. Biol. Med. 122* : 321

PETERSON, P.J. & SPEDDING, D.J. (1963) - *N.Z. Jl agric. Res. 6* : 13

PULSFORD, M.F., RAC, R. & IRVING, E.A. (1966) - *Aust. vet. J. 42* : 165

QUARTERMAN, J., MILLS, C.F. & DALGARNO, A.C. (1966) - *Proc. Nutr. Soc. 25* : xxiii

ROBERTSON, T.G. & DURING, C. (1961) - *N.Z. Jl Agric. 103* : 306

ROTRUCK, J.T., HOEKSTRA, W.G. & POPE, A.L. (1971) - *Nature, Lond. 231* : 223

ROTRUCK, J.T., POPE, A.L., BAUMANN, C.A., HOEKSTRA, W.G. & PAULSON, G.D. (1969) - *J. Anim. Sci. 29* : 170

ROTRUCK, J.T., POPE, A.L., GANTHER, H.E. & HOEKSTRA, W.G. (1972) - *J. Nutr. 102* : 689

ROTRUCK, J.T., POPE, A.L., GANTHER, H.E., SWANSON, A.B., HAFEMAN, D.G. & HOEKSTRA, W.G. (1973) - *Science 179* : 588

SCHUBERT, J.R., MUTH, O.H., OLDFIELD, J.E. & REMMERT, L.F. (1961) - *Fedn Proc. 20* : 689

SETCHELL, B.P. (1962) - *Aust. vet. J. 38* : 392

SETCHELL, B.P., HARRIS, A.N.A., FARLEIGH, E.A. & CLARK, F.L. (1962) - *Aust. vet. J. 38* : 62

SHARMAN, G.A.M. (1954) - *Vet. Rec. 66* : 274

SHERIFF, D., RAC, R. & RANKIN, G.J. (1973) - *Vet. Rec. 92* : 89

SKERMAN, D.D. (1962) - *Proc. Aust. Soc. Anim. Prod. 4* : 22

SLEN, S.B., DEMIRUREN, A.S. & SMITH, A.D. (1961) - *Can. J. Anim. Sci. 41* : 263

SMITH, W.S. (1962) - *Aust. vet. J. 38* : 394

THOMSON, G.E. & LAWSON, B.M. (1970) - *N.Z. vet. J. 18* : 79

TOLLERSRUD, S. (1969) - *Acta vet. scand. 10* : 359

TRINDER, N., HALL, R.J. & RENTON, C.P. (1973) - *Vet. Rec. 93* : 641

TRINDER, N., WOODHOUSE, C.D. & RENTON, C.P. (1969) - *Vet. Rec. 85* : 550

UNDERWOOD, E.J. (1971) - "Trace Elements in Human and Animal Nutrition". Academic Press Inc. : New York

WALKER, D.J., HARRIS, A.N.A., FARLEIGH, E.A., SETCHELL, B.P. & LITTLEJOHNS, I.R. (1961) - *Aust. vet. J. 37* : 172

WESTERMARCK, H. (1964) - *Nord. VetMed. 16* : 264

WESWIG, P.H., ROFFLER, S.A., ARNOLD, M.A., MUTH, O.H. & OLDFIELD, J.E. (1966) - *Am. J. vet. Res. 27* : 27

WHANGER, P.D., MUTH, O.H., OLDFIELD, J.E. & WESWIG, P.H. (1969) - *J. Nutr. 97* : 553

WHANGER, P.D., WESWIG, P.H., MUTH, O.H. & OLDFIELD, J.E. (1970) - *Am. J. vet. Res. 31* : 965

WISE, W.R., WESWIG, P.H., MUTH, O.H. & OLDFIELD, J.E. (1968) - *J. Anim. Sci. 27* : 1462

WOLF, E., KOLLONITSCH, V. & KLINE, C.H. (1963) - *J. agric. Fd Chem. 11* : 355

WRIGHT, P.L. & BELL, M.C. (1963) - *Proc. Soc. exp. Biol. Med. 114* : 379

YOUNG, S. & KEELER, R.F. (1962*a*) - *Am. J. vet. Res. 23* : 955

YOUNG, S. & KEELER, R.F. (1962*b*) - *Am. J. vet. Res. 23* : 966

EPILOGUE

Emeritus Professor E.J. Underwood
University of Western Australia

Some 300 years ago Francis Bacon wrote "reading maketh the full man, conference the ready man and writing the exact man". During the last two days I have done some reading, attended a conference and attempted some writing, but I feel neither full, ready nor exact. Despite these formidable handicaps I am both pleased and honoured to be given the task of summarizing a symposium which I found notable and personally rewarding for several reasons.

In the first place it is clear that the organizers have recognized that trace elements in soil, plant and animal systems cannot be considered properly in isolation from each other. They form part of an interdependent and interacting total ecosystem of considerable, indeed daunting, complexity. The papers we have heard from soil scientists, plant nutritionists, biochemists, agronomists and animal nutritionists have revealed an appreciation of this interdependence and this complexity. This has been good for us all as specialists in one field or another and gives great hope for the future of trace element research and its application to the land.

A further notable feature of the Symposium is the way it has, rightly and properly on this Jubilee occasion, highlighted the very substantial contributions that South Australian scientists, both from the Waite Institute and elsewhere, have made from the beginnings of the trace element era more than 50 years ago to the present time. These contributions have ranged from practical studies involving the successful solution of trace element problems of crops and stock in the field to sophisticated investigations of the biochemical mechanisms, metabolic movements and modes of action of trace elements in the cells and tissues of plant and animal bodies.

The prominent part played by South Australian scientists was immediately apparent from the first two papers given by Professors Donald and Prescott and by John Lee, each of whom has had a long and distinguished association with trace element research in this State. These speakers placed the whole proceedings in a proper historical perspective and made it clear that the early interest in trace elements stemmed from the need to prevent and control a number of naturally occurring drastic disease conditions of crops and stock which were limiting land development. The remarkable success of trace element research as a basis for large-scale land development in Australia, compared with any other country in the world, was emphasized by Professor Donald at the beginning of the Symposium. It was not until much later in the period under review that research emphasis shifted more towards attempts to understand the underlying mechanisms involved and the ways in which trace elements interact significantly with each other and with other elements and compounds within the soil, the plant and the animal. The emphasis also shifted inevitably to investigations of less severe manifestations of trace element disorders, without well-marked and characteristic signs of disease, but affecting larger areas and more animals. This was very apparent from the paper by Dr. Egan.

It would be impossible for me to mention every speaker and every one of the 22 admirable papers presented over the last two days — nor would you want me to do this and merely give an indigestible, potted version of all that has been said and which you have already heard. I will therefore draw your attention to some of the points which have emerged from the Symposium which seem to me to have particular relevance and significance.

For example, in the first of the sectional review papers by Dr. Norrish, we were told of the recent success of the electron probe microanalyser and similar instruments in illuminating our understanding of the mineral associations and forms of retention of many trace elements in soils, about which we were previously largely ignorant. This dependence for progress on the development and use of new analytical tools and techniques was explicit or implicit in many of the papers that followed. It suggests that all laboratories concerned with trace element research, whether with soils, plants or animals, should include a group of scientists specifically responsible for testing, developing and adapting new and improved analytical techniques to the needs of their colleagues and examining the opportunities for advances that such techniques possess. The experience with Cr, which I mentioned in my paper, is particularly pertinent. The normal concentrations of Cr in blood, milk and liver have been reduced by one to two orders of magnitude in the last ten years since the contamination from stainless steel instruments was recognized and the atomic absorption analytical technique that was used was improved.

In the following section, Professor Loneragan tackled the concept and determination of 'availability' of nutrients to plants from soils. The development of reliable techniques for meaningfully measuring the availability of trace elements, as well as the major elements, has proved frustrating and tantalizingly elusive to a whole generation of soil scientists and plant nutritionists. Perhaps this is inevitable because of the inherent difficulty in using 'static' chemical procedures for determining availability for what is essentially a dynamic process — plant growth. Professor Loneragan gave examples from his own work of the vital importance of knowing about the mobility of elements within the plant when attempting to use plant concentrations in diagnosing deficiencies or excesses. He went on to remind us that the rate of absorption of trace elements growing on soils depends upon both soil and plant factors. The importance of soil factors, as well as plant, in influencing absorption was made clear also in the papers by Quirk, McKenzie, Tiller, Nambiar and Johnson. In this connection you will recall how Mr. McKenzie told us of the remarkable occlusion of Co on Mn oxide nodules and Dr. Johnson of the soils that form insoluble compounds of Se and ferric hydroxide, so that high soil Se levels do not result in Se levels in the herbage that are toxic to stock. The similarity of this situation with that which pertains in animals became particularly apparent later from the notable and attractively presented paper on 'Trace Element Interactions in Animals' given by Dr. Suttle, himself an important contributor to knowledge in this area. Dr. Suttle emphasized that the availability and absorption of trace elements by animals can be greatly affected by interactions with other elements and compounds within the feed and within the animal. To use his own words, "they (i.e. the interactions) can occur from the moment the diet is mixed to the final excretion of metabolites" and "are a major factor contributing to the poor correlation between trace element intake and animal health". One could equally say that such interactions and the forms of the elements in the soil and the plant contribute similarly to the poor correlation between total trace element levels in soils, their uptake by plants and plant health and requirements. This point was brought out again later in the papers by King and Alston and by Reuter of the South Australian Department of Agriculture. The former workers stressed the need to recognize the differences between soils before soil analysis can be used effectively for diagnosing deficiencies in wheat.

In the sections of the Symposium dealing with functions of the trace elements, particularly those by Professor Nicholas and Dr. Boardman concerned with plants, and those of Drs. Smith and Gawthorne and of Dr. Godwin and myself concerned with animals, two points emerged of great importance. The first is that the trace elements, especially the transition metals, lie at the very root of a wide range of metabolic processes including processes of such overwhelming significance to the whole of life on earth as photosynthesis, nitrogen fixation and DNA synthesis. The second point that emerged is that there are still several trace elements, known to be essential for plants or animals or both, whose specific functions are unknown. These include Na, Cl and B in plants and Sn, Ni, F and V in animals. Our continued ignorance of the site of action of B in plants is remarkable because it has been known to be essential for growth since 1923, a year before the Waite began, as Professor Prescott reminded us in his opening paper. Research aimed at identifying the biochemical and physiological sites of action of these elements surely represents the supreme challenge of the future in the field of endeavour which we have been discussing for the last two days.

Finally, it is a pleasure to refer to the concluding sessions on the distribution, diagnosis and correction of trace element disorders in plants and animals, which brought out so clearly the difficulties and the successes in applying efficiently and economically the more basic findings to field problems — problems frequently complicated by immense variability, by the coexistence of two or more deficiencies and interacting factors and often by the absence of well-marked clinical or pathological changes specific to a single deficiency, imbalance or toxicity, thus leading to difficulties in diagnosis. This last point was well brought in the papers by Mr. Reuter and Drs. Judson and Obst, who also emphasized in an impressive way the importance of climatic, soil and plant factors in the occurrence, recognition and correction of trace element disorders in animals and the need for more research on residual effects and into field diagnostic tests. Workers in these more applied aspects of trace element research in many ways have the most difficult and at the same time the most important task, although they rarely receive the recognition they deserve. These are the 'applicators', the people who seem to me to be most obviously motivated by those other words of Francis Bacon when he wrote "the real and legitimate goal of the sciences is the endowment of human life with new inventions and riches".

In conclusion, I would again like to congratulate all concerned on a highly successful conference which has given scientists of diverse backgrounds a better overall awareness of the problems, progress and interdependence of trace element research on soils, plants and animals, and of the gaps in that research that most need to be filled. In the foyer of the great Sloan-Kettering Cancer Research Institute in New York there is a plaque containing the moving words 'Within these walls a few labor that many may live'. If world population and food and fibre needs continue to grow at the present rate, as seems inevitable, if trace element research at the Waite Institute and elsewhere continues to yield the rich dividends in food and fibre productivity in the future as it has in the past, as seems likely, then long before the Waite is celebrating its Centenary, which I doubt if I will be attending, I suggest it would not be inappropriate to place a plaque in the foyer of this Institute which similarly reads 'Within these walls a few labor that many may live'.

Subject Index

A

B

C

Z

A 5
B 6
C 7
D 8
E 9
F 0
G 1
H 2
I 3
J 4